Soulless Matter, Seats of Energy

Companion Volumes

Charming Beauties and Frightful Beasts:
Non-Human Animals in South Asian Myth, Ritual and Folklore
Edited by Fabrizio M. Ferrari and Thomas W.P. Dähnhardt

Roots of Wisdom, Branches of Devotion:
Plant Life in South Asian Traditions
Edited by Fabrizio M. Ferrari and Thomas W.P. Dähnhardt

Soulless Matter, Seats of Energy

Metals, Gems and Minerals in South Asian Traditions

Edited by
Fabrizio M. Ferrari and Thomas W.P. Dähnhardt

equinox

SHEFFIELD UK BRISTOL CT

Published by Equinox Publishing Ltd.
UK: Office 415, The Workstation, 15 Paternoster Row, Sheffield, South Yorkshire S1 2BX
USA: ISD, 70 Enterprise Drive, Bristol, CT 06010

www.equinoxpub.com

First published 2016

British Library Cataloguing-in-Publication Data
A catalogue record for this book is available from the British Library.

ISBN-13 978 1 78179 128 8 (hardback)
978 1 78179 129 5 (paperback)

Library of Congress Cataloging-in-Publication Data
Names: Ferrari, Fabrizio M., editor. | Dähnhardt, Thomas W.P., 1964- editor.
Title: Soulless matter, seats of energy : metals, gems and minerals in South Asian traditions / Edited by Fabrizio M. Ferrari and Thomas W.P. Dähnhardt.
Description: Bristol, CT : Equinox Publishing Ltd, 2016. | Includes bibliographical references and index.
Identifiers: LCCN 2016002486 (print) | LCCN 2016006626 (ebook) | ISBN 9781781791288 (hb) | ISBN 9781781791295 (pb) | ISBN 9781781794364 (e-PDF) | ISBN 9781781794371 (e-epub)
Subjects: LCSH: South Asia—Religion. | South Asia—Religious life and customs. | Metals—Religious aspects. | Metals—Folklore. | Metals—Mythology—South Asia. | Minerals—Religious aspects. | Minerals—Folklore. | Minerals—Mythology—South Asia.
Classification: LCC BL1055 .S64 2016 (print) | LCC BL1055 (ebook) | DDC 202/.12—dc23
LC record available at http://lccn.loc.gov/2016002486

Typeset by S.J.I. Services, New Delhi
Printed and bound by Lightning Source Inc. (La Vergne, TN), Lightning Source UK Ltd. (Milton Keynes), Lightning Source AU Pty. (Scoresby, Victoria).

Contents

List of Figures vii

List of Tables viii

Introduction ix

Section One: Myth and Ritual

1 Five Stones – Four Rivers – One Town: The Hindu *Pañcāyatanapūjā* 3
Mikael Aktor

2 A 'Sulfurous' Śakti: The Worship of Goddess Hiṅgulā in Baluchistan 28
Francesco Brighenti

3 From Iron to Sapphire: Indian Myths and Rituals about Saturn, the Implacable Lord of Celestial Spheres 51
Monia Marchetto and Manuel Martin Hoefer

Section Two: Science and Health

4 Mineral Healing: Gemstone Remedies in Astrological and Medical Traditions 73
Anthony Cerulli and Caterina Guenzi

5 Mercury Tonics (*Rasāyana*) in Sanskrit Medical Literature 94
Dagmar Wujastyk

6 When *Ngülchu* Is Not Mercury: Tibetan Taxonomies of 'Metals' 116
Barbara Gerke

Section Three: Power and Devotion

7 In Search of the *Sādhu*'s Stone: Metals and Gems as Therapeutic Technologies of Transformation in Vernacular Asceticism in North India 143
Antoinette E. DeNapoli

8 *'Deg Tegh Fateh!'*: Metal as Material and Metaphor in Sikh Tradition 174
Eleanor Nesbitt

Section Four: Body and Embodiment

9 A Little Lipstick Goes a Long Way: Chit-Chatting with Women in the *Rāmāyaṇa* and *Mahābhārata* 201
Deeksha Sivakumar

10 *Ratna*: A Buddhist World of Precious Things 219
Mattia Salvini

11 When Earth Comes Alive: Earth-Bodied Beings in Jain Tradition 255
Ana Bajželj

Index 275

List of Figures

1.1. A *pañcāyatanapūjā* set in the home of Mr Ramachandran Balasundara Athreya, Chennai — 5

1.2. A *bāṇaliṅga* from the workshop of Mangal Kewat, Bakawa, Madhya Pradesh, acquired on 22 December 2012 — 13

1.3. Five *śālagrāmas* representing the five deities of the *pañcāyatanapūjā*, acquired from Himālay Rudrākṣa Bhaṇḍār, Paśupatināth, Kathmandu, on 2 September 2014 — 17

1.4. A *pañcāyatanapūjā* set bought from Giri Trading, Mylapore, Chennai, on 29 September 2014 — 21

2.1. Naturally(?) formed crescent-shaped figure on the rock wall opposite the shrine of Hiṅglāj Devī, Hinglaj, Pakistan, 2010 — 40

4.1. *Navaratna* jewel — 90

6.1. The illustrated section on *ngülchu* in Jampal Dorje's *Stainless Rosary* — 128

7.1. Guru Ma teaches about *doṣa* and complete living on *Śarad Pūrṇimā* — 155

7.2. Members of Guru Ma's *bhakti* community listening to the *Śarad Pūrṇimā kathā* — 155

7.3. While holding a *bhakt*'s child, Kailash Das talks about her success with stone therapy — 166

7.4. Kailash Das displays the silver toe-rings she had made to heal her dog-bite — 166

7.5. Tulsi Giri explains the health and healing benefits of stone therapy for her navel condition. Photograph by the author — 168

7.6. Copper ring on Tulsi Giri's right big toe — 168

List of Tables

3.1. The features of the nine planets 54
4.1. Planets (*grahas*) and associated gems 77
4.2. The placement of the nine planetary gemstones according to the spatial directions 90

Introduction

FABRIZIO M. FERRARI AND THOMAS W.P. DÄHNHARDT[1]

In view of the title chosen for the present collection,[2] and in response to stimulating conversations with some contributors, the editors feel that it will be useful to outline here some early Indian philosophical ideas on what is generally addressed as the inanimate world. In particular we will survey how earth, minerals, gems and metals have been conceptualized in relation to life and sentience.

As it will become clear in the chapters to follow, the phenomenal world in Indian traditions cannot be merely reduced to *either* animate *or* inanimate. Earth, stones and metals can be infused with life, or have natural potency[3] which permits them to escape such a rigid classification. One further problem emerges when we try to ontologically define what determines 'life,' and the lack thereof.

In the famous *puruṣasūktam* from the tenth *maṇḍala* of RV, a later hymn, man dominates the world by transforming it into food. In so doing, he is able to tell the animate, what eats (*sāśana*), from the inanimate, what eats not (*anaśana*):

1 Fabrizio M. Ferrari is Professor of Indology and South Asian Religions in the Department of Theology and Religious Studies at the University of Chester (UK).
Thomas W.P. Dähnhardt is Assistant Professor in Hindi and Urdu Languages and Literatures in the Department of Asian and North African Studies at the Ca' Foscari University of Venice (Italy).

2 The volume concludes our excursus on nature in South Asian traditions. See Ferrari and Dähnhardt (2013) on non-human animals; Ferrari and Dähnhardt (2016) on plant life.

3 Such is the case of minerals and metals in South Asian medical and alchemical traditions. See Meulenbeld (2002, vol. I: 104 *et passim*) on early Āyurvedic sources.

From his three quarters the Man (*puruṣa*) went upward, but a quarter of him came to be here again
From there he strode out in different directions toward what eats and what does not eat.[4] (RV_2 x.90.4; tr. Jamison and Brereton)

Food (*ánna-*) is a core theme in early Vedic literature. Agni is *annavat* (provided with food), *annāda* (he who eats food) and *annapati* (lord of food = provider of food) (Gonda 1987: 81). It is no surprise that life was associated with food, its consumption and the manifold ways to produce it. This brings us to the theme of this volume. Raw materials of organic and inorganic origin served for the crafting of tools for domestic and more specialized labor (e.g. tilling and farming, weaponry, carpentry, masonry as well as handicraft, jewelry, et cetera) as well as for commercial purposes.[5] Stones, minerals and metals of various qualities were also employed in a number of ritual contexts, including the *aśvamedha* and the pressing of *soma* (Caland 1990: 364–67).

Vedic Saṃhitās and the Brāhmaṇas reflect a sacrificial culture and do not provide sufficient evidence to ascertain whether philosophical arguments on matter were current. Some isolated passages can be found where living beings are grouped together in view of possessing different forms of *ātman* (AiĀ II.3.2). Similarly, the Upaniṣads – with their tendency to interiorize sacrifice by means of asceticism – provide a basis to reflect on categories of being and existence, the origin of the world and the place of man in it. See, for instance, the following excerpt from the *Khilakāṇḍa* of *Bṛhadāraṇyaka Upaniṣad*:

Prajapati then thought to himself: 'Now, why don't I prepare a base for that semen?' So he created woman and, after creating her, had intercourse with her. A man, therefore, should have intercourse with a woman. Prajapati stretched out from himself the elongated stone for pressing Soma and

4 *tripā́d ūrdhvá úd ait púruṣaḥ pā́do 'syehā́bhavat púnaḥ* |
táto víṣvaṅ ví akrāmat sāśanānaśané abhí || RV_1 x.90.4.
Translators are indicated at all times. If no indication is given, the translation is by the editors.

5 Mining and metallurgy must have been important aspects of Vedic economy. See Rau (1974) and Chakrabarti (1992) for a study of early metallurgy in India. As for mining, there is scant evidence in Vedic sources. Rau (1974: 26) only mentions MaiU VI.28, where the term 'mine' (*avaṭa*) is used in a symbolic fashion. More detailed evidence on techniques of mining and metallurgy as well as the social and economic background appears in Kauṭilya's *Arthaśātra*. See the study by Olivelle (2012).

> impregnated her with it. [2] Her vulva is the sacrificial ground; her pubic hair is the sacred grass; her labia majora are the Soma-press; and her labia minora are the fire blazing at the center. A man who engages in sexual intercourse with this knowledge obtains as great a world as a man who performs a Soma sacrifice, and he appropriates to himself the merits of the women with whom he has sex. The women, on the other hand, appropriate to themselves the merits of a man who engages in sexual intercourse with them without this knowledge. [3] (BĀU VI.4.2–3; tr. Olivelle)

Human anatomical particulars are associated with the technical terminology of one of the most important sacrifices, viz. the pressing of *soma*. This evinces the importance of generating offspring in Vedic culture (hence the emphasis on food; see above) to the extent that conceiving is considered an act as meritorious as sacrificing.[6] Yet, what about *grāvan*, the stone for pressing? Is *grāvan* animate, infused with life, or inanimate? Or is it just a metaphorical extension of Prajapati's power?

In Vedic texts, when stones and metals are mentioned, they generally serve as figures of speech (e.g. similes, hyperboles, metonyms) or are evoked in relation to their practical use. See for instance the following passages from *Atharvaveda*, possibly the earliest (c. 1200 BCE) attested evidence of the use of iron in India (Witzel 1989: 138):

> Cut along this skin with the dark [metal], O slaughterer, joint by joint with the knife (*así*) ; do not plot against [him]; do not be hostile to [him]; prepare him joint-wise; set him up apart in the third firmament.[7] (AV^{Ś} IX.5.4; tr. Whitney)

> Dark metal [= iron] [is] the flesh of [*odaná*, a rice dish], red [metal = copper] its blood.[8] (AV^{Ś} xi.3.7; tr. Whitney, modified)

One noteworthy exception is the earth, which is described as pervaded by energy and therefore capable of acting. AV^{Ś} XII.1 (= AV^{P} XVII) is especially revealing. The earth (*bhū́mi*) is 'mistress of what is and what is to be' (AV^{Ś} XX.1.1b; tr. Whitney). She is mother (*mātā́*) (XII.1.12) and, pervaded by Agni (XII.19) (= infused with energy/life?), provides humans with plenty of resources to live, including precious materials.[9]

6 Cf. BĀU VI.2.13 and its parallel in ChU V.8.

7 *ánuchya śyāména tvácam etā́ṃ viśastar yathāparv àsínā mā́bhí maṃsthāḥ* | *mā́bhí druhaḥ paruśáḥ kalpayainaṃ tṛtī́ye nā́ke ádhi ví śrayainam* ||

8 *śyāmám áyo 'sya māṃsā́ni lóhitam asya lóhitam* || AV^{Ś} XI.3.7.

9 *nidhíṃ bíbhratī bahudhā́ gúhā vásu maṇíṃ híraṇyaṃ pṛthivī́ dadātu me* | (XII.1.44a-b).

The fact that most Vedic texts seem not concerned with the issue at stake does not exclude the possibility that an intellectual debate was actually current. Early Buddhist and particularly Jain sources indicate that disquisitions on the sentience/insentience of matter, especially in relation to karmic retribution, were central.[10] As for Brahmanical circles, in verse Upaniṣads, and more convincingly in post-Buddhist theistic verse Upaniṣads, we notice important changes in the way the world is perceived. When examining the origin of the universe, matter (*pradhāna* or *prakṛti*) is described as the matrix of everything that is manifested and inevitably subject to decay. Such process is regulated by Īśvara, the Lord, who presides over both physical realities and individual souls (*ātman*) (ŚU I.10). With the dissemination of the theory of karmic retribution and various notions of liberation from rebirth, a number of philosophical traditions investigate where souls/selves can be found and what defines life. We will survey some of these ideas over the next pages.

In the first century BCE, Sarvāstivāda Buddhist scholasticism develops the idea that everything – animate and inanimate – exists in aggregates of particles called *dhammas* (Skt. *dharmas*) which respond to a principle of causality according to the so-called law of 'dependent origination' (P. *pratītyasamutpāda*; Skt. *pratītyasamutpāda*).[11] According to this theory, what existed in the past and will exist in the future is as real as everything existing in the present. The Sarvāstivādins make reference to five aggregates (*skandha*) as categories for *dhammas*, namely: (1) body (Skt., P. *rūpa*), (2) sensation (Skt., P. *vedanā*), (3) perception (Skt. *saṃjñā*; P. *saññā*), (4) mental formations (Skt. *saṃskāra*; P. *saṅkhāra*) and (5) consciousness (Skt. *vijñāna*; P. *viññāṇa*). While insentient objects (plants, rocks, metals, earth, et cetera) are composed *only* of physical *dhammas* and their nature is impermanent (*anitya*), human and non-human sentient beings are composed of all five aggregates. It is so evinced that it is not the self (Skt. *ātman*; P. *atta*) but consciousness (5) conditioned by action (Skt. *karma*; P. *kamma*) which determines one's rebirth (1), mental formations (4) and all successive experiences (2 and 3). As for the self, the Sarvāstivādins argue that it constantly changes, hence the expression 'no-self' (Skt. *anātman*; Pkt. *anatta*).

Possibly in response to such arguments, a coherent Brahmanical ontology emerges, namely Vaiśeṣika. Within this system, it is argued that

'Bearing treasure [and] good in many hidden places, let the earth give me jewels and gold' (tr. Whitney, modified).

10 See Introduction in Ferrari and Dähnhardt (2016).

11 For a summary of Sarvāstivāda philosophy, see: Willemen et al. (1998: 19–35).

everything that exists (*bhava*; *sat*) is both knowable and nameable.[12] Such awareness brings about the emancipation of the self. In order to investigate existence, Vaiśeṣikas refer to six fundamental categories (*padārtha*): (1) substance (*dravya*), (2) quality (*guṇa*), (3) activity (karma), (4) sameness (*sāmānya*), (5) difference, individuality (*viśeṣa*) and (6) inherence (*samavāya*) (VS I.1.4).[13] A seventh one, absence (*abhāva*), will be added later. Nyāya logicians too propose their own categories[14] but according to them matter – both animate and inanimate – exists as a combination of single and eternal atoms (*aṇu*) which represent the limit of minuteness (*aṇutva*). In order to form gross matter, atoms combine in dyads (*dvyaṇuka*). The union of three dyads makes a triad (*truṭi*; *trasareṇu*), which is thought of as the limit of 'diminishing grossness' (*mahattvāpakarṣa*) (Bhaduri 1947: 71). Nyāya thus proposes that the universe is real, whether we perceive it or not.

Since Nayāyikas and Vaiśeṣikas are both interested in cognition, the *padārthas* are designed to discuss: (a) subjects of thought, (b) objects of thought and (c) the process of thought itself. Both schools converge in their contention that material substances (*dravya*) result from: earth (*pṛthvī*), water (*jala*; *āpas*), fire (*tejas*), wind (*vāyu*), air/wind (*ākāśa*), time (*kāla*), space/direction/(*diś*), self (*ātman*) and mind (*manas*) (Mishra 1936).[15]

12 This is summarized by the commentator Praśastapāda (fifth century CE): *ṣaṇṇām api padarthānām astitvabhidheyatvajñeyatvāni* (PBh II.3, p. 16). See Balcerowicz 2010 for an in-depth study of existence and non-existence in Vaiśeṣika.

13 The same categories can be found in *Carakasaṃhitā* (first century CE) along with various ideas from Sāṃkhya-Yoga, Nyāya and Vaiśeṣika. See Comba (2001).

14 The sixteen *padārthas* of Nyāya appear as a combination of epistemological and logical concepts which have little to do with ontology. They are listed in NS I.1.1: *pramāṇa* (means of valid cognition), *prameya* (object of valid cognition), *saṃśaya* (doubt), *prayojana* (purpose), *dṛṣṭānta* (acknowledged matters), *siddhānta* (presuppositions), *avayava* (elements of argumentation), *tarka* (reasoning), *nirṇaya* (decision), *vāda* (discussion), *jalpa* (wrangling), *vitaṇḍa* (quibble), *hetvābhāsa* (fallacy), *chala* (distortion), *jāti* (futility) and *nigrahasthāna* (point of defeat). While in early Vaiśeṣika texts an independent ontology is developed and there seems to be no awareness of Nyāya, Nayāyikas do borrow from Vaiśeṣika's ontology.

15 This is rejected by the Buddhist philosopher Dignāga (480–540 CE) who in his *Pramāṇasamuccaya* argues that categories of Nyāya-Vaiśeṣika are naturally imposed by the activity of the mind. Reality, animate and inanimate, is made of groups and combinations of instantaneous particulars (*svalakṣaṇa*). Dharmakīrti (600–660 CE) agrees with Dignāga on particulars. For him too it is unimportant a distinction between animate and inanimate. Since the only valid means of knowledge (*pramāṇa*) are inference and sensory perception, to be real means to be causally effective (*arthakriya*), ergo impermanent. This has

Living organisms are featured by a body which is the site of conscious activity, organs of senses and the capacity to experience pleasure and pain (NS I.1.11). As a consequence, agency (*kartṛtva*) is regarded as a specific feature of animate/sentient beings.[16] Conversely, inanimate bodies are the mere object of action, feelings and perception. All material substances – animate/sentient or inanimate/insentient – can be distinguished into: (1) earthy objects: clay, stones and immobile objects such as plants and trees; (2) watery objects: rivers, lakes, seas but also rain, ice, snow, et cetera; (3) fire objects: terrestrial (flames, embers, etc.), aerial (sunlight, flashes, etc.); gastric (digestive fire); underground (the component of metals); (4) airy objects. The last are derived from wind and the atmosphere, and include *prāṇa*. Usually rendered as 'vital breath', *prāṇa* is acknowledged as a property of *ātman*, the soul/self of animate beings:

> The ascending life-breath, the descending life-breath, the closing of the eye-lids, the opening of the eye-lids, life, the movement of the Mind, and the affections of the other senses, and also Pleasure, Pain, Desire, Aversion, and Volition are marks (of the existence) of the Soul. (VS III.2.4; tr. Sinha)

While *prāṇa* does not appear in Nyāya foundational texts such as NS, the sixth century champion of Nyāya, Uddyotakara writes that:

> This living body is not soul-less (= that which is endowed with soul [*nirātmakam*]). Because [if it is soul-less,] the contradiction would follow that it should have no breathing (*aprāṇa*), etc. (NV 294.10–11, tr. Kanō 2001: 410)

The definition of a living being as one possessing *prāṇa* is not new as it is attested in earlier sources.[17] Five vital breaths, or winds, are indicated as evidence of life in animate beings: (1) *prāṇa* (out breath), (2) *vyāna* (inter breath), (3) *apāna* (in breath), (4) *udāna* (up breath) and (5) *samāna* (link breath) (TaitU I.7; ChU I.3; MaiU II.6).[18] In *Manusmṛti*, an animate being is called *prāṇin* whereas an inanimate object is *aprāṇa* (lit. 'without

---

major implications, as it leads to conclude that what is permanent and cannot change (e.g. gods and Veda) is unreal.

16 Cf. Pāṇinian grammarians below.

17 RV I.48.10, 101.5; $AV^{Ś}$ XI.4.14; AV^{P} V.15.8, V.22.9. In the Upaniṣads, *prāṇa* is directly associated to *ātman* (its source) and *brahman* (its seat). See Zysk (1993: 204n60). Similarly, the lack of breath is identified with death. See relevant sources from AV in Zysk (1993: 200n4).

18 See Bodewitz (1986).

breath/wind') (Manu ix.223; see also v. 28–30). Early medical compendia too investigate breath in human beings. According to the *tridoṣa* theory which features prominently in Āyurveda, diseases are believed to result from imbalance between three humors, wind (*vāta* or *vāyu*), bile (*pitta*) and phlegm (*kapha* or *śleṣman*). Wind is naturally associated to *prāṇa*, the source of life in all animate beings (CaS.Śā i.70; SuS.Śā vii.4).

Coming back to the Brahmanical *darśanas*, Mīmāṃsā, with its focus on Vedic sacrifice (*yajña*) and its exegesis, maintains the universe is real, and so are the gods and everything is described in Veda. Eight fundamental categories are listed,[19] the first of which, *drāvya*, includes the same substances of Nyāya-Vaiśeṣika. According to Mīmāṃsakas, these can be known by means of the language of Veda (and the rules of its grammar), which is looked at as an indispensable tool to distinguish by means of reason (*hetu*) the established (*siddha*) from what is to be established (*sādhya*) and, among the others, the animate from the inanimate (*acetana*) (MS i.2.35).[20] Let us see how this works.

In discussing liturgical speech, Jaimini holds that *dharma* is directly expressed by verbs, rather than by nouns and adjectives (MS ii.1.1–4). This is exemplified in a response to the objection (*pūrvapakṣa*) that the sacrifice of the bones (*asthoyajña*) is actually performed by the bones themselves (MS x.2.47). Śabara (before fifth century CE) comments that on similar instances sentence structure clearly indicates that inanimate objects (here, the bones) cannot be confused with the agent as they are *made to perform* by the direct injunction of the sacrificer (ŚBh part iv, pp. 1848–49). Śabara then moves to linguistics and observes that the (ending of a) verb, the active component of a sentence which he calls *bhāvanā*, serves the purpose of producing: (a) an effect (object) by means of an instrument (sacrifice); and (b) 'a procedure explaining how the object can be acquired through that instrument [rites composing sacrifice]' (Freschi 2013: 153–54). Since Mīmāṃsā is a system primarily concerned with the science of sacrifice, it is not surprising that its major schools, the Bhaṭṭas and Prābhākaras, have both investigated and agreed upon such terms. Disagreement however

19 The first six are the same as those listed by Vaiśeṣika plus potency (*śakti*) and number (*sāṃkhya*). 'Inherence' is called *paratantra* and rather than having 'difference' (*viśeṣa*) we have 'similarity' (*sādṛśya*). Non-existence (*abhāva*) is an addition of the commentator Kumārilabhaṭṭa.

20 This serves Jaimini to contrast the objection that Vedic mantras are absurd as they address inanimate objects as if they were alive. It is in fact man, capable of reasoning, who exalts *yajña* through them (MS i.1.32 and ŚBh on the same).

emerges on the conditions of *ātman* and the subject. For the Bhaṭṭas, the self is both self-conscious and the substratum of consciousness. In his *Tantravārttika*, Kumārilabhaṭṭa (seventh century) reflects on *arthavāda* ('explanation of the meaning') and *bhāvanā* ('causing to be'), and expounds that each *bhāvanā* is incited by a subject (*prayojakakriyā* and *prayojaka-vyāpāra*, TV_2 23 and 25; cf. the analysis of Śabara's comment to MS II.1.1[21]). This grammatical subject, the 'I' (*aham*), is ultimately understood to be the sacrificial subject precisely because an act of understanding (*pratyaya*) is required. For the Prabhākaras, *ātman* is neither sentient nor intelligent, but it is the substratum of knowledge as well as of all feelings. By means of reflexivity, every action permits the subject to be not just an agent but a knower and a moral agent (PrP, *Tattvāloka*, p. 329).[22]

Eventually one can tell an animate subject (e.g. a warrior) from an inanimate object (e.g. his sword) not just because of the presence of breath (*prāṇa*) or the self (*ātman*)[23] in the former but in view of recognition (*pratyabhijña*), a property of the 'I' (*aham*)[24] which indicates knowledge and ability to conduct sacrificial activity. The body is thus an instrument of experience whereas Vedic injunctions (language) are instruments of knowledge. Desire, i.e. active awareness of producing consequences, gives the sacrificial subject self-awareness. As Śabara says: 'if we leave cognition aside, we cannot indicate anything at all.'[25]

The grammatical tradition (*vyākaraṇa*) inaugurated with Pāṇini (400 BCE) and developed with the commentators of his *Aṣṭādhyāyī*, such as Kātyāyaṇa and Patañjali (250–100 BCE) and Bhartṛhari (fourth/fifth century CE), has equally influenced philosophical approaches to the question discussed in the present introduction. A good starting point, as suggested in a series of studies by George Cardona (1974, 2014), is the analysis of 'the direct participant in the accomplishment of action' (2014: 85), or *kāraka* (from A

21 *bhāvārthāḥ karmaśabdāstebhyaḥ kriyā pratīyetaiṣa hyartho vidhīyate* ||

22 Cited in Freschi 2014: 72. See Verpoorten (1987: 34) for a summary of discrepancies between Bhaṭṭās and Prābhākaras.

23 Moving from an objection on MS I.1.5d (*tat pramāṇam, anapekṣatvāt*) in ŚBh ('We perceive it [*ātman*] with the help of breath [*prāṇa*],' tr. Verpoorten), Prābhākara and Śālikanātha investigate whether *prāṇa* can be an indicator of *ātman*. Due to it being non-coeval with the body, vital breath is eventually not regarded as 'the logical mark of the presence of *ātman*' but is identified as one of its proprieties (Verpoorten 2012: 279). See also Biardeau 1968.

24 See: Śabara on MS I.1.5; ŚV *ātmavāda* 125, 135.

25 *tasmān na vijñānaṃ pratyākhyāya kasyacid rūpaṃ nidarśayituṃ śakyam* | ŚBh (vol. I: 85) on MS I.1.5, tr. Verpoorten (2012: 286).

I.4.23). We will focus specifically on A I.4.54 (*svatantraḥ karttā*) and A I.4.55 (*tatprayojako hetuś ca*):

> The independent [*kāraka*] is called *kartṛ* (agent) | 54
> Its instigator, the *hetu*, [is called *kartṛ*] too | 55

Kārakas should not be confused with grammatical subjects. A *kāraka* is always expressed in relation to action (*kriyā*) which can be active (the semantic role of the 'subject' is expressed through the nominative case whereas that of the 'object' is expressed through the accusative) or passive (the semantic role of the 'subject' is expressed through the instrumental case and that of the 'object' through the nominative). Irrespective of their status of animate/inanimate or sentient/insentient, all *kārakas* are featured by agency (*kartṛtva*) although only the *kartṛ* is the lead (= independent) agent whose action permits other *kārakas* to 'exist' (Cardona 1974: 239). The *kartṛ* is seen as a typology of *kāraka*[26] whose independent (*svatantra*) action permits animate beings and inanimate objects to be/come agents. These however are all signified by their dependent (*paratantrā*) function with respect to the *kartṛ*. While this will lead to major theological conclusions,[27] it is noteworthy that as early as Bhartṛhari, knowledge of grammar is looked at as the way to achieve *mokṣa*, or absolute knowledge. Since reality is Veda and the only means to understand Veda is the study of its language, words are seen as composite entities. They are more real than parts of words (roots, stems, etc.) which in turn are paralleled to the constituents of reality (Bronkhorst 1995).

Another relevant point is the concept of desire (*icchā*) of the subject in relation to a most desired (*iṣṭatamaṃ*) object.[28] This rule has a series of implications. Among these it seems relevant here to note that only animate/sentient beings do wish. Later commentators discuss whether this Pāṇinian rule is valid, since there are sentences where the agent appears to be inanimate. For Kātyāyana, when an agent is inanimate/insentient, the sentence must be a simile (*upamānādvā siddham*; *vārttika* 14 in Mbhā

26 The *kārakas*, amongst other things, permit the formulation of cases and other morphological elements. Pāṇini lists six *kārakas*: *apādāna* ('ablative', A I.4.24–31), *sampradāna* ('dative', A I.4.32–37, 39–41, 44), *karman* ('accusative', A I.4.38, 43, 46–53), *karaṇa* ('instrumental', A I.4.42), *adhikaraṇa* ('locative', A I.4.45) and *kartṛ* (A I.4.54–55).

27 See below, Utpaladeva and Kashmir Śaivism.

28 A I.4.49 (*karturīpsitataṃ karma*); A III.1.7 (*dhatoḥ karmaṇaḥ samānakarttṛkādic-chāyāṃ vā*).

vol. II: 14). Alternatively, he notes that everything that acts is intelligent and animate (*sarvasya vā cetanāvattvāt*; *vārttika* 15, ibid.: 15). Patañjali objects to Kātyāyana's *vārttika* 12 (*āśaṅkāyām acetaneṣu upasaṅkhyānam*) and is categorical: animate beings only are able to wish.[29] Therefore a sentence like '*aśmā* [stone] *luluṭhiṣate* [third pers. sing. desiderative of √*luṭh* (to roll)]' should be intended as '*I think/believe/infer that* the stone is about to roll' (Mbhā on A III.1.7.2). Bhartṛhari observes that the typology of a *kāraka* is not defined by its potential to participate in action (*sādhana*). In fact, its existence (*sat*) or non-existence (*asat*) depends upon reason (*buddhi*), or the way in which it is thought and spoken of (VkP III.7.3; cf. III.7.90).

This permits us to trace a parallel with Madhyamaka Buddhism (from c. 150 CE), a school that probably influenced Bhartṛhari. The highest reality, which Bhartṛhari calls *brahman*, is divided by means of language into the phenomenal world. As claimed by the Madhyamaks, the manifest world is not real. It is thus of no avail to argue what is animate and what is inanimate. Rather it is of the utmost importance to know the rules that govern language since such knowledge enables one to discern the real from the unreal, and therefore to achieve liberation (Nakamura 1972).

Nyāya-Vaiśeṣika and Mīmāṃsā hold that an absolute difference (*atyanta bheda*) exists between material cause (*upādāna kāraṇa*) and effect (*hetu*). This is of great relevance to the present discussion as it helps us to ascertain that early Brahmanical systems, just like grammarians, were not primarily concerned with a rigid classification of the manifest world in living/sentient creatures (humans, animals and plants) and inanimate/insentient matter (rocks, stones, earth and metals).

The Sāṃkhya system moves from entirely contrasting premises. Sāṃkhyas reduce reality to two co-fundamental principles, *puruṣa* (self; = *ātman*) and *prakṛti* (material nature; = *pradhāna*). The latter is constituted by three *guṇas* (qualities, strands), viz. *sattva*, *rajas* and *tamas* (SK 12–13; cf. MBh XII.194). The *guṇas* become bound to the inactive *puruṣa*, whom they continuously transform (*parināma*). It is by means of their combination (*saṃghāta*) (SK 21) that the hierarchy of elements – from the most subtle to the grossest one[30] – becomes manifest in the world (SK 52–54). The preoccupation of Sāṃkhya, however, is not just to demonstrate the existence of the self (*puruṣa*) and its relation with nature by telling the animate (human beings, animals, plants) from the inanimate (rocks, minerals, metals).

29 *cetanāvataḥ etat bhavati icchā iti* | (MBhā vol. II: 14)

30 On the qualities of matter, cf. Bhartṛhari's *Mahābhāṣyadīpika* I.23.21–23 (in Bronkhorst 1994: 309).

Followers of Sāṃkhya develop a system in which beings and objects are classified with respect to their capacities to *either* obtain *or* help in obtaining valid (liberating) knowledge, i.e. awareness of the *difference* (*vivkekakhyāti*) between *puruṣa* and *prakṛti*.

Nature, *prakṛti*, is the substantial source of manifestation (*upādānakāraṇa*) of the universe. Although between *puruṣa* and *prakṛti* there is both 'logical absence of identity' and 'material absence of relation' (Larson 1980: 309), nature in all its forms – animate/inanimate, sentient/insentient – is experienced by the 'I' (*ahaṃkāra*) through intentional consciousness (*buddhi*) in a process that involves the five *buddhīndriyas* (sight, hearing, smelling, tasting and touching), the five *karmendriya* (speaking, grasping, walking, excreting and sexual pleasure), the five *tanmātras* (sound, tact, form, flavor and odor) and the five *bhūtas* (earth, water, fire, air and space).[31] The manifest world (both animate and inanimate) is believed to be present in a subtle form in *mūlaprakṛti*, the uncreated (*avikṛti*) and fundamental source of productivity (SK 3). Its essence is thus non-different (*abheda*) from its causal conditions. Knowledge and awareness that the self is not part of it produces emancipation (*kaivalya*) from causality, or *mokṣa*. It so appears that the distinction between the animate and the inanimate is worth investigating only insofar it is inscribed in the study of the distinction between existent and non-existent. See how this is elaborated in chapter II of Vijñānabhikṣu's *Sāṃkhyasārā* (second half of the sixteenth century):

> [m]ateriality can be eschewed with the help of the knowledge of the fundamental principles. This knowledge consists in recognizing the difference between animate (*cetana*) and the inanimate (*acetana*) entities, beginning with the unmanifest materiality and ending with the five gross elements. (Larson and Bhattacharya 1987: 402)
>
> [...] [t]his is why all inanimate entities are called nonexistent *from the transcendental standpoint*. (ibid.: 406; emphasis added)

The position of Sāṃkhya confirms that philosophical investigation in early India was not primarily concerned with categories such as animate and inanimate. This is not so in Ājīvikism and Jainism.

31 Cf. MBh XII.184.1–2 and XII.194.6. Cf. Also VisM XX.73, where the fifth century Theravāda commentator Buddhaghosa explains that minerals, gems, stones and the soil (but also grass, trees and creepers) are insentient in that they are devoid of sense faculties (P. *anindriyabaddha*).

Since both traditions promote a soteriological system grounded on karmic retribution, it becomes imperative not to cause harm to other living beings. In relation to abstinence from violence, the *Āyāraṁgasutta* informs us that the faculty of knowing the doctrine, and therefore to observe all appropriate vows, is proper for living beings of whom *ātman* (Pkt. *atta*) is a feature:

> The Self is the knower (or experiencer), and the knower is the Self. That through which one knows, is the Self. (Āyār i.5.5.171; tr. Jacobi)

This and similar reflections contribute to the development of a distinctive Jain epistemology, namely *anekāntavāda*, the doctrine of complexity of reality. According to this method, anything can be viewed from multiple and equally valid perspectives, yet only a Jina is able to grasp the all-encompassing truth of that particular aspect of reality. Traditionally this important epistemological advancement has been attributed to Mahāvīra Vardhamāna, the last Jina. Yet as recently detailed by Balcerowicz (2015: 186–204), in the commentary (*cūrṇi*) to *Nandīsūtra* (Pkt. *Naṁdīsutta*) of Jinadāsagaṇi Mahattara (seventh century)[32] and in the *vṛtti* of Haribhadra (eighth century), a tripartite computational system (*parikarma*) which anticipates some aspects of the seven forms of modal judgment (*syādvāda* or *saptabhaṅgī*; see below) is attributed to the Trairāśikas, a denomination, or an order, of the Ājīvikas. According to this method, the world is made of: (1) living elements (*jīva*), (2) lifeless elements (*ajīva*) and (3) living elements with lifeless elements (*jīvājīva*).[33] The ability to discern between the three permits the ascetic to observe one of the five great vows (Pkt. *paṁcamahavvaya*; Skt. *pañcamahāvrata*; Āyār ii.15.1–5) and a distinctive feature of Jainism, namely *ahiṃsā* (non-injury). Consequently, one is able to progress on the path towards liberation from rebirth.

Since souls (*jīva*) can be embodied in stationary (*sthāvara*) one-sensed (earth-, water-, fire-, wind-, and tree-)forms of life as well as two-sensed, three-sensed, four-sensed and fifth-sensed moving (*trasa*) forms (humans, animals, gods, etc.),[34] to tell the animate from the inanimate is important to Jain monks. With time, the tripartite model illustrated above would be

32 See Flügel (2012: 127n41).

33 Other examples include: (1) the world (*loka*); (2) non-world (*aloka*); and (3) the world with non-world (*lokāloka*). See also: (1) the existent (*sat*); (2) the non-existent (*asat*); and (3) the existent with the non-existent (*sad-asat*). *Nandīsūtracūrṇi* 105 and *Nandīsūtravṛtti* 107, in Balcerowicz (2015: 186–89).

34 See Āyār i.1.2–6. See also DrS 11.

expanded in the method of the seven forms of modal judgment (*saptabhaṅgī*) introduced by the particle *syāt* (*syādvāda*):

1. in a certain sense something is (*syādasti*),
2. in a certain sense something is not (*syānnasti*),
3. in a certain sense something is and is not (*syādastinasti*),
4. in a certain sense something is inexpressible (*syādavaktavyam*),
5. in a certain sense something is and yet it is inexpressible (*syādastyavaktavyam*),
6. in a certain sense something is not and it is inexpressible (*syādnastyavaktavyam*),
7. in a certain sense something is and is not, and it is also inexpressible (*syādastinastyavaktavyam*). (NA 30)[35]

This method would be firmly opposed by Brahmanical systems. Vedāntins, for instance, would agree with Jains that forms of perception, cognition and knowledge are attributes of the self/soul, which in turn is wanting in inanimate things (*jaḍa*).[36] Yet Advaita Vedānta cannot accept *anekāntavāda*. In particular, Śaṅkara (between eight and ninth centuries) objects to the *saptabhaṅgīs* in BS II.2:33–36. Moving from a strict non-dual stance, he interprets *syāt* as 'maybe' (rather than with the acceptation given by Jains: *kathaṃcit* = in some way/in a certain sense) and denounces *syādvāda* as self-contradictory possibilism. This appears clearly when one claims that an object is:

1. *maybe* animate
2. *maybe* inanimate
3. *maybe* animate and inanimate
4. *maybe* inexpressible
5. *maybe* animate and inexpressible
6. *maybe* inanimate and inexpressible
7. *maybe* animate, inanimate and inexpressible.

35 This frame begins to appear since the end of the fifth century but it will take few centuries more to fully develop (Balcerowicz 2011: 16).

36 In one of the oldest doxographic documents, Haribhadra says of Jainism: 'Since existence is featured by birth (*utpāda*), decay (*vyaya*) and permanence (*dhrauvya*), an entity (*vastu*) is said to have infinite qualities (*anantadharmakaṃ*) and is the field of conception (*mānagocaraḥ*).' (SDS IV.57) Ācārya Nemicandra (tenth century) defines so a living soul: '*Jīva* is characterized by consciousness (*cetanā*) that is concomitant with *upayoga* – perception (*darśana*) and knowledge (*jñāna*), is incorporeal (*amūrta*), a causal agent (*kartā*), coextensive with the body, enjoyer of the fruits of karmas (*bhoktā*), having the world as its abode, emancipated (*Siddha*), and of the nature of darting upwards.' (DrS 2, tr. Jain).

The debate changes radically with the penetration of devotional and theological elements. Within the Viśiṣṭādvaita Vedānta tradition, Rāmānuja (eleventh century) observes that both *cit* (consciousness) and *acit* (non-consciousness) are different but at the same time they are qualities of God:

> Brahman is in the causal state, when its body consists of the individual selves and physical nature, in their subtle condition not distinguishable by differentiations of name and form. The passage of the world to this phase of existence is what is termed 'dissolution'. Brahman, having as its body, the individual selves and nature, in their gross manifested condition distinguished by differentiations of names and forms is in the state of the effect. The assumption of this manifestation and grossness of aspect is described as 'creation'. (VAS 93, tr. Raghavachar)

In the case of the Kashmir Tantric tradition, which incorporates both Vedāntic stances and elements from Bhartṛhari's philosophy of language, it is argued that just like the lack of a *kartṛ* in a sentence invalidates all other *kārakas*, so Īśvara is the supreme agent whose mere presence make it possible for everything – animate and inanimate – to exist. So Utpaladeva (tenth century) comments:

> *Kārika*: Even if the unity of consciousness is maintained to be the only ultimate reality, there cannot be action, for two entities divided as regards the nature of their manifestation (*ābhāsabhinnayoḥ*), without a preliminary act of thought which grasps and establishes the unity (*ekatvaparāmarśaṃ*), characterized by the desire to act. (IPK II.4.20)
> *Vṛtti*: [...] an insentient reality cannot even be the agent of the action of being – 'it exists, is' – since it does not possess the freedom that is manifested through 'wanting to be' (*bubhūṣāyogena*). Thus the ultimate truth in this regard is that the knowing subject, and he alone, 'causes' the insentient reality 'to be' (*bhāvayati*), or, in other words, appears in various forms such as mount Himācala and so on. (tr. Torella)

At an ontological level, as noted by Bronkhorst (2011: 69), 'even in the case of the verb "to exist, to be," the subject, or the agent of the action in question, is God, the supreme "knower."' See again Utpaladeva:

> *Kārika:* Indeed, the foundation of insentient realities rests on the living being; knowledge and action are considered the life of the living being. (IPK I.1.4)
> *Vṛtti:* There are two kinds of reality: insentient and sentient. The establishment of an insentient nature rests on the living being; the being such of the living, i.e. life, is represented precisely by knowledge and action. (tr. Torella)

Devotional and ritualistic culture is an extremely successful aspect of religious culture from late Vedic times. Icon worship is core to the most widespread religious modes of the subcontinent, namely Tantrism and *bhakti* (cf. RgV III.31). These however have been the object of criticism from a range of philosophical traditions which find it difficult to reconcile knowledge and liberation with the practice of worshiping inanimate matter. The Jain polymath Hemacandra (eleventh century) presented these ideas in the mouth of sympathizers of materialism (Cārvākas?) so as to defend the doctrine of karmic retribution:

> What merit has been acquired by one stone that it is worshipped by bathing, ointment, wreaths, clothes, and ornaments? What evil has been acquired by another stone that it is polluted? (TŚPC, vol. I: 37; tr. Johnson)

Despite criticism, man-made icons such as *maṇḍalas*, *yantras* and images (*mūrtis*) made of stone, clay, metals, et cetera are regularly worshiped across the subcontinent and this worship (*pūjā*) is the basis of daily religious life in a number of traditions. The object of worship may be considered a symbol of a deity but more often it is the deity ipso facto. This, however, is possible only after the *mūrti* is infused with life (= *prāṇa*, vital breath) during the *prāṇapratiṣṭhā* ritual.[37] It can be thus inferred that the notion of breath (*prāṇa*) as an indicator of life and sentience is a ubiquitous one. Further ideas, such as the ways in which life can be infused, the potency of certain matter and the practical applications of these ideas in a variety of traditional settings, are discussed in the chapters of this volume.

The first section specifically investigates myths and rituals in which minerals and metals, otherwise insentient and inanimate, become seats of power. In Chapter 1 Mikael Aktor examines the worship of Śiva, Viṣṇu, Sūrya, Gaṇeśa and Devī in the context of the *pañcāyatanapūjā*. The deities are worshiped not in their anthropomorphic forms but in their aniconic forms as five stones, each originating from a particular location. The chapter examines individual stones and their mythological associations with their respective deities, their collection, manufacture and distribution, as well as their contemporary ritual use. The next contribution is a historical and ethnographic study of the westernmost among the Śākta-*pīṭhas*, that of goddess Hiṅgulā ('She of cinnabar') in Pakistan's Baluchistan province. Francesco Brighenti surveys the ritual use of cinnabar in South Asian traditions, with particular reference to alchemy and Tantric culture. The chapter is a first attempt at discussing the polarity of sulfur versus mercury, the

37 See Bühnemann (1988: 191–95) for details about *prāṇapratiṣṭhā*.

symbolism of cinnabar, and the worship of volcanic and gas-jet phenomena against the background of the cult of Hiṅgulā. In the third chapter, Monia Marchetto and Manuel Hoefer investigate astronomical and mythological aspects of Śani (Saturn) through textual sources and popular beliefs. The chapter particularly investigates the special relation of Śani with a mineral, iron, and a precious stone, blue sapphire. Both elements are discussed against the context of myth, ritual and the sacred geography of India.

The next section deals with science and health. Anthony Cerulli and Caterina Guenzi explore the science of 'external' remedies in India. Drawing on textual references and ethnographic case-studies, the chapter relates how gemstones (e.g. rubies, pearls and sapphires) and amulets (assembled from materials such as devil tree, wild asparagus, seashells and other substances) have been used in preventative therapies as part of a person's daily regimen and as specific remedies for disorders during pregnancy, planetary-based afflictions and madness, among other things. The preparation, testing and use of minerals and amulets as means to treat the body and mind reveal a worldview in which the human person is entwined within multiple and intersecting ecosystems. In Chapter 5, Dagmar Wujastyk examines uses of mercury in Sanskrit medical treatises from about the seventh century, when they are first mentioned in *Aṣṭāṅgahṛdayasaṃhitā*. Mercury preparations were also widely used in rejuvenation therapy (*rasāyana*) – one of the eight classical areas of Indian medicine. From about the eleventh century, mercury became a significant ingredient in *rasāyana* therapy with works such as Cakrapāṇidatta's *Cakradatta* (eleventh century) and Vaṅgasena's *Cikitsāsārasaṃgraha* (eleventh/twelfth century) recording multiple recipes for mercurial tonics. The use of mercury in *rasāyana* recipes reflects a profound change in Indian medicine during this period, namely the increasing importance of iatrochemistry. The following chapter focuses on mercury too, but draws attention to Tibetan sources. Barbara Gerke analyses Tibetan classifications and sources of *ngülchu* (*dngul chu*), 'silver-water,' often translated as mercury, which in its processed form is described as the 'king of essences' in Tibetan pharmacopeias since at least the twelfth century CE. Based on textual analysis and ethnographic fieldwork among contemporary Tibetan pharmacologists in India and Nepal, the chapter argues that due to the varied sourcing across history, *ngülchu* cannot simply be equated with metallic mercury, as certain shiny-silvery substances procured from leaves, animals and other objects are also labelled *ngülchu,* share similar characteristics, and go beyond what popular 'Western' science would classify as a 'metal.'

The next section, titled Power and Devotion, opens with an ethnographic study of female *sādhus* in contemporary Rajasthan. Antoinette DeNapoli investigates gendered representations of metals, minerals and gems, and the ways that their practices shape and reconfigure the more standardized parameters for defining what *saṁnyāsa* is all about in contemporary India. The chapter first analyzes the 'rhetoric of renunciation,' i.e. the stories and songs about the earth, its properties and humans' responsibility to the planet, including ideas about ecological sustainability. It then proceeds to examine the *sādhus*' use and classification of metals, minerals and gems, the deities associated with these substances, the problems they are thought to cure and/or prevent, and personal experiences of illness. The following chapter explores metal-related traditions in Sikhism. Eleanor Nesbitt discusses the imagery of *Gurū Granth Sāhib* (or *Ādi Granth*) on metals, minting and gemstones, as well as the equation of the Guru with the philosopher's stone. Khālsā traditions are then investigated. This includes an analysis of the evolution of the five Ks and of the *niśān sahib*, the flagpole and pennant bearing the Khālsā's insignia. Finally, the chapter elucidates Sikhs' emphasis on iron in the context of Gurū Gobind Siṅgh's epithets for Akāl (Timeless One; God) as *sarab loh* (all iron; pure iron) and Sikh interpretations of the invocation of Bhagautī in the congregational prayer (*ardās*) as a prayer to the sword rather than to Bhagavatī (the Hindu goddess).

The last section of the volume revolves around concepts of body and embodiment. In the first chapter, Deeksha Sivakumar analyzes two conversations from *Rāmāyaṇa* and *Mahābhārata*. The former is between the wife of *ṛṣi* Atri, the ascetic Anasūyā, and young Sītā, after her marriage to Rāma. The latter is an exchange between the wife of the five Pāṇḍava brothers, Draupadī, and Satyabhāmā, Kṛṣṇa's favorite wife, at a time when the former's exile in the forest was about to finish. On both occasions, it is argued, cosmetics and makeup techniques – always divulged in confidence – help a woman to acquire and keep her partner, or to ensure her husband's affection and obedience. By rereading the traditional concept of *pativratā*, cosmetic knowledge demonstrates the conscious participation of women in actively controlling their social and marital life. In Chapter 10, Mattia Salvini engages with the concept and the symbolism of *ratna* (jewel) in Buddhist literature and culture. By definition a Buddhist is 'one who takes refuge in the Three Jewels.' A successful Buddhist practitioner may accumulate merit to reach Indra's divine realm inlaid with four precious gems; he or she might reach 'the meditative absorption which is like a diamond' to become awakened. If on the other hand one focuses on

Amitābha's name, this will bring about rebirth in a pleasant realm where the landscape is made of precious substances and gems. Those of greater capacity may attempt Buddhahood in one lifetime, in which case they will have to rely on the swifter 'Diamond Vehicle.' From the name of textual collections like the 'Heap of Jewels' to the contemporary Burmese and Thai Buddhist practice of covering statues with layers of gold-leaf, hardly any aspect of Buddhist religious life is untouched by imagined or actual gems and precious metals. Moving from these and other examples, the chapter intends to show the importance of a certain part of the mineral world for the larger universe of Buddhist religious culture. The last chapter explores the notion of earth-beings as proposed by the Jain tradition. Ana Bajželj discusses textual sources on the attribute of consciousness (*caitanya*) in living beings. Earth-beings represent one kind of possible embodiment in which living substances are bound by bodies that have the nature of earth. The interaction between living substances and non-living (*ajīva*) and non-conscious (*acetana*) earth-matter, as well as the distinction between earth utilized as a body and earth without a current embodying function, is discussed in Jain literature. Jainism does not put forward a notion of earth as such but instead refers to a multitude of earth-beings such as raw soil, particles of dust, sand, raw minerals, pebbles, sand, salt, iron, copper, lead, silver, gold, diamonds, et cetera. The chapter looks at various accounts of these subcategories, their status within the broader ontological-cosmological doctrine, the character of their interactions with other living beings, the karmic contexts within which embodiments in the form of earth-beings are possible, the lifespan of these embodiments and their possible future rebirths. Additionally, the chapter points out and critically examines the implications of these doctrines in the practical life of Jain laity and mendicants.

The editors auspicate that this volume, along with the preceding two, will serve as a useful contribution to the study of nature and environment in South Asian traditions. We are aware much has been left out. Some traditions have been minimally addressed, whereas others, like Indian materialism, and non-Indic religions (Zoroastrianism, Judaism, Christianity, the Bahá'í faith) are not included at all. A further serious limitation of the present series is that the contributions do not cover all regions of South Asia. Despite our best efforts, we were not able to include enough contributions relating to Pakistan, Bangladesh, Nepal, Sri Lanka, Bhutan and the Maldives. This calls for reflection, and indicates the necessity to promote more actively research on the history of religions in a variety of South Asian regional contexts beyond India.

ABBREVIATIONS AND REFERENCES

P. = Pali   Pkt. = Prakrit   Skt. = Sanskrit

A	Pāṇini: *Aṣṭādhyāyī*. Sharma, R.N. (tr.) (2002). *The Aṣṭadhyāyī of Pāṇini.* 6 vols. Second revised and enlarged edition with index of sutras (translated and explained). Delhi: Munshiram Manoharlal.
AiĀ	*Aitareya Āraṇyaka*. Keith, A.B. (tr.) (1909). *Aitareya Āraṇyaka*. London: Oxford University Press.
AV^{P}	*Atharvaveda*, Paippalāda recension. Bhattacharya, D. (ed.) (1997). *The Paippalāda-Saṃhitā of the Atharva-veda, critically edited from palmleaf manuscripts in the Oriya script discovered by Durgamohan Bhattacharyya and one śāradā manuscript*. Vol. 1, consisting of the first fifteen *kāṇḍas*. Calcutta: The Asiatic Society.
AV^{Ś}	*Atharvaveda*, Śaunaka recension. (1) GRETIL e-text based on the ed.: *Gli inni dell' Atharvaveda (Saunaka)*, trasliterazione a cura di Chatia Orlandi, Pisa 1991, collated with the ed. R. Roth and W.D. Whitney: *Atharva Veda Sanhita*, Berlin 1856: <http://gretil.sub.uni-goettingen.de/gretil/1_sanskr/1_veda/1_sam/avs_acu.htm> (accessed 15 October 2015). Input by: Vladimir Petr and Petr Vavrousek. (2) Whitney, W.D. (tr.) (1905). *Atharva-Veda Saṃhitā*. Harvard Oriental Series, vol. VII. Cambridge, Mass.: Harvard University Press.
Āyār	*Āyāraṁgasutta* (Pkt.) = *Ācārāṅgasūtra* (Skt.). Jacobi, H. (tr.) (1884). *Gaina Sûtras, Part I. Âcârâṅga Sûtra and Kalpa Sûtra*, pp. 1–213. Oxford: Clarendon Press.
Balcerowicz 2010	Balcerowicz, P. (2010). 'What exists for the Vaiśeṣika?' In: Balcerowicz, P. (ed.), *Logic and Belief in Indian Philosophy*, pp. 249–360. Warsaw Indological Series Vol. 3. Delhi: Motilal Banarsidass.
Balcerowicz 2011	Balcerowicz, P. (2011). 'Dharmakīrti's Criticism of the Jaina Doctrine of Multiplexity of Reality (*Anekāntavāda*).' In Krasser, H., H. Lasic, E. Franco and B. Kellner (eds), *Religion and Logic in Buddhist Philosophical Analysis. Proceedings of the Fourth International Dharmakīrti Conference Vienna, August 23–27, 2005*, pp. 1–32. Wien: Verlag der Österreichischen Akademie der Wissenschaften.
Balcerowicz 2015	Balcerowicz, P. (2015). *Early Asceticism in India. Ājīvikism and Jainism*. London: Routledge.
BĀU	*Bṛhadāranyaka Upaniṣad*. In Olivelle (1998: 29–165).
Bhaduri 1947	Bhaduri, S. (1947). *Studies in Nyāya-Vaiśeṣika Metaphysics*. Poona: Bhandarkar Oriental Research Institute.
Biardeau 1968	Biardeau, M. (1968). 'L'ātman dans le Commentaire de Sabarasvāmin.' *Mélanges d'Indianisme à la mémoire de L. Renou*, pp. 109–25. Publications de l'Institut de Civilisation Indienne, Série in-8°, fasc. 28. Paris: E. de Boccard.

Bodewitz 1986 — Bodewitz, H.W. (1986). '*Prāṇa*, *Apāna* and Other *Prāṇa*-s in Vedic Literature.' *The Adyar Library Bulletin*, Golden Jubilee Volume, pp. 326–48.

Bronkhorst 1994 — Bronkhorst, J. (1994). 'The Qualities of Sāṅkhya.' *Wiener Zeitschrift für die Kunde Südasiens und Archiv für Indische Philosophie*, xxxviii: 309–22.

Bronkhorst 1995 — Bronkhorst, J. (1995). 'Grammar as the Door to Liberation.' *Annals of the Bhandarkar Oriental Research Institute*, 76(1/4): 97–106.

Bronkhorst 2011 — Bronkhorst, J. (2011). *Language and Reality. On an Episode in Indian Thought*. Leiden: Brill.

BS — Vārahamihira: *Bṛhatsaṃhitā* (Version 4.3, May 8, 1998). GRETIL e-text based on the edition of A.V. Tripathi (Sarasvati Bhavan Granthamala Edition) with reference to H. Kern's text and his translation, <http://gretil.sub.uni-goettingen.de/gretil/1_sanskr/6_sastra/8_jyot/brhats_u.htm> (accessed 15 February 2016). Input by: Michio Yano and Mizue Sugita.

Bühnemann 1988 — Bühnemann, G. (1988). *Pūjā. A Study in Smārta Ritual*. Vienna: Institut für Indologie der Universität Wien, Sammlung De Nobili.

Caland 1990 — Caland, W. (1990). *Kleine Schriften*. Ed. M. Witzel. Wiesbaden: Otto Harrassowitz.

Cardona 1974 — Cardona, G. (1974). 'Pāṇini's *Kārakas*: Agency, Animation and Identity: *śabda-pramāṇakā vayam/yac chabda āha tad asmākaṃ pramāṇam* Mahābhāṣya paspaśā (i.11.1–2), ad 2.1.1 (i.366.12–13).' *Journal of Indian Philosophy*, 2(3/4): 213–306.

Cardona 2014 — Cardona, G. (2014). 'Pāṇinian Grammarians on Agency and Independence.' In Dasti, M.R. and E.F. Bryant (eds.), *Free Will, Agency, and Selfhood in Indian Philosophy*, pp. 85–111. New York: Oxford University Press.

CaS — *Carakasaṃhitā*. Sharma, R.K. and B. Dash (tr.) (2011). *Carakasaṃhitā. Text with English Translation and Critical Exposition Based on Cakrapāṇi Datta's Āyurveda Dīpikā*. 7 vols. Reprint 1977 edition. Varanasi: Chowkhamba Sanskrit Series Office.

Chakrabarti 1992 — Chakrabarti, D.K. (1992). *The Early Use of Iron in India*. New Delhi: Oxford University Press.

ChU — *Chāndogya Upaniṣad*. In Olivelle (1998: 166–287).

Comba 2001 — Comba, A. (2001). 'Carakasaṃhitā, Śārīrasthānai and Vaiśeṣika Philosophy. In Meulenbeld, G.J. and D. Wujastyk (eds.), *Studies on Indian Medical History*, pp. 39–56. Delhi: Motilal Banarsidass.

DrS — Nemicandra: *Dravyasaṁgraha*. Jain, V.K. (ed., tr.) (2012). *Ācārya Nemichandra's Dravyasaṃgraha With Authentic Explanatory Notes*. Dehradun: Vikalp Printers.

Ferrari and Dähnhardt 2013 — Ferrari, F.M. and T.W.P. Dähnhardt (eds.) (2013). *Charming Beauties and Frightful Beasts. Non-Human Animals in South Asian Myth, Ritual and Folklore*. London: Equinox.

Ferrari and Dähnhardt 2016 — Ferrari, F.M. and T.W.P. Dähnhardt (eds.) (2016). *Roots of Wisdom, Branches of Devotion. Plant Life in South Asian Traditions*. London: Equinox.

Flügel 2012	Flügel, P. (2012). 'Sacred Matter: Reflections on the Relationship of Karmic and Natural Causality in Jaina Philosophy.' *Journal of Indian Philosophy*, 40(2): 119–76.
Freschi 2013	Freschi, E. (2013). 'Did Mīmāṃsā Authors Formulate a Theory of Action?' In Mirnig, N., P.D. Szántó and M. Williams. (eds.), *Puṣpikā: Tracing Ancient India Through Texts and Traditions. Contributions to Current Research in Indology*. Vol. I, pp. 151–72. Oxford: Oxbow Books Press.
Freschi 2014	Freschi, E. (2014). 'Does the Subject Have Desires? The Prābhākara Mīmāṃsā Answer.' In Ciotti, G., A. Gornell and P. Visigalli (eds.), *Puṣpikā 2: Tracing Ancient India Through Texts and Traditions. Contributions to Current Research in Indology*, pp. 55–86. Oxford: Oxbow Books.
Gonda 1987	Gonda, J. (1987). *Rice and Barley Offerings in the Veda*. Leiden: E.J. Brill.
IPK	Utpaladeva: *Īśvarapratyabijñākārikā*. Torella, R. (ed., tr.) (1994). *The Īśvarapratyabijñākārikā of Utpaladeva with the Author's Vṛtti*. Serie Orientale Roma LXXI. Roma: Istituto Italiano per il Medio ed Estremo Oriente.
Kanō 2001	Kanō, K. (2001). '*Pariśeṣa, Prasaṅga*, and *Kevalavyatirekin* – The Logical Structure of the Proof of *Ātman*.' *Journal of Indian Philosophy*, 29(4): 405–22.
Larson 1980	Larson, G.J. (1980). 'Karma as a "Sociology of Knowledge" or "Social Psychology of Process/Praxis."' In Doniger, W. (ed.), *Karma and Rebirth in Classical Indian Traditions*, pp. 303–16. Berkeley: University of California Press.
Larson and Bhattacharya 1987	Larson, G.J. and J.Sh. Bhattacharya (eds.) (1987). *Encyclopedia of Indian Philosophy. Vol. 4: Sāṃkhya. A Dualist Tradition in Indian Philosophy*. Delhi: Motilal Banarsidass.
MaiU	*Maitrī (Maitrāyaṇī) Upaniṣad*. Van Buitenen, J.A.B. (1962). *The Maitrāyaṇīya Upaniṣad. A Critical Essay, with Text, Translation and Commentary*. The Hague: Mouton.
Manu	*Manusmṛti* or *Mānavadharmaśāstra*. Olivelle, P. (ed. and tr.) (2005). *Manu's Code of Law. A Critical Edition and Translation of the Mānava-Dharmaśāstra*. New York: Oxford University Press.
MBh	Vyāsa: *Mahābhārata*. Sukthankar, V.S. et al. (eds.) (1933–66). *The Mahābhārata for the First Time Critically Edited*. 19 vols. Poona: Bhandarkar Oriental Research Institute.
Mbhā	Patañjali: *Mahābhāṣyā*. Kielhorn, F. (ed.) (1962–1972). *The Vyākaraṇa-mahābhāṣyā of Patañjali*. Revised by K.V. Abhyankar. 3 vols., third edition. Poona: Bhandarkar Oriental Research Institute.
Meulenbeld 2002	Meulenbeld, G.J. (2002) *A History of Indian Medical Literature*. 5 vols. Leiden: E.J. Brill.
Mishra 1936	Mishra, U. (1936). *Conception of Matter according to Nyāya-Vaiçeṣika*. With a foreword by Ganganatha Jha and an introduction by Gopinath Kaviraj. Allahabad: Law Journal Press.

MS	Jaimini: *Mīmāṃsāsūtra*. Sandal, M.L. (tr.) (1923). *The Mîmâmsâ Sûtras of Jaiminî*. 2 vols. Allahabad: S.N. Basu at the Panini Office.
NA	Siddhasena Divākara: *Nyāyāvatāra*. Balcerowicz, P. (ed., tr.) (2001). *Jaina Epistemology in Historical and Comparative Perspective. Critical Edition and English Translation of Logical-Epistemological Treatises: Nyāyâvatāra, Nyāyâvatāra-vivṛti and Nyāyâvatāra-ṭippana with Introduction and Notes*. Stuttgart: Franz Steiner Verlag.
Nakamura 1972	Nakamura, H. (1972). 'Bhartṛhari and Buddhism.' *Journal of the Ganganatha Jha Kendriya Sanskrit Vidyapeetha*, 28: 395–405.
NS	Gotama: *Nyāyasūtra*. Vidyâbhuṣana, S.C. (tr.) (1913). *The Nyâya Sutras of Gotama*. Allahabad: The Pâṇini Office.
NV	Uddyotakara: *Nyāyavārttika*. In Thakur, A. (ed.) (1967). *Nyāyadarśana, with Vātsyāyana's Bhāṣya, Uddyotakara's Vārttika, Vācaspati Miśra's Tātparyaṭīkā and Udayana's Nyāyavārttika-tātparya-pariśuddhi*. Mithila: Mithila Institute Series 20.
Olivelle 1998	Olivelle, P. (ed. and tr.). (1998). *The Early Upaniṣads. Annotated Text and Translation*. New York: Oxford University Press.
Olivelle 2012	Olivelle, P. (2012). 'Material Culture and Philology: Semantics of Mining in Ancient India.' *Journal of the American Oriental Society*, 132(1): 23–30.
PBh	Praśastapāda: *Padārthadharmasaṁgraha*. Dvivedin, V.P. (ed.) (1895). *The Praśastapādabhāṣya with Commentary Nyāyakandalī of Śrīdhara*. Sri Garbi Dass Oriental Series no. 13. Delhi: Sri Satguru Publications. GRETIL e-text available at: <http://gretil.sub.uni-goettingen.de/gretil/1_sanskr/6_sastra/3_phil/vaisesik/paddhs_u.htm> (accessed 5 November 2015). Input by: Muneo Tokunaga and Yasutaka Muroya.
PrP	Śālikanāthamiśra: *Prakaraṇapañcikā*. Śāstrī, A.S. (ed.). *Prakaraṇapañcika with the Commentary Nyāyasiddhi of Jayapuri Nārāyaṇabhaṭṭa*. Kāśī: Kāśī Viśvavidyālaya.
Rau 1974	Rau, W. (1974). *Metalle und Metallgeräte im vedischen Indien*. Wiesbaden: Steiner.
RgV	Śaunaka: *Ṛgvidhāna*. Bhat, M.S. (ed.) (1987). *Vedic Tantrism. A Study of Ṛgvidhāna of Śaunaka with Text and Translation*. Delhi: Motilal Banarsidass.
RV	*Ṛgvedasaṃhitā*. (1) Thomson, K. and J. Slocum (last updated: Tuesday, 13 May 2014, 12:56). *The Rigveda. Metrically Restored Text*. University of Texas at Austin. E-text available at: <http://www.utexas.edu/cola/centers/lrc/RV/> (accessed 18 October 2015). (2) Jamison, S.W. and J.P. Brereton (trs.) (2014). *The Rigveda: the Earliest Religious Poetry of India*. 3 vols. New York: Oxford University Press.
Śā	*Śārīrasthāṇa*
ŚBh	Śabara: *Śābarabhāṣya*. In: Abhyaṅkar, K.V.Ś. and G.Ś. Jośī (eds.) (1971–1980). *Śrīmajjaiminipraṇītaṁ Mīmāṃsādarśana*. Vols. 1–3.

Ānandāśrama Saṃskṛtagranthāvaliḥ 97. Pune: Ānandāśrama Ānandāśramamudraṇālaya.

SDS Haribhadra: *Ṣaḍdarśanasamuccaya.* Pullè, F.L. (ed.) (1887). 'Shaṭdarçanasamuccayasutraṃ', *Giornale della Società Asiatica Italiana*, 1: 47–73.

SK Īśvarakṛṣṇa: *Sāṃkhyakārikā.* Śarmā, V.P. (ed.) (1970). *Śrīmadīśvarakṛṣṇaviracitā Sāṃkhya-kārikā: Māṭharācāryaviracita-Māṭharavṛtti'-sahitā.* CSS 296. Vārāṇasī: Caukhambā Saṃskṛta Sīrīj Āphis.

SuS *Suśrutasaṃhitā.* Bhishagratna, K. (tr.) (2011). *Suśruta Saṃhitā.* 3 vols. Reprint. Varanasi: Chowkhambha Sanskrit Series Office.

ŚU *Śvetāśvatara Upaniṣad.* In Olivelle (1998: 413–33).

ŚV Kumārilabhaṭṭa: *Ślokavārttika.* Jha, G. (tr.) (1983). *Ślovavārtika. Translated from the Original Sanskrit with Extracts from Commentaries of Sucarita Miśra (the Kāśikā) and Pārthasārathi Miśra (the Nyāyaratnākara).* Reprint. Delhi: Sri Garib Das Oriental Series, No. 8.

TaitU *Taittirīya Upaniṣad.* In: Olivelle (1998: 288–314).

TŚPC Hemacandra: *Triṣaṣṭiśalākāpuruṣacaritra.* In Johnson, H.M. (1931–1962). *The Lives of Sixty-three Illustrious Persons.* 6 vols. Baroda: Oriental Institute.

TV Kumārilabhaṭṭa: *Tantravārttika.* (1) Abhyaṅkar, K.V.Ś. and G.Ś. Jośī (eds.) (1971–1980). *Mīmāṃsādarśana.* Ānandāśrama Saṃskṛtagranthāvaliḥ 97. Pune: Ānandāśramamudraṇālaya. (2) GRETIL e-text based on: Kei Kataoka (2004). *Koten Indo no Saishiki Kôi-ron: Śābarabhāṣya & Tantravārttika ad 2.1.1-4 Genten Kôtei Yakuchû Kenkyû* [The Theory of Ritual Action in Mīmāṃsā: Critical Edition and Annotated Japanese Translation of Śabarabhaṣya & Tantravarttika ad 2.1.1–4.]. Tokyo: Sankibo Press, <http://gretil.sub.uni-goettingen.de/gretil/1_sanskr/6_sastra/3_phil/mimamsa/ktv_2-1u.htm> (accessed 23 October 2015). Input by: Kei Kataoka.

VAS Rāmānuja: *Vedārthasaṁgraha.* (1) Śāstrī, R.M. (ed.) (1864). *Vedārtha Saṁgraha with the Gloss Called Tātparyadipikā by Sudarśanasūri Edited with a Commentary Called Snehapūrtti.* Kāśī: Meḍikal Hal Namakamudrālaya. (2) Raghavachar, S.S. (tr.) (1978). *Vedartha Sangraha of Sri Ramanujacarya, with a Foreword by Swami Adidevananda.* Mysore: Sri Ramakrishna Ashrama.

Verpoorten 1987 Verpoorten, J.-M. (1987). *Mīmāṃsā Literature.* Wiesbaden: Otto Harrassowitz.

Verpoorten 2012 Verpoorten, J.-M. (2012). 'Some Aspects of Ātman according to Prabhākara and Śālikanātha.' In: Balcerowicz, P. (ed.), *Worldview and Theory in Indian Philosophy*, pp. 277–98. Warsaw Indological Series Vol. 5. Delhi: Motilal Banarsidass.

VisM	Buddhaghosa: *Visuddhimagga*. Warren, H.C. (ed.) (1989). *Visuddhimagga of Buddhaghosācariya*. Edition revised by Dh. Kosambi. Delhi: Motilal Banarsidass.
VkP	Bhartṛhari: *Vākyapadīya*. Rau, W. (ed.) (2002). *Bhartṛharis Vākyapadīya. Versuch einer vollständigen deutschen Übersetzung nach der kritischen Edition der Mūla-Kārikās*. AAWL, Einzelveröffentlichung Nr. 8. Stuttgart: Franz Steiner.
VS	Kaṇāda: *Vaiśeṣikasūtra*. Sinha, N. (tr.) (1923). *The Vaiśeṣika Sûtras of Kaṇâda with the Commentary of Śaṅkara Miśra and Extracts from the Gloss of Jayanârâyaṇa*. Second edition revised and enlarged. Allahabad: The Panini Office.
Willemen et al. 1998	Willemen, C., B. Dassein and C. Cox. (1998). *Sarvāstivāda Buddhist Scholasticism*. Leiden: E.J. Brill.
Witzel 1989	Witzel, M. (1989). 'Tracing the Vedic Dialects.' In Caillat, C. (ed.), *Dialectes dans les Littératures Indo-Aryennes*, pp. 97–266. Paris: Institut de Civilisation Indienne.
Zysk 1993	Zysk, K.G. (1993). 'The Science of Respiration and the Doctrine of the Bodily Winds in Ancient India.' *Journal of the American Oriental Society*, 113(2): 198–213.

Section One

Myth and Ritual

Chapter 1

Five Stones – Four Rivers – One Town: The Hindu *Pañcāyatanapūjā*

MIKAEL AKTOR[1]

In 2014, on 15 August – Indian Independence Day – a ceremony took place in Chennai whereby 108 brahmin couples were initiated into the worship known as the *pañcāyatanapūjā*. In this *pūjā* five deities – Śiva, Viṣṇu, Sūrya, Gaṇeśa and Devī – are worshiped in the form of five natural stones traditionally related to five different locations in South Asia ranging from northern Nepal to the Indian state of Tamil Nadu in the south. The five stones are (in the order of the five deities just mentioned): the *bāṇaliṅga*, an oval-shaped stone from the Narmada river near Omkareshwar, Madhya Pradesh; the *śālagrāma*, a mostly black fossil from the Kali Gandaki gorge near Muktinath, Nepal; the *sphaṭika*, a colorless and transparent crystal traditionally related to the town of Vallam, Tamil Nadu; the *śoṇabhadra*, a red, rounded stone traditionally related to the Son river in Bihar; and the *suvarṇamukhi*, an angular metallic-like stone traditionally related to the Swarnamukhi river at Srikalahasti, Andhra Pradesh.

The initiation ceremony, announced as *samaṣṭipañcāyatanapūjāvaibhavaṃ* (Skt. 'the glory of the worship of the five seats of god for the common good'), was arranged and promoted by Sarma Sastrigal, a leading figure of the *smārta* tradition of Chennai, and sanctioned by the Kāñcī Maṭha, whose junior Śaṅkarācārya had blessed all 108 *pañcāyatanapūjā* sets of stones before the initiation. According to Sarma Sastrigal, the *pañcāyatanapūjā*, formerly a key ritual among *smārta* brahmin householders, has gradually been neglected over the course of the last 100 years to the level of near

1 Mikael Aktor is Associate Professor in the Department of History at the University of Southern Denmark.

extinction today.[2] In many ways this ritual epitomizes the anti-sectarian ideology of Smārtism, related as it is to the theology of Advaita Vedānta (Jackson 2013), but its revival is also a traditionalist attempt to preserve a fragile brahmin identity in the southern Indian context of modernization and globalization.

In this chapter I will consider each of the following points: the ritual, its structure and performance; the stones, their physical and ritual properties; the theological meanings attached to the ritual; and the modern context of *smārta* traditionalism.

PAÑCĀYATANAPŪJĀ: STRUCTURE AND PRACTICE

The Sanskrit word *āyatana* has several meanings most of which are related to the notion of a dwelling place, a seat, a base or similar object. In ritual contexts the word may refer to a temple, an altar and an idol. In the more narrow context of the five gods of *smārta* ritual, Jan Gonda (1975: 198–99) renders *āyatanas* as 'those gods (...) in whom Brahman is supposed to be present or to reside and who are, represented by their images, the object of worship in sanctuaries.' How this is understood in terms of the relationship between materiality and gods, whether as dwelling places or direct manifestations, and what is supposed to be the relation between these forms and the ultimate formless nature of *brahman* as understood by *smārtas* will be discussed at the end of this chapter.

The five stones of the *pañcāyatanapūjā* are normally placed on a round or square plate of bronze or brass. Often it is furnished with legs and upwardly-bent edges, except on one side where the plate has a spout to allow water and milk poured over the stones to flow down into a pot or tray placed beneath the primary plate (see figure 1.1).

2 Personal communication, 9 October 2014. In addition to Sarma Sastrigal, Chennai, I wish to thank the following individuals each of whom have supplied important information and advice during the research undertaken during the course of the preparation of this chapter which took place in Nepal and India from September to November 2014: Peter Carsten Ilsøe, Geological Museum, Copenhagen; Shiva Shankar (Stig Lundgren) and Seshadri Seetharaman, Stockholm; Martin Gansten, Malmö; Ankit and Keshab Dulal, Kathmandu; Rabin Subedi and Krishna Prasad Subedi, Muktinath; Ramesh Shukla, Department of Geology, Patna University; Shyam Prasad Sharma, Srikalahasti; E.V. Sri Shankar, Ramachandran Balasundara Athreya and Rajalakshmi Athreya, Chennai; V. Rammohan, Department of Geology, University of Madras, Guindy Campus, Chennai.

Figure 1.1. A *pañcāyatanapūjā* set in the home of Mr Ramachandran Balasundara Athreya, Chennai. Photograph by the author (9 October 2014).

Any one of the five stones can be placed at the center according to the situation and individual preference of the devotee. Figure 1.1 depicts a *pañcāyatanapūjā* set at a performance a few days after the Navarātri festival. Devī is therefore at the centre of the set, but is represented here not by the *suvarṇamukhi* stone, but by the *śrīyantra* engraved on a silver plate. In this arrangement, the four other deities are placed around Devī/ *śrīyantra* in the following order (starting from the lower right corner from the perspective of the devotee): Viṣṇu as the *śālagrāma* (due to the flower arrangement it has been placed a bit too high; it should have been placed at the corner of the square bronze plate like the other three); Śiva as the *bāṇaliṅga* (but here coupled with the *suvarṇamukhi* so as to form a couple: Śiva-Śakti) in the lower left corner; Gaṇeśa as the *śoṇabhadra* in the upper left corner; and Sūrya as the *sphaṭika* in the upper right corner. Behind the *pañcāyatanapūjā* set with the five stones and the *śrīyantra* we see the anthropomorphic form of the goddess.

However, this arrangement is not the most common. According to a more prevalent interpretation of the scriptural sources, the set of five stones is rotated. In this case the parallel arrangement with Devī/*suvarṇamukhi* at the center should comprise Viṣṇu/*śālagrāma* in the upper left corner (again from the perspective of the devotee), Śiva/*bāṇaliṅga* in the upper right, Gaṇeśa/*śoṇabhadra* in the lower right and Sūrya/*sphaṭika* in the lower left. According to this classical arrangement but with Śiva/*bāṇaliṅga* at the centre (that is, the *śivapañcāyatanapūjā*, the standard), Viṣṇu/*śālagrāma* should be in the upper left corner, Sūrya/*sphaṭika* in the upper right, Gaṇeśa/*śoṇabhadra* in the lower right and Devī/*suvarṇamukhi* in the lower left. Hence the order given at the start of this chapter: Śiva, Viṣṇu, Sūrya, Gaṇeśa and Devī.[3]

The *pañcāyatanapūjā* is performed as a series of many minor rituals roughly divided into two sections. First there is a series of rituals aimed at the purification of the worshiper and the utensils; subsequently there is a series of rituals formed as services to the deities. The latter generally follows the classical series of 16 services (*ṣoḍaśopacāra*) that constitute the orthoprax *pūjā* ritual (Bühnemann 1988: 101–82; Valpey 2012). All the rituals consist of a combination of bodily activities and mantra recitation. A normal structure takes the following form.[4]

The first twelve acts are of purification:

1. Sipping water (*ācamana*).
2. Touching different parts of one's own body (*aṅgavandana*) with the various names of Viṣṇu and Viṣṇu *avatāras*.
3. Meditating on Gaṇeśa as the destroyer of obstacles (*gaṇapatidhyāna*).
4. Controlling the breath (*prāṇāyāma*) with the *gāyatrīmantra*.
5. Declaring the purpose (*saṃkalpa*) of the performance of this *pañcāyatanapūjā*.
6. Worshiping one's seat (*āsanapūjā*).
7. Worshiping the bell (*ghaṇṭāpūjā*); the bell is rung at several points during the ritual.

3 The various arrangements according to which deity sits in the center are prescribed in a verse attributed to Bobadeva and quoted in a number of Sanskrit texts, e.g. in chapter three of the *Śālagrāmaparīkṣā* by Rāmabhaṭṭa (Shapiro 1987: 431; see also Bühnemann 1988: 37, 50–51). In present-day practice there are variations of the standard arrangement, as can be seen in local publications, e.g. Giri Trading (2014). See also Rao (2009, vol. 1: 4) where Viṣṇu and Devī are interchanged in the *śivapañcāyatanapūjā* (Viṣṇu in the northwest, Devī in the northeast) compared to the standard arrangement given by Bühnemann.

4 This list is based on Śāstri (2003: 41–67); Giri Trading (2014).

8. Worshiping the water pot (*kalaśapūjā*) whilst invoking seven holy rivers; water, used for sprinkling on the devotee's own body and the utensils and for bathing the deities, is a major element of the ritual.
9. Worshiping the conch (*śaṅkhapūjā*); the conch is not used as a trumpet (as it is in the temple) but for pouring water taken from the water pot over the deities.
10. Worshiping one's own Self (*ātmapūjā*); the body is praised as the abode of the gods.
11. Worshiping the seats of the deities (*pīṭhapūjā*); the 'seats' are the plates on which the stones are placed, but some may also be placed in a small metal cylinder.
12. Praising the teacher (*gurudhyāna*); this may be done at different places in the ritual.

Sixteen services for the deities follow:

13. [1] Invocation (*āvāhana*) of the deities in each of the five stones.
14. [2] Offering a seat (*āsana*).
15. [3] Offering water for washing the feet of the deities (*pādya*).
16. [4] Offering water as for a respected guest (the deities) (*arghya*).
17. [5] Offering water for the deities to sip (*ācamana*).
18. [6] Bathing the deities (*snāna*); this may be an elaborate part of the ritual. Water, milk and other liquids are poured over the deities accompanied by a series of mantras. The liquids flow from the plate down to a pot or tray below. In conclusion, each stone is dried with a piece of cloth.
19. [7] Offering garments (*vastra*); a small piece of cloth is put around each stone.
20. [8] Offering the sacred thread (*yajñopavīta*), but only to the male deities.
21. [9] Offering anointments (*anulepana, ābharaṇa*); small amounts of sandalwood and turmeric paste are smeared on the stones.
22. [10] Offering leaves and flowers (*puṣpa*); again this part of the ritual may be very elaborate with separate praise to each deity.
23. [11] Offering incense (*dhūpa*); incense sticks are waved in front of the plate and around the deities.
24. [12] Offering light in the form of a small butter lamp (*dīpa*).
25. [13] Offering food (*naivedya*), mostly in the form of grains, sweets and fruit.
26. Offering betel nut to refresh the mouth.
27. Offering a lamp (*pañcamukhadīpa*).

28. Waving the camphor lamp (*kapūranīrājana*) in front of the deities; the worshiper and perhaps others present at the ritual touch the flames with their fingertips and subsequently touch their eyelids.
29. Offering flowers together with Vedic mantras (*mantrapuṣpa*).
30. [14] Circumambulation (*pradakṣiṇa*); the worshiper stands up and turns around.
31. [15] Prostration (*namaskāra*); the worshiper lies flat in prostration before the deities.
32. [16] Offering gifts and entertainments (*dāna*); a mental offering of a parasol, a fan, a mirror, a chariot and other items is made, as well as dance and music.

The ritual is concluded by:

33. Sprinkling water on one's head (*tīrthaprokṣaṇa*) and sipping water; this is done while praying that the purposes of the worship (item 5 above) are fulfilled and while dedicating the worship to *brahman* alone.

The procedure may be prolonged or shortened according to specific needs and situations. A full worship will last about one and a half hours.

A particular point in the procedure deserves special mention. Whereas most mantras during the ritual address the deities rather than the stones that constitute their material forms, the invocations (the first of the sixteen *upacāras*) explicitly mention these stone forms. Each deity is invoked first by a Vedic verse and subsequently by a direct invocation of the deity in each stone:

> ... I focus my thought on, I invoke Mahāgaṇapati [Gaṇeśa] in this material image [*asmin bimbe*][5] ... I focus my thought on, I invoke Śiva united with his consort Umā and his son Skanda in this *liṅga* [*asmin liṅge*]. ... I focus my thought on, I invoke Ambikā [Devī] in this material image [*asmin bimbe*]. ... I focus my thought on, I invoke Mahāviṣṇu united with his consort Bhūmī in this *śālagrāma* discus [*asmin sālagrāmacakre*].[6] ... I focus my thought on, I invoke Sūryanārāyaṇa united with his consorts Chāyā and Saṃjñā in the material image of this crystal [*asmin sphaṭikabimbe*]. (Śāstri 2003: 48–49)

5 Skt. *bimba* is often translated as 'image' in the sense that an object is a material reflection of something (Colas 2013; MWSED 1899: 731, s.v. *bimba*). If we assume gods to be anthropomorphic beings, however, the word 'image' applied to aniconic objects like the five stones in the *pañcāyatanapūjā* seems awkward. This will be discussed in more detail below.

6 The *śālagrāma* has a pattern that resembles Viṣṇu's discus.

But the invocations are somewhat contradictory. According to medieval sources, neither the *śālagrāma* nor the *bāṇaliṅga* requires invocation. Both are considered direct and permanent manifestations (*vibhūti*) of their respective gods, unlike temporary forms of, say, clay or earth, and also unlike artefacts made by humans such as images in the temples or those on the domestic altar which have to go through a formal consecration (*prāṇapratiṣṭhā*) and invocation (*āvāhana*) at the time of establishment (Bühnemann 1988: 136; Hikita 2005: 246n15; Rao 2009, vol. 1: 7, vol. 2: 179; Shapiro 1987: 22–23).

THE FIVE STONES

There exists an unequal amount of textual material about each of the five stones used in the *pañcāyatanapūjā*. For instance, while we have much Purāṇic and Śāstric material on the *śālagrāma* and a good deal on the *bāṇaliṅga*, there is almost nothing about the three other stones. The preface to the second volume of S.K. Ramachandra Rao's two-volume work (2009), which is a compendium of textual material relating mainly to the *śālagrāma* and the *bāṇaliṅga*, states that while the first volume deals only with the *śālagrāma*, the second volume 'focuses attention on other details of formal worship like Bana-linga (for Siva), Sona-sila (for Ganesha), and Dhatu-patra and Yantra (for Devi).' Here the 'Sona-sila' is equal to the *śoṇabhadra* and the 'Dhatu-patra and Yantra,' described elsewhere in the work as 'metallic ore,' are most probably equal to the *suvarṇamukhi* (Rao 2009, vol. 1: 7–8). But of these three objects, volume 2 actually deals only with the *bāṇaliṅga*. Whilst we do not know the precise reason for the omission of the promised material on the *śoṇabhadra* and the *suvarṇamukhi*, it might be that an original plan to include scriptural sources about these two stones had to be dropped due to a lack of textual material. Rao's work also does not deal with the *sphaṭika*, although it is mentioned in passing (ibid.: 8).

This unequal distribution of textual material was confirmed during the course of an interview I had with a Chennai-based scholar. He claimed to possess a vast knowledge of the classical textual material relevant to the *śālagrāma* and *bāṇaliṅga*, but had to admit that the material on the three other stones is sparse, except for the prescription that these stones must be worshiped as the manifestations of their respective deities.[7]

7 Interview with Dr E.V. Sri Sankar, Chennai, 29 September 2014.

The unequal distribution of textual material might indicate that the *pañcāyatanapūjā* was grafted on to an already existing cultic use of the *śālagrāma* and *bāṇaliṅga*. Of the five stones, these two are the most extraordinary: one a fossil with characteristic marks not found on other stones and the other originating from, in topographic terms, a very special source.

The cultural awareness and cultic use of extraordinary stones are acknowledged as elements of prehistoric practice (e.g. Bednarik 2002; see also Shapiro 1987: 199–204), and, even without archeological finds, it seems reasonable to hypothesize that these two stones would have been used ritually by local people from the two respective areas long before the creation of any textual or epigraphic evidence of such use. Frits Staal, according to Shapiro (1987: 44–51), suggests a possible link between the use of *śālagrāma* stones among indigenous people and the Vedic rite of constructing a fire altar (*agnicayana*), and Banerjea (1956: 82–83) argues that *śālagrāma*s and *bāṇaliṅga*s were worshiped before they became associated with pan-Indian gods like Viṣṇu and Śiva; Shapiro (1987: 64) takes this argument as an indication of the 'archaic sanctity' of these two objects.

Temple architecture indicates that an arrangement with one main deity in the center and four others in each surrounding corner of a square platform – the so-called *pañcāyatana* temple type (Kramrisch 1976: 200, 255) – started to appear after the decline of the Gupta Empire, the earliest preserved example being the ruins of the Daśāvatāra Viṣṇu Temple in Deogarh dating to the sixth century (Bühnemann 1988: 49). This reorganization of temple architecture and worship was a *smārta* response to the collapse of the old order and the social disruption that followed; it was an attempt to create an inclusive base for pan-Indian religious unity in the midst of growing sectarian fragmentation (Jackson 2013). The idea was further systematized and emphasized by Śaṅkara, who flourished around 700 CE (Fortsthoefel 2012), and, in more recent times, it has been a consistent idea put forth by *smārta*s that he was the first to promote the *pañcāyatanapūjā* (Farquhar 1920: 179; Jackson 2013). At the very least, it is in Śaṅkara's work that we find the first clear textual reference to the *śālagrāma* stone as the seat of Viṣṇu (BSB I.2.7, I.2.14, I.3.14; Rao 2009, vol. 1: 4; Narayanan 2012). With the *pañcāyatana* structure already in place in temple architecture and ritual, and one of the five stones explicitly mentioned in his work, it is possible that the *pañcāyatanapūjā* became established as a *smārta* ritual during the time of Śaṅkara. However, we lack definitive evidence. The tradition may have been established centuries later, but with reference to Śaṅkara as a means of legitimation.

Regarding the geological identity of the five stones, the Purāṇic and later Sanskrit texts do not provide any precise information. However, I have seen family *pañcāyatanapūjā* sets which have been passed down through the generations and which are of generally the same type as those I have seen in some of the larger shops in Chennai that sell such sets today, although there may be some variation.[8] Today there is a consistent tradition about where each of the stones should be collected, but, apart from the *bāṇaliṅga*s coming from the Narmada river and the *śālagrāma*s coming from the Gandaki river, I have not come across any Sanskrit texts that mention the original locations of the stones.[9] Many sources on the internet, however, consistently associate the *sphaṭika* with the town of Vallam (Tamil Nadu), the *śoṇabhadra* with the Son river (Bihar) and the *suvarṇamukhi* with the Swarnamukhi river (Andhra Pradesh).[10] The fieldwork which was undertaken in Nepal and India as preparation for the writing of this chapter was planned to include travel to all five sites in order to understand more about the precise locations and how the stones are collected and supplied to the shops that sell them. However, it turned out that, apart from the sites related to the *bāṇaliṅga* and the *śālagrāma*, present realities do not accord with the traditions about the locations as handed down (see below). In this sense, the pattern accords with the hypothetical history of the *pañcāyatanapūjā*: because the *pañcāyatanapūjā* as a *smārta* ritual was developed late and grafted onto pre-existing traditions within Śaivism (for the *bāṇaliṅga*) and Vaiṣṇavism (for the *śālagrāma*), there is not the same richness of information about the three other stones. The fact that the ritual had almost disappeared until recently and is regularly performed only

8 For instance in Giri Trading (2015) and Bakthi Today Pavithra Saamagri Parisodhana Nilayam (see Vellampalli 2013).

9 Apart from the verses by Bopadeva on the arrangements of the five deities according to the points of the compass and similar verses in some other texts (see Bühnemann 1988: 51), there are only very few classical Sanskrit texts that contain a coherent description of the *pañcāyatanapūjā*. Bühnemann (1988: 37) mentions Nīlakaṇṭha's *Bhagavantabhāskara* (seventeenth century), in which the *Ācāramayūkha* has a section on 'How to maintain the five objects of worship' (*pañcāyatanasthāpanaprakāra*) (Śeṇḍe 1985: *Ācāramayūkha* 80–99). Of the five stones of the *pañcāyatanapūjā*, however, the text mentions only the *śālagrāma* and the *bāṇaliṅga*. The text discusses who is entitled to perform the ritual and who is not (in the context of the *śālagrāma*); it describes what flowers and leaves are offered and other similar details.

10 See, for instance, the talk given by the former pontiff of Kāñcī Maṭha (Kāñcī Kāmakoṭī Pīṭha) (Varadarajan 2006) or the similar information on the website of the Maṭha (Kanchi Kamakoti Peetham 2015).

among relatively small groups of southern Indian *smārta* brahmins further explains the limited amount of information.

THE BĀṆALIṄGA

Bāṇaliṅgas, the manifestations of Śiva, are stones of various sizes, colors and patterns, but ideally of a uniform oval shape and originating from the Narmada river in Madhya Pradesh. More specifically, they are stones that, due to the particular natural conditions in the river at a certain location, are naturally shaped as ovals. As they are characterized only by the river in which they are found and the unique natural conditions that shape them, they do not share a single geological type. It is the combination of a waterfall and an underwater cave at a specific spot, the village of Dharaji about 15 kilometers upstream from Omkareshwar, that gives – or rather gave – the stones their even oval shape as a consequence of the whirling forces of the waterfall on the stones in the cave (Aktor 2015: 25–31). The underwater cave, known as the Dhāvṛīkuṇḍ, is mentioned in manuals on the *Narmadāparikramā* pilgrimage (the full circumambulation of the Narmada river), in which it is indicated that people from this village dive into the cave under the waterfall and collect these sacred stones for which they receive a donation (Neuß 2012: 153). However, since 2007, when the Omkareshwar dam was opened, the waterfall has been submerged and it is no longer possible to collect such stones. Instead, stones of a suitable shape are collected at other locations along the river and processed mechanically to gain the even oval shape of the original *bāṇaliṅgas*. In fact, the mechanical procedure enables the workshops to produce stones of a much more symmetrical shape and with a smooth polished surface. One such *bāṇaliṅga* that I acquired during the course of fieldwork in the Omkareshwar area in 2012 is shown in figure 1.2.[11]

These workshops started operating in an attempt to create a new business venture even before the submergence of the Dharaji waterfall, but since 2007 the demand for their products has grown substantially.

The manufacturing process involves the use of chisels, electric angle grinders, rotating steel drums and polishing machines.[12] This transition

11 I acquired the stone on my visit to the workshop owned by Mangal Kewat in the village of Bakawa, Khargone district, Madhya Pradesh, on 22 December 2012 (NSLS).

12 Fieldwork observation and interviews with workshop owners in the village of Bakawa, 22 December 2012.

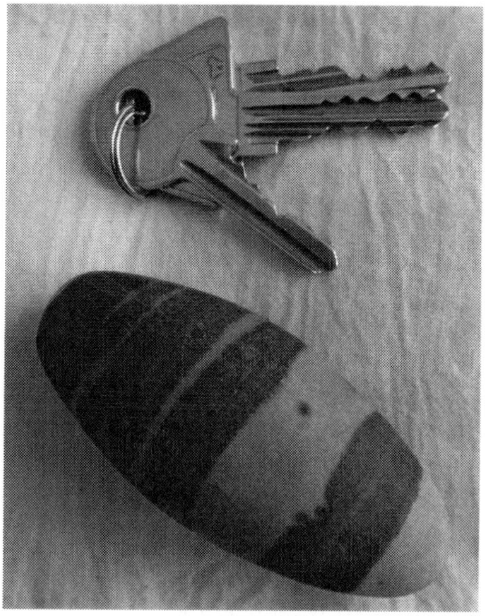

Figure 1.2. A *bāṇaliṅga* from the workshop of Mangal Kewat, Bakawa, Madhya Pradesh, acquired on 22 December 2012. Photograph by the author.

from a natural to a mechanical process of production has been brought about by the need to adapt to the large-scale landscape changes caused by modern hydroelectric river engineering, but, from a scriptural point of view, this is not entirely uncontroversial. *Śivaliṅgas* are divided into two main groups: manmade artefacts (*mānuṣaliṅga*) and natural stones, often rounded in shape, which have traditionally been assigned the status of 'self-manifested *liṅgas*' (*svayambhūliṅga*). Of these, the latter category is regarded as the most prestigious, and thus *bāṇaliṅgas* are often claimed as belonging to this group (Rao 1914: 82; Rao 2009, vol. 2: 173). As *svayambhūliṅgas* they should not undergo manufacturing of any sort. Nonetheless, in agreement with some scriptures that regard *all* stones from the Narmada river as *bāṇaliṅgas*, not only those from the Dhāvṛīkuṇḍ underwater cave (Rao 2009: 170–71), the modern practice is not seen as a process that erodes the *svayambhū*-status of the modern workshop-produced stones. Thus the stones which are collected from the river for processing are already *bāṇaliṅgas* although they have not yet achieved (naturally) the ideal shape – at least this seems to be the understanding.

The new workshop production has also created a much larger market. Prior to the emergence of these workshops and the opening of the

Omkareshwar dam, *bāṇaliṅgas* were not considered as items to be bought and sold. People who wanted to acquire the original, natural, *bāṇaliṅgas* would typically have been temple trustees or others with a professional interest in the stones. They would have to approach the one family in Dharaji whose privilege it was to dive into the Dhāvṛīkuṇḍ. The work was not without risks, but the divers considered their work as a service to Śiva and felt themselves protected directly by Mother Narmadā. On average, not more than one stone was collected each day. For this service the divers would receive a fair 'donation.' Today, without this addition to its income, the family is visibly poor.[13]

The most precious *bāṇaliṅgas* are white, but that does not exclude stones of other colors being used and valued. Typically, the different colors as well as lines and dots on individual stones are assigned particular significance in medieval texts. A stone which is half white and half red is considered an *ardhanarīśvaraliṅga*, that is, associated with Śiva in his aspect as half god (white) and half goddess (red), and a stone which is multicolored with marks like Śiva's matted hair is seen as a *mṛtyuṃjayaliṅga*, associated with Śiva as the death-conquering god.[14] It is not unusual that the visual and other qualities of aniconic religious objects are read in this way. It constitutes a kind of anthropomorphization which connects the aniconic object with a well-known anthropomorphic iconography.

THE ŚĀLAGRĀMA

Śālagrāmas, the manifestations of Viṣṇu, are stones from the pilgrimage site known as Śālagrāmakṣetra or Viṣṇukṣetra, which is the area along and around the Kali-Gandaki gorge in Mustang, Nepal, with the pilgrimage town of Muktinath at its centre. According to some myths, all stones from this area can be worshiped as *śālagrāmas*; although this is the case in theory, what is generally meant by *śālagrāmas*, and certainly those regarded as the most sacred, are stones from the area that contain

13 Interview conducted on 30 December 2012 with two brothers, Nam Singh and Ram Singh, who used to dive for *bāṇaliṅgas* prior to the submergence of the Dharaji waterfall in 2007. For further details, see Aktor (2015: 30–31, 33n22).

14 'The Mṛtyuñjaya-liṅga is multicolored, and has marks of matted hair and spear; it is worshipped by gods and titans alike [...] The Ardha-nārīśvara-liṅga is distinguished by half of it being white, and the other half red; there will also be marks of trident and hand-drum' (Hemādri's *Caturvargacintāmaṇi* [*Lakṣaṇakāṇḍa*] quoted in Rao 2009: 185, 187).

ammonite fossils (Shapiro 1987: 5, 25, 252). These have been found in unusually large numbers in and around the rivers and lakes of this area (Enay and Cariou 1997: 3, 15).

Ammonites are the fossils of extinct marine *Ammonoidea* cephalopods. These mollusks created a chambered shell for protection, and it is the marks of this shell which typically distinguish their fossils. They are present in the Himalayan region as a result of the formation processes of these mountains. Prior to the geological formation of the Himalaya, this area of the planet was part of the open Tethys Ocean. When the Indo-Australian tectonic plate, which had moved away from the Gondwana supercontinent of the southern hemisphere, collided with the Eurasian plate at the transition from the Upper Cretaceous to the Paleogene geological period (around 55 million years ago), sediments from the Tethys Ocean containing the remnants of its former animals were pushed up gradually to form the present-day mountain range.

It is the typical grooved impressions left by the chambered shell which form the patterns that in the Hindu tradition are seen as the marks of Viṣṇu's discus, the *cakra*. These stones can be of various shapes, but the fossils are typically spirals with the shell impression forming a coil of slanting lines. It is this '*cakra*' pattern that more than anything else contributes to the ritual value of the stones, and the fuller the *cakra* the more precious the *śālagrāma*. According to medieval texts, even broken or damaged *śālagrāmas* are ritually efficacious if they have the marks of the *cakra* (Rao 2009: 20–21).

In addition to the *cakra* pattern, many *śālagrāma* stones have cavities, so-called 'mouths' (*vaktra*) or 'doors' (*dvāra*), which may contain smaller *cakras*, each a fossil in its own right. Purāṇic texts describe them as having been created by 'adamantine worms' (*vajrakīṭa*) with extraordinarily sharp teeth which are able to cut through hard stone (Rao 2009: 25–26).[15] According to *śālagrāma* myths, which are found with some variation in the texts, these worms were the gods themselves. For example, Gandakī was a pious woman who did hard penance in the Himalaya. As a result, Brahmā, Viṣṇu and Śiva appeared in order to grant her a boon, but her only wish was that the three gods would enter her womb and be born of her. The gods were upset by this idea and refused her wish. This angered Gandakī,

15 Rao (2009: 25) is not specific as to the exact Purāṇic source. Shapiro (1987: 87) mentions a similar story with reference to the *Varāhapurāṇa*, but he also warns of the 'instability' of this literature where 'the same material is now attributed to one text and then to another, or not found at all' (Shapiro 1987: 79).

who cursed the gods to become worms. They made a counter curse that Gandakī would become a dark river. As a consequence, Viṣṇu entered the stones of the river with his *cakra*, and the other gods became worms able to carve out *cakras* from the stones. Thus Viṣṇu was born from the womb of the Gandaki River as *śālagrāmas* and the other gods as worms (Rao 2009: 29–31, with reference to the 'Āsura-khaṇḍa section of Skandapurāṇa' on p. 31). Shapiro (1987: 80) notices the non-sectarian plot of these myths in which Viṣṇu and Śiva are forced to cooperate.

According to the distribution of cavities, the number and position of *cakras*, the colors and other particularities, *śālagrāmas* are distinguished as one of the innumerable forms of Viṣṇu, his *avatāras* as well as his many epithets. But some are also seen as manifestations of the other four deities of the *pañcāyatanapūjā*. These *śālagrāmas* are without the *cakra* marks specific to Viṣṇu. The *Śālagrāmaparīkṣā*, a text from the late seventeenth century (Shapiro 1987: unpaginated abstract), mentions Śiva's five *brahmamantra* forms (*sadyojāta*, *vāmadeva*, *aghora*, *tatpuruṣa* and *īśāna*), one *śālagrāma* specific to Gaṇeśa, a few specific to Sūrya and some to 'Śakti' (Devī).[16] The last are characterized by patterns like a sleeping *kuṇḍalinī* snake (Shapiro 1987: 464–66, 468–72). With these other *śālagrāmas* it is possible to perform the *pañcāyatanapūjā* with five *śālagrāmas* alone, instead of the usual set of five stones from different locations (*bāṇaliṅga* etc.).

Today, the streets around the Paśupatinātha temple complex in Kathmandu are the centre of the *śālagrāma* trade, as they were in the early 1980s when Shapiro did his fieldwork (Shapiro 1987: 6, 133). I acquired a set of five *śālagrāmas* from one of the shops in this area;[17] each of the stones is considered a manifestation of one of the five *pañcāyatana* deities. Unlike the prescriptions in the *Śālagrāmaparīkṣā*, however, these stones do all possess the *cakra* marks, except for the one that represents Śiva. The family that owns this particular shop has been collecting *śālagrāmas* for more than thirty years with several journeys having been made to the Muktinath and Damodar areas of Nepal. The family members are also well trained in the interpretation of the stones according to Purāṇic sources or manuals like

16 The five *brahmamantra* aspects of Śiva are related to five fundamental activities of the god: emission, maintenance, reabsorption, veiling and grace. See Davis 1991: 458–51.

17 Himālay Rudrākṣa Bhaṇḍār, Paśupatināth, Kathmandu (see Nepalrudraksha n.d.). I met the Dulal family in their shop on 2 and 14 September 2014. Keshab Dulal, the father of the shop-owner, followed me to Muktinath and was my guide there. We went for river walks between Muktinath and Kagbeni and found ammonite fossils in both the riverbed and porous parts of the rocks.

Figure 1.3. Five *śālagrāmas* representing the five deities of the *pañcāyatanapūjā*, acquired from Himālay Rudrākṣa Bhaṇḍār, Paśupatināth, Kathmandu, on 2 September 2014. Photograph by the author.

the *Śālagrāmaparīkṣā*, that is, in how to associate a certain specimen with a particular aspect of Viṣṇu or another god. Figure 1.3 shows this set of five *śālagrāmas*.

In the center is Śiva, without a *cakra* mark but with an embedded round ball resembling a *svayambhūliṅga*. In the upper left corner the oblong *śālagrāma* with *cakra* marks at both ends is considered to be *viṣṇupada*, Viṣṇu's footprint. The round *śālagrāma* in the upper right corner is Sūrya (the sun god) with the *cakra* marks as his rays. Gaṇeśa is represented by the *śālagrāma* in the bottom right corner; it is identified primarily by its chubby shape. The *śālagrāma* in the bottom-left corner is Ambikā (Devī), distinguished as such by its shape, which suggests a female torso. However, when I asked the family about the identities of other *śālagrāmas* that I had photographed in the shop, I noticed that their identifications would sometimes vary.

Although the population of the Mustang area is predominantly Buddhist, the Muktinath temple is primarily a pilgrimage site for Viṣṇu devotees from both Nepal and India, and over the centuries *śālagrāmas* have primarily been sought after by Vaiṣṇavas. Probably, *śālagrāmas* have become less common as a result of such demand, but they are in no way impossible to find. Local people still collect them and sell them to trekkers and tourists. Undoubtedly, fake *śālagrāmas* are also in circulation. The specimens that I saw on display in the souvenir shop at Jomsom Airport were clearly carved from ordinary black stones.

THE SPHAṬIKA

Sphaṭikas, the manifestations of Sūrya, the Sun God, are roundly-cut quartz crystals traditionally assigned to the area around the town of Vallam, near Thanjavur, Tamil Nadu. Quartz crystal is sold in shops all over India cut into various shapes, such as pyramids, *śivaliṅgas*, *śrīyantras* and others. Geologists from Madras University told me that the crystal can be found in many locations across India, but it seems that Rajasthan is a particularly rich source today.[18] In the past, the area around Vallam may have been one such source, but it is not one today.

In the section about Vallam in the *Imperial Gazetteer of India* published in 1908 we read that '[t]he quartz crystals found in the neighbourhood, known as Vallam stones, are made into spectacles and ornaments' (Frowde 1908: 297). The text also mentions a 'formerly strong fort' of which '[f]ew traces of the defences now survive, except the moat. A sacred tank within the fort is hewn in the solid rock and unusually deep. It is called Vajratirtham ('the diamond pool'), and the popular tradition is that it was dug by Indra' (ibid.: 297). Remnants of the moat can still be seen today, but only a few people know about the earlier fort. I talked to a variety of people in the area, such as owners of shops that sell gemstones or *pūjā* items, both in Thanjavur and in Vallam, but no one knew anything about quartz crystals from Vallam or the concept that the *sphaṭika* stone of the *pañcāyatanapūjā* is attributed to Vallam. I am not the first seeker of the *sphaṭika* with this experience. Commenting on the content of a webpage about the *pañcāyatanapūjā* one person writes: 'I live in Trichy near to Tanjavur, where can i find lord Surya's crystal stone in Trichy, Vallam and Thanjavur areas or are still present in the Vallam river and I asked many people with no had knowledge of this' [sic] (Brahmanda 2012). However, the online database of minerals, mindat.org, mentions the 'Vallum Diamond' as a quartz crystal located in the Tanjore (Thanjavur) district, but it has no information about actual mines or industries (mindat.org$_c$).

18 Interview with Professor V. Rammohan and his colleagues, Department of Geology, University of Madras, Guindy Campus, 8 October 2014.

As indicated in the *Imperial Gazetteer*, the area around Vallam may very well have been a former source of quartz crystals and other gemstones, and the political and religious significance of nearby Thanjavur, with its palace and famous Bṛhadīśvara temple, may have been a further reason to include this location in the system of the *pañcāyatanapūjā*.

THE ŚOṆABHADRA

Śoṇabhadra stones, the manifestations of Gaṇeśa, are described as red stones from the Son river in Bihar. Those I have seen in homes where the *pañcāyatanapūjā* is performed and in the shops in Chennai are red jasper; but whether they must be jasper is uncertain. I have also seen red laterite collected from the Son river near its source at Amarkantak, Madhya Pradesh, presented as *śoṇabhadra* stones. The jasper specimens may be more or less polished but typically they are smooth and with a rounded shape. Laterite examples are more rough, porous and with an uneven shape. According to one informant from Chennai, what connects this stone to Gaṇeśa is its shape. The shape must be pyramidal, but also rounded and chubby like the contours of the sitting Gaṇeśa.[19]

Some sources on the internet are a little more precise about the location from where the stones should be collected, namely from the riverbed where the Son flows into the Ganga from the south, which is some 40 kilometers west of Patna city along the Ganga (see Krishnamurthy n.d.). However, geologists from Patna University told me that it is unlikely that specimens of jasper as big as those I saw in the Chennai shops could be found in the river at this site.[20] If they are supposed to have been transported a long distance by the river to near its outflow into the Ganga they will, as a consequence of this journey, end up as small particles. Again, not a single person I met in the area or in nearby Patna had any knowledge about the *śoṇabhadra* and its supposed origin in the Son river.

This lack of information does not mean that jasper was not found in the area in earlier times. Various objects made of jasper have been found at archeological sites in this part of Bihar (Vishnu 1993: 18; Singh 2008: 120), although it is not certain that pieces of jasper were frequently collected from the river. It is possible that the river location was selected in

19 Interview with Dr E.V. Sri Sankar, Chennai, 29 September 2014.

20 Interview with Professor Ramesh Shukla and his colleagues, Department of Geology, Patna University, 16 October 2014.

the *pañcāyatana* tradition in order to connect the *śoṇabhadra* to the river mythology of the *śālagrāma* and the *bāṇaliṅga* and their connection to the Ganga. Patna is not only the site where the Son flows into the Ganga from the south but it is also the location where the Gandaki flows into the river from the north (just north of Patna city). In addition, the Son has its source at Amarkantak, Madhya Pradesh, the same area from where the Narmada river originates. Thus, all three north-central Indian rivers that are related to the *pañcāyatanapūjā* – the Gandaki, the Son and the Narmada – are connected geographically, and the first two of these rivers flow into the Ganga.

THE SUVARṆAMUKHI

Lastly, the *suvarṇamukhi* stones, manifestations of Devī, are described as stones 'with a streak of gold' from the Swarnamukhi river at Srikalahasti, Andhra Pradesh (Varadarajan 2006). The geologists I consulted all identified those that I had photographed in private homes as well as those bought in a Chennai shop as specimens of pyrite.[21] This is a cubic iron sulfide crystal, which, according to the same geologists, is not commonly found in India.[22] One priest from the famous Śrīkālahastīśvara Śiva temple whom I interviewed knew about the *pañcāyatanapūjā* and the *suvarṇamukhi* stone, but positively rejected the notion that it is collected in the Srikalahasti area.[23] Shop-owners in the pilgrim bazars who sell small pyramids or *liṅga*s allegedly cut from quartz crystal knew nothing of the round *suvarṇamukhi* stone or the *pañcāyatanapūjā*.

However, an email message that I received from a member of the Advaita-L mailing list[24] added new information:

21 Interviews with Professor V. Rammohan and his colleagues, Department of Geology, University of Madras, Guindy Campus, 8 October 2014; and with Professor Ramesh Shukla and his colleagues, Department of Geology, Patna University, 16 October 2014.

22 However, mindat.org has two records of pyrite in Andhra Pradesh: one in Jonnagiri, Kurnool district, and one in Chigargunta, Chittoor district. Neither of them, however, is located near the Swarnamukhi river. See mindat.org$_a$ and mindat.org$_b$.

23 Interview with Shyam Prasad Sharma, Śrīkālahastīśvara temple, Srikalahasti, 5 October 2014.

24 Advaita-L (http://www.advaita-vedanta.org/lists/) is a mailing list which aims 'to discuss advaita-vedānta as taught by Śrī Śaṅkara and the smārta sampradāya.'

> It is 'swarṇamākṣika śilā' [*suvarṇamākṣika* stone] and not swarṇamukhi śilā. This is not found in Srikalahasti. Moreover, the Swarnamukhi River is completely dried up. Nothing can be found there now as everything is dried up. Once my Guruji told me that these swarṇamākṣika śilās are found in Rāyalaseema region of Andhra Pradesh (near Kadapa) and these are natural formations of rocks in the mountains. Some shine like silver hence called 'rājatamākṣika śilā' and a few shine like gold and hence called 'swarṇamākṣika śilā. (Advaita-L, 5 October 2014).

The *suvarṇamākṣika* stones are used in Āyurvedic medicine: the ground stone, mixed with various ingredients, is purified by being exposed to prolonged and repeated heat and then powdered (Devanathan 2013). Of the two stones mentioned in the message on Advaita-L, *rājatamākṣika* is pyrite (iron sulfide crystal) and *suvarṇamākṣika* with the golden sheen is chalcopyrite (copper iron sulfide crystal) (Savalgi et al. 2011: 1649). The *suvarṇamukhi* stones I have seen in the context of *pañcāyatanapūjā* are all pyrite specimens.

Formerly, and ideally, the *pañcāyatanapūjā* sets consisting of these five stones – the *bāṇaliṅga*, the *śālagrāma*, the *sphaṭika*, the *śoṇabhadra* and the *suvarṇamukhi* – were handed down from one generation to the next. But today, when the tradition has been almost forgotten and has, therefore, become part of a revitalization process, people have to buy new sets of

Figure 1.4. A *pañcāyatanapūjā* set bought from Giri Trading, Mylapore, Chennai, on 29 September 2014. Photograph by the author.

stones from a shop. Chennai in particular has become a center for the distribution of *pañcāyatanapūjā* sets, although this is quite a small-scale business. A few shops also sell the stones online. Figure 1.4 shows a set that was bought in the large Giri Trading shop in Mylapore, Chennai, on 29 September 2014.

The price for this set was ₹620 (€8.30). The *bāṇaliṅga* in the center is a polished black hematite and very different from those I saw in the Omkareshwar area in 2012. The stone sitting upper left is a *śālagrāma* (although not an ammonite fossil), upper right is a *sphaṭika* (quartz crystal), lower right is a *śoṇabhadra* (red jasper) and lower left sits a *suvarṇamukhi* (pyrite).

With the professionalization of the distribution of these sets of stones via shops like Giri Trading, it is no wonder that supply of the individual stones has also become professionalized. Rather than the stones being collected from the sites stipulated by tradition (where they may well be extinct), shops may find it easier to deal with existing companies specialized in sourcing gemstones and minerals from wherever it is most feasible.

WHY *PAÑCĀYATANAPŪJĀ*? WHY STONES?

Is this kind of professionalization and disregard of traditional ways a problem within the group of *pañcāyatanapūjā* practitioners? It does not seem to be. The basic idea of image worship among Advaita Vedāntins is that *brahman* is one and ultimately formless but that it can be worshiped in many diverse forms according to the taste and tradition of the individual devotee.[25] This notion was sharply expressed in one of the interviews I conducted in Chennai: 'Brahman is everywhere. You might worship God in a dog or a donkey as well as in these stones. Or I might start worship Dr Aktor.'[26] In this sense, the *pañcāyatanapūjā* set is first of all a devotional device for purifying and preparing the mind for contemplation of *brahman*. It was also made clear to me that aniconic forms are seen as more fitting visual representations of the formless *brahman* than anthropomorphic forms. Also the natural occurrence of the stones suggests the all-pervasiveness of *brahman*.

25 Notice how Jan Gonda captured this idea precisely when he (as already mentioned) identified *āyatana* in the *smārta* context as 'those gods (...) in whom Brahman is supposed to be present or to reside and who are, represented by their images, the object of worship in sanctuaries' (1975: 198–99).

26 Interview with Dr E.V. Sri Sankar, Chennai, 29 September 2014.

The non-sectarian and inclusive purpose of the ritual was also frequently expressed. The email from the member of Advaita-L, quoted above, continues:

> These pañcāyatana śilā vyavasthā were introduced by Ācārya Śaṅkara to demonstrate abheda-dṛṣṭi in upāsana. It was this noble idea of abheda darśana and also to unite Bhārata Deśa in those days Bhagavatpāda introduced pañcāyatana worship and established 4 mutts in four directions (Advaita-L, 5 October 2014).

It is claimed here that Śaṅkara introduced the worship of the five different *pañcāyatana* stones in order to stress the goal of undivided unity in worship. By worshiping all five deities, and not just one exclusively, the concept that they are ultimately one is suggested. By selecting stones from sacred rivers and locations across South Asia, the Hindu unity of that territory is also stressed, just as in the establishment of the four centers of Advaita Vedānta (Dwarka in the west, Puri in the east, Joshimath in the north and Sringeri in the south).

The *advaita* orientation may also affect the view of the relationship between gods and material. For *śaiva*s as well as for *vaiṣṇava*s this relationship is one of identity. The *bāṇaliṅga*s and *śālagrāma*s are direct manifestations of Śiva and Viṣṇu respectively. As for the *śālagrāma*, Shapiro (1987: 22–23) writes: 'The signifier, the śālagrāma, is not different from the signified; the śālagrāma *is* Viṣṇu or any other deity one wishes it to be, and no difference is recognized.'[27] But Śaṅkara does not subscribe to the same idea of identity; for him, the relationship between god and image is more one of superimposition, and I do not think it makes a difference in Śaṅkara's view that the *śālagrāma* is not an 'image' in the same way as the image in the temple. In the words of Gérard Colas (2004: 159):

> [W]hile Śaṅkara is eager to assert the notion of the embodiment of god, he clearly distinguishes concrete images from the gods they represent in passages where he mentions images. A divine image has no value by itself. It receives value only through a superimposition, a mental application of the idea of a god to it.

27 Narayanan elaborates: 'For the Śrīvaiṣṇava the *śālagrāma* is not merely a petrified fossil; it is fully God, made of a pure, transcendental, non-material substance called *śuddha sattva* – the material of Viṣṇu's own body. ... We are confronted with a paradox: what appears to non-Hindu eyes as the most gross and material representation of the deity is understood by the Śrīvaiṣṇava to be a divine, auspicious form, composed of a non-material substance that exists only in heaven and in the Śrīvaiṣṇava temple on earth' (1985: 60, 62).

This more psychological understanding of objects of worship also seems to form the background to the relaxed attitudes towards the *pañcāyatana* stones that I encountered among those who perform the ritual. Yes, this *pūjā* is important as a devotional purification of the mind and as a statement of *smārta* values, but in principle the objects of worship could take any form, because, ultimately, these stones, like other images, are only the superimposed seats of the one *brahman*.

THE PRESENT CONTEXT

The *pañcāyatanapūjā* initiation ceremony that took place in Chennai on 15 August 2014 was announced as *samaṣṭi*. In a *vedānta* context, *samaṣṭi* denotes a practice that pertains to the community or the common good, in contrast to one that is *vyaṣṭi*, pertaining to the individual. By utilizing natural stones from five different locations spread across South Asia, the *pañcāyatanapūjā* has always been a *samaṣṭi* practice with strong references to Bhāratavarṣa as a religious unit. But how far does the notion of community extend today? Sarma Sastrigal, on whose initiative the ceremony was arranged, wrote about the event on his Facebook page: 'This puja done by smartas has a strong prayer for universal wellbeing in addition to the individual's spiritual progress.'[28] While this universalistic aim is not uncommon in modern forms of Advaita Vedānta, still, in this particular context, and as suggested by the simultaneous celebration of the Indian Independence Day, the *samaṣṭi* in this case is probably related closely to India and, more specifically, the revitalization of *smārta* identity in the south. Underlying the event is a wish to promote traditional religious values and customs, and a critique of the westernization of Indian society. Some groups of *smārta* brahmins see themselves as role models to wider society in their attempt to revitalize cultural and religious practices that are dying out under the pressure of globalization. There was a strong emphasis during several of the interviews I conducted on the glorious Indian past, the exceptional antiquity of the Vedas, Sanskrit as the mother of all languages, scientific inventions already being anticipated in the *Mahābhārata* and similar traditionalistic ideas. In Sarma Sastrigal's book, *The Great Hindu Tradition*, this traditionalism is expressed in many ways, not least with respect to practical matters, such as, for instance, how to bind the traditional *dhoti* in the *pancha kaccham* style (2011: 52–58).[29] Today, tradition is in need of how-to

28 Sarma Sastrigal on Facebook, 17 August 2014.

29 For an illustration, see Thapas (2012).

books, it cannot be taken for granted. Under these circumstances the *pañcāyatanapūjā* also has to adapt: to hydroelectric projects that change river flows; to market mechanisms that change supply patterns; and to new media that globalize communications.

ABBREVIATIONS AND REFERENCES

Skt. = Sanskrit

Aktor 2015 — Aktor, M. (2015). 'The Śivaliṅga Between Artifact and Nature: The Ghṛṣṇeśvaraliṅga in Varanasi and the Bāṇaliṅgas from the Narmada River'. In K.A. Jacobsen, M. Aktor and K. Myrvold (eds.), *Objects of Worship in South Asian Religions: Forms, Practices and Meanings*, pp. 14–34. London: Routledge.

Banerjea 1956 — Banerjea, J.N. (1956). *The Development of Hindu Iconography*. 2nd ed. New Delhi: Munshiram Manoharlal.

Bednarik 2002 — Bednarik, R. (2002). 'Mauports and very early palaeoart.' *Auranet*, <http://www.ifrao.com/manuports-and-very-early-palaeoart/> (accessed 25 August 2015).

BEH — Jacobsen, K.A., H. Basu, A. Malinar and V. Narayanan (eds.) (2009–2014). *Brill's Encyclopedia of Hinduism*. Leiden: Brill Online.

Brahmanda 2012 — Brahmanda, D. (2012). 'Panchayatana Pooja,' *Divine Brahmanda*, <http://www.divinebrahmanda.com/2012/03/panchayatana-pooja.html> (accessed 25 August 2015).

BSB — Śaṅkara: *Brahmasūtrabhāṣya. Brahmasūtra with Śāṅkarabhāṣya* (Works of Śaṅkarācārya, Vol. III) (1964). Delhi: Motilal Banarsidass.

Bühnemann 1988 — Bühnemann, G. (1988). *Pūjā. A Study in Smārta Ritual*. Leiden: Brill.

Colas 2004 — Colas, G. (2004). 'The Competing Hermeneutics of Image Worship in Hinduism (Fifth to Eleventh Century AD).' In Granoff, P. and K. Shinohara (eds.), *Images in Asian Religions. Texts and Contexts*, pp. 149–79. Vancouver: The University of British Columbia Press.

Colas 2013 — Colas, G. (2013). 'Iconography and Images (Murtī): Ancient Concepts.' In BEH.

Davis 1991 — Davis, R.H. (1991). *Ritual in an Oscillating Universe. Worshipping Śiva in Medieval India*. Princeton: Princeton University Press.

Devanathan 2013 — Devanathan, R. (2013). 'Pharmaceutical and Analytical Studies on Swarna Makshika Bhasma: An Ayurvedic Formulation.' *Asian Journal of Pharmaceutical and Clinical Research*, 6 (1): 26–29.

Enay and Cariou 1997 — Enay, R. and E. Cariou. (1997). 'Ammonites Faunas and Palaeobiogeography of the Himalayan Belt During the Jurassic: Initiation of a Late Jurassic Austral Ammonite Fauna.' *Palaeogeography, Palaeoclimatology, Palaeoecology*, 134: 1–38.

Farquhar 1920 — Farquhar, J.N. (1920). *An Outline of the Religious Literature of India*. London: Oxford University Press.

Forsthoefel 2012 — Forsthoefel, T. (2012). 'Śaṅkara.' In BEH.

Frowde 1908 — Frowde, H. (1908). *The Imperial Gazetteer of India*, vol. 24. Oxford: Clarendon Press.

Giri Trading 2014 — Giri Trading. (2014). *Nitya Pañcāyatana Pūjā*. Chennai: Giri Trading.

Giri Trading 2015 — Giri Trading (2015). 'Giri Trading Shop.' <http://www.giri.in> (accessed 25 August 2015).

Gonda 1975 — Gonda, J. (1975 [1969]). 'Āyatana.' In Gonda, J., *Selected Studies*, Vol. 2, pp. 178–256. Leiden: E.J. Brill.

Hikita 2005 — Hikita, H. (2005). 'Liṅga Worship as Prescribed by the Śivapurāṇa.' In Einoo, S. and J. Takashima (eds.), *From Material to Deity. Indian Rituals of Consecration*, pp. 241–82. New Delhi: Manohar.

Jackson 2013 — Jackson, W.J. (2013). 'Smārta.' In BEH.

Kanchi Kamakoti Peetham 2015 — Shri Kanchi Kamakoti Peetham (2015). 'Puja' (*Hindu Dharma*, part 22, chapter 5), *Kamakoti.org*, <http://www.kamakoti.org/hindudharma/part22/chap5.htm> (accessed 25 August 2015).

Kramrisch 1976 — Kramrisch, S. (1976 [1946]). *The Hindu Temple*. 2 vols. Delhi: Motilal Barnarsidass.

Krishnamurthy n.d. — Krishnamurthy, V. (n.d.). 'My Pancayatana Puja.' <http://www.krishnamurthys.com/profvk/PancAyat.html> (last accessed 20 February 2015).

mindat.org$_a$ — mindat.org. (1993–2015). 'Chigargunta.' *mindat.org and the Hudson Institute of Mineralogy*, <http://www.mindat.org/locentry-639120.html> (accessed 8 September 2015).

mindat.org$_b$ — mindat.org. (1993–2015). 'Jonnagiri.' *mindat.org and the Hudson Institute of Mineralogy*, <http://www.mindat.org/locentry-639123.html> (accessed 8 September 2015).

mindat.org$_c$ — mindat.org. (1993–2015). 'Vallum Diamond.' *mindat.org and the Hudson Institute of Mineralogy*, <http://www.mindat.org/min-39104.html> (accessed 8 September 2015).

MWSED — Monier-Williams, M. (1899). *A Sanskrit-English Dictionary*. Oxford: Clarendon Press.

Narayanan 1985 — Narayanan, V. (1985). 'Arcāvatāra: On Earth as He Is in Heaven.' In J.P. Waghorne and N. Cutler (eds.), *Gods of Flesh Gods of Stone: The Embodiment of Divinity in India*, pp. 53–66. Chambersburg, Pennsylvania: Anima Publications.

Narayanan 2012 — Narayanan, V. (2012). 'Śālagrāma.' In: BEH.

Nepalrudraksha n.d. — Nepalrudraksha. (n.d.). 'Himalaya Rudraksha Bhandar.' <http://nepalrudraksha.com/> (accessed 25 August 2015).

Neuß 2012 — Neuß, J. (2012). *Narmadāparikramā – Circumambulation of the Narmadā River. On the Tradition of a Unique Hindu Pilgrimage*. Leiden: E.J. Brill.

NSLS — Narmadeshwar Shiv Ling Stone. At <http://www.narmadeshwar-shivlingstone.com> (accessed 20 February 2015 but no longer visible in September 2015).

Rao 1914	Rao, T.A.G. (1914). *Elements of Hindu Iconography*, Vol. 2, reprint 1997. Delhi: Motilal Banarsidass.
Rao 2009	Rao, S.K.R. (2009 [1996]). *Śālagrāma-Kosha* (2 vols.). Delhi: Sri Satguru Publications.
Śāstri 2003	Śāstri, L.A. (2003). *Nityāhnikam. The Daily Routine of Every Brahmin.* Kalpathi: R.S. Vadhyar & Sons.
Sastrigal 2011	Sastrigal, S. (2011). *The Great Hindu Tradition. An Insight into Vedic Principles, Sastras and Heritage*. 2nd edition. Chennai: Sri Sarma Sastrigal.
Savalgi et al. 2011	Savalgi, P.B., B.J. Patgiri and V.J. Shukla. (2011). 'Characterization of Different Samples of Swarna Makshika.' *International Journal of Research in Ayurveda & Pharmacy*, 2(6): 1648–50.
Śeṇḍe 1985	Śeṇḍe, N. (1985). *Bhagavantabhāskara of Śrī Nīlakaṇṭha Bhaṭṭa.* Reprint. Delhi: Chaukhamba Sanskrit Pratishthan.
Shapiro 1987	Shapiro, A.A. (1987). *Śālagrāmaśilā. A Study of Śālagrāma Stones with Text and Translation of Śālagrāmaparīkṣā*. PhD Dissertaion, Columbia University.
Singh 2008	Singh, S. (2008). *A History of Ancient and Early Medieval India. From the Stone Age to the 12th Century*. New Delhi: Pearson Education.
Thapas 2012	Thapas (2012). 'How to Wear a Pancha Kachcham.' <https://thapas.wordpress.com/2012/10/26/how-to-wear-a-pancha-kachcham> (accessed 8 September 2015).
Valpey 2012	Valpey, K. (2012). 'Pūjā and Darśana.' In BEH.
Varadarajan 2006	Varadarajan, A. (2006). 'Deivathin Kural Series – 57.' <http://advaitham.blogspot.in/2006/09/deivathin-kural-series-57.html> (accessed 8 September 2015).
Vellampalli 2013	Vellampalli, S.H. (2013). 'Pavithra Saamagri Parisodhana Nilayam.' *Bakthi Today*, <http://bakthitoday.blogspot.dk/2013/08/panchayatan.html> (accessed 25 August 2015).
Vishnu 1993	Vishnu, A. (1993). *Material Life of Northern India*. New Delhi: Mittal Publications.

Chapter 2

A 'Sulfurous' Śakti: The Worship of Goddess Hiṅgulā in Baluchistan

FRANCESCO BRIGHENTI[1]

THE SHRINE OF HIṄGULĀ IN MAKRAN

The cave temple of the Hindu goddess Hiṅgulā (Old Indo-Aryan form) or Hiṅglāj (New Indo-Aryan form) is situated at the eastern extremity of the range of mountains in Pakistani Baluchistan called by her name – the Hinglaj Range, being the highest mountain chain to extend to the coastal zone of Makran. The site of the shrine, a gorge between rocky cliffs, is also called Hinglaj.[2] The site is accessed from the deserted Makran coastland through the valley of the Hingol River (also known as Aghor near its mouth), which likewise takes its name from goddess Hiṅgulā. The cave-like open-air sanctuary of the goddess in the Hinglaj water gorge consists of a walled enclosure lying under a slightly overhanging high wall of sandstone.

The cult image of the goddess is aniconic, being represented by a group of connected stones, narrow in the middle and globular at both ends, that rest on an oblong rock (apparently part of the natural cave floor) around which a raised mud platform has been built and later renovated at various times. The assemblage overall suggests the shape of a recumbent anthropomorphic figure, and is regarded by the Hindus as a *svayambhūmūrti*

1 Francesco Brighenti is an independent scholar based in Venice (Italy) and a Research Associate of the University of Hyderabad (India).

2 The transcription 'Hinglaj,' without diacritics, is used in this chapter to refer to the geographical site of the shrine of goddess Hiṅglāj for consistency with the transcription system employed for all other geographical names.

(self-generated image) of the goddess. A low and narrow semicircular tunnel beneath the image platform, which may have been originally – at least in part – a natural crevice in the rock, is used by pilgrims, who crawl on their knees to perform the *pradakṣiṇā* or ritual circumambulation of the image itself in a rite of purification and rebirth. This underground passage is known as the *garbha* ('womb') of Hiṅglāj Devī (Stein 1943: 202; Schaflechner 2014: 201–17).

The shrine is the westernmost among the fifty-one or fifty-two Śākta *pīṭhas* or holy seats of the Goddess as the divine active power (Śakti), conceived of in a well-known Hindu myth as the places where the dissevered limbs of Śiva's spouse Satī, an earlier incarnation of Pārvatī, fell from the sky and gave rise to as many great places of pilgrimage.[3] Satī's *brahmarandhra* (the fontanel at the top of the head) is generally said to have fallen at this spot, though another tradition holds it is her navel that fell here (see below). Hinglaj is the best known Hindu place of pilgrimage in Baluchistan, one of the *vāmācāra* ('left-hand', i.e. Tantric esoteric) type. For many centuries it has been a site of pilgrimage for yogis from all over north India, as far as Bengal, despite the fact it has long been in the custody of Muslims, who regulate the pilgrimage rites in that place (Crooke 1913: 716; Briggs 1938: 106–07). Even a celebrated Ṣūfī saint of Sindh like Śāh 'Abdul Laṭīf (1689–1752) once made a pilgrimage to Hinglaj in the company of a band of Hindu yogis (Advani 1970: 16–18). Ṣūfī faqirs, too, e.g. Quṭb 'Alī Śāh (1808–1910), conferred with Hindu yogis at Hinglaj (Gajwani 2000: 39), and the shrine is often referred to as 'the goal' in Ṣūfī mystical poetry of Sindh (Schimmel 1975: 390).

Because in modern times the stone block worshiped as Hiṅglāj Devī by the Hindus has been taken by Muslim Baluchs, particularly from the heterodox Zikrī sect,[4] as marking the grave of the female Muslim saint Bībī Nānī ('Lady Grandmother'), legions of nineteenth and twentieth century scholars have traced it back to the cult of goddess Nanā (Gr. Nanaia) fostered by the Parthians, Scythians and Kuṣāṇas in the centuries straddling the beginning of the common era. The hypothesis, first advanced by Masson (1834: 172), is that 'the Muhammedans in Nanni [i.e. Bībī Nānī], may have preserved the Greek name Nanaia.' This assumption was

3 For a synthetic presentation of the mythologies about the Śākta *pīṭhas*, see Kinsley (1986: 184–87).

4 Schaflechner (2014: 54–55). This source states that the Zikrīs were in charge of the sanctuary until the late-twentieth century, when the organization was taken over by a Hindu committee.

accepted acritically, and with little additional investigation, by most of the authors who have referred to the cave temple at Hinglaj in their works after Masson. At any rate, since the life story of 'Bībī Nānī' (one more tomb of whom is venerated near the Bolan Pass in northern Baluchistan, see below) is virtually unknown in Muslim sources, and since this justifies the legitimate suspicion that the Baluch name 'Bībī Nānī' may point to some earlier, non-Islamic cult prevalent in Baluchistan in the pre-Islamic age, the possibility that goddess Nanā was worshiped at Hinglaj will be discussed in this chapter.

A GODDESS NAMED AFTER CINNABAR (OR VERMILION)

The name Hiṅgulā/Hiṅglāj clearly derives from the Sanskrit word for cinnabar/vermilion, *hiṅgula*. Hiṅglāj is from Hiṅgula-ja, where the suffix *-ja* stands for 'born from, produced by, belonging to.' Long ago Mazumdar speculatively proposed the derivation of the goddess's name Hiṅgulā from a set of Munda words for 'fire' (Mun. *sɛŋgɛl*; Ku. *ʃiŋgal*, etc.), yet his linguistic hypothesis has been apparently ignored by the scholars (Mazumdar 1921: xxviii). Eggermont, from a stray reference to the shrine of Hiṅgulā in Makran as 'Hiṅgu-pīṭha' found in the eighteenth century CE Tantric work entitled *Śaktisaṅgama Tantra* (Sircar 1960: 77), elaborates a theory that Hiṅgulā was the goddess of the asafetida plant (*Ferula assa-foetida* L., Skt. *hiṅgu*), native of Afghanistan and Iran (Eggermont 1975: 59–61). The latex from this plant, having a pungent, alliaceous odor due to the presence of sulfur compounds, is traditionally used in South Asia as a flavoring agent attributed with strong medicinal properties (Bakhru 2001: 6–11). In spite of such an intriguing link with sulfur (note that the Hiṅgulā cult seemingly has alchemical implications) Eggermont's theory, resting on an isolated geographical reference taken from a very late work, must be rejected.[5]

Cinnabar is a bright red mineral form of mercury sulfide (HgS). The formation of natural cinnabar deposits is tied to comparatively recent volcanic activity near the surface and the presence of hot springs. Mercury sulfide is denominated cinnabar when processed, by pounding, from the mineral itself, and vermilion when prepared synthetically by heating mercury with sulfur in a closed vessel. Cinnabar is the most common ore of

5 Moreover, etymologically the word *hiṅgula* 'cinnabar' and its derivatives are to be kept separate from the word *hiṅgu* '*Ferula assa-foetida* L.' See Mayrhofer (2001: 538 s.v. *hiṅgula*).

mercury. The metal is extracted from this mineral by a heating process that separates it from sulfur. Recombining mercury and sulfur to obtain vermilion by sublimation and subsequent condensation can chemically reverse this process.

It is not entirely clear whether the Sanskrit term *hiṅgula*, first attested in the *Mahābhārata* in the form *hiṅgulaka*,[6] originally indicated natural cinnabar or synthetic vermilion. The eminent linguist Manfred Mayrhofer attributes the first meaning to it, and furthermore adds that the word's etymology is unclear. A specific word for vermilion, *sindūra*, is only attested in Classical Sanskrit.[7]

Cinnabar is not available as a natural resource in India but is widely found in Nepal, Tibet and Central Asia (including Afghanistan). As to vermilion, both the Chinese and the Indians had the technology to synthetize it from ancient times (Lo Bue 1981: 60–61; White 1996: 66). In KauAŚ (II.12.2; II.13.46; II.14.34 and 48; II.22.6) *hiṅguluka*, a variant form of the word *hiṅgula* or mercury sulfide, is an article of commerce, although it is not clear whether by this term Kauṭilya meant cinnabar or vermilion. If in Kauṭilya's time *hiṅguluka* was being imported into India from foreign lands (Saletore 1973: 158–59),[8] then it was most probably cinnabar, the most expensive red pigment of the time.

Natural cinnabar, or else the less expensive, chemically produced vermilion, has been used in South Asia since ancient times to besmear both aniconic (stones, trees, etc.) and anthropomorphic cult images of goddesses and *yakṣa/yakṣī* deities, most probably as a mineral substitute for the offering of meat and blood (Skt. *bali*) (Eck 1982: 51–52), as well as the foreheads of 'sacred men (or women)' such as oracles, diviners, certain classes of ascetics, etc., and to mark the parting of the hair of married women to symbolize the happy state of a wife who enjoys her husband's protection (despite HgS being a toxic substance). Both the use of cinnabar or vermilion paste to besmear the cult images of Hindu deities and its use to mark the foreheads of married women may have originally determined the choice of the name Hiṅgulā for the goddess worshiped at Hinglaj in Makran. As regards the latter interpretation, it is traditionally said that

6 MWSED (1298 s.v. *hiṅgulaka*). The term occurs in MBh III.267.11 in the *bahuvrīhi* compound *piṣṭahiṅgulakānana* 'having one's face smeared with powdered vermilion' (or, rather, 'cinnabar'?). See Scharf (2003: 533).

7 Mayrhofer (2001: 538 s.v. *hiṅgula*; 512 s.v. *sindūra*).

8 Whereas Saletore (1973) translates *hiṅguluka* as 'vermilion', Chatterjee (1990: 56) translates it as 'cinnabar.'

goddess Satī, being Śiva's wife, wore vermilion in the parting of her hair and hence, after the upper part of her head fell in Makran, she became known by the name Hiṅgulā. However, the notion that Satī's *brahmarandhra* had fallen at Hinglaj is attested at such a late date (see below) that it is possibly irrelevant for establishing the origin of the theonym Hiṅgulā (Schaflechner 2014: 26).

According to the ancient principles of Tantric alchemy, sulfur symbolizes the Goddess or feminine principle (Śakti) and woman's uterine blood, while mercury symbolizes Śiva and man's seed. Cinnabar/vermilion, being the alchemical union of mercury and sulfur, the primordial pair of opposites, is considered a most powerful panacea in Tantric alchemical treatises. Some Tantric authors maintain that, to be technically successful, the alchemical processes of separation and recombination of mercury and sulfur should be accompanied by the performance of erotic rites. The goddess's name Hiṅgulā, 'She of Cinnabar/Vermilion,' may possibly bear some connection with these alchemical traditions (Brighenti 2001: 202). White has explored in some detail the Tantric alchemical significance of the cult and shrine of Hiṅglāj Devī and defines the latter as an 'alchemical hierophany.'[9] Yet, White's alchemical hypothesis, as far as this Hindu goddess is concerned, remains to be verified.

EARLY LITERARY REFERENCES TO GODDESS HIṄGULĀ AND THE SPREAD OF HER CULT

According to Agrawala (1964: 136–40), the earliest literary mention of the shrine of Hiṅgulā in Makran would be found in the *Vāmanapurāṇa*.[10] The critical edition of this text edited by Anand Swarup Gupta (VāmP) only states that a youthful goddess named Carcikā, born from Śiva's sweat on his forehead during his combat with the demon Andhaka, settled in 'the best of places,' a mountain called Haiṅgulata (*haiṅgulatādrimuttamam*, XLIV.47). The adjective *haiṅgula* means 'coming or derived from [goddess] Hiṅgulā,'[11] and in some later Tantric texts the *pīṭha* of Hiṅgulā is alternatively termed as Hiṅgulāṭa (Sircar 1973: 35; Bhattacharyya 2002: 60). Such

9 See especially White (1996: 196, 205–06, 245–46, 311–12).

10 Although no scholarly consensus has been reached regarding the date of composition of VāmP, this work is generally assigned to the ninth–tenth centuries CE. See Rocher (1986: 240–41).

11 MWSED (1305 s.v. *haiṅgula*).

lexical data may support Agrawala's claim that the Haiṅgulata mountain referred to in VāmP is identical with the Hinglaj mountain range near the Makran coast, where the *pīṭha* of Hiṅgulā is situated. As a cautionary note, however, it should be mentioned that in the Buddhist *Kuṇālajātaka*, an apparently very early Pāli text, Mount Hiṅgula (P. Hiṅgulapabbata) is placed in the Himalaya (Himavanta), not at the west end of India.[12]

In the list of sacred sites of the Goddess included in the DBhP (vii.38.6), an important Śākta work which is generally held to be slightly later than VāmP,[13] the shrine of Hiṅgulā in Makran is paired with that of Jvālāmukhī in the Kangra valley in one verse. This appears significant in view of the close connection in cult between these two Śākta *pīṭhas* (see below). The late medieval *Brahmavaivarta Purāṇa* (*Kr̥ṣṇajanmakhaṇḍa* lxxvi.21),[14] in describing pilgrimage to certain holy places, mentions the 'image of Durgā' (possibly aniconic like the present one) at the shrine of Hiṅgulā.

Some medieval Tantric texts, e.g. the *Kulārṇava* and *Kubjikā Tantras*, mention Hiṅgulā in their lists of *pīṭhas* (Sircar 1973: 18–19). However, it is only in the *Pīṭhanirṇaya*, a small Tantric work dated by Sircar to c. 1690–1720, that we find for the first time the shrine of Hiṅgulā at Hinglaj listed as the first – and, consequently, foremost – among the fifty-one Śākta *pīṭhas*, most likely on account of its being identified, as likewise stated for the first time in that work, as the place where Satī's *brahmarandhra* had fallen (Schaflechner 2014: 25–26).[15] A competing but less influential tradition, reflected in the sixteenth century Bengali poem *Caṇḍīmaṅgala*, states that it was Satī's navel – not her *brahmarandhra* – that fell at Hinglaj (Sircar 1973: 32–33).

In agreement with the dates proposed by scholars for early references to goddess Hiṅgulā in Purāṇic and Tantric literature, an inscribed Yoginī image representing Durgā as Mahiṣāsuramardinī, found from the principal cell of the Sixty-four Yoginī temple in Khajuraho (Madhya Pradesh) from

12 Fausbøll (1891: 415). See also Norman (1983: 81–82) for the early dating of the *Kuṇālajātaka*.

13 See Rocher (1986: 172). The dates generally proposed by scholars for the DBhP span the tenth to the fourteenth centuries ce. The list of Devī *tīrthas* given in DBhP vii.38.5–30 is, however, thought to be a late medieval interpolation – see Sircar (1973: 107).

14 BrVP is generally recognized as one of the more recent Purāṇas. Its present text is dated to the fifteenth–sixteenth centuries ce. See Rocher (1986: 163).

15 For the dating of the *Pīṭhanirṇaya* and the debunking of its traditional ascription to the earlier *Tantracūḍāmaṇi*, see Sircar (1973: 3–4, 24). For the list of fifty-one *pīṭhas* given in the same work, see ibid.: 35–37.

the tenth century CE, is labelled 'Hiṅghalāja' (Dehejia 1986: 117). The label inscription on the pedestal, showing paleographical development from '*ga*' to '*gha*', is said to have been added by worshipers of goddess Hiṅgulā a century or more after the image was carved (Desai 2013: 112–13). This implies that the original theonym Hiṅgalāja, a variant form of the name Hiṅgulā, must have been in use in at least the tenth century CE.

The spread of the cult of Hiṅgulā to Gujarat, Rajasthan and the northern Deccan in the medieval period appears to have been mainly due to the migration or periodic movements from Baluchistan-Sindh of communities who venerated her as their clan or lineage goddess (*kuladevī*) or, at any rate, as an important deity. These include the Lambāḍi/Banjārā nomadic traders, who worshiped her for protection of their travels along the caravan routes (Deshpande 1995: 173–74); the Cāraṇa pastoral caste of Rajasthan and Gujarat, some of whose communities claim to have come from Sindh, and a section of whom evolved into the royal bards of their Rājput patrons, who in turn started to worship Hiṅglāj Devī (Kamphorst 2008: 231–38, 249, 253–55);[16] the Bhāvasāras, a caste of dyers and tailors of Rajasthan, Gujarat and Maharashtra, who also possibly migrated from Sindh (Deshpande 1995: 174); and several others. In both Gujarat and Rajasthan there are many temples dedicated to Hiṅglāj Devī which, in the case of Rajasthan, are often situated along the trade routes crossing the Thar desert (Tambs-Lyche 2004: 123; Kamphorst 2008: 238, 253–54). The most revered temple of Hiṅglāj Devī in Rajasthan is the one in the hill fort of Hinglajgarh, situated just across the present border with Madhya Pradesh.

It is, in any event, apparent that there is no definite evidence for the existence of a goddess named Hiṅgulā (or the like) in India before the ninth–tenth centuries CE. This notwithstanding, there are reasons to believe the site of Hinglaj in Makran could have been a center of goddess worship since at least one thousand years before that, even though it might not yet have been dedicated to a Hindu goddess in this early phase. This working hypothesis is based on two arguments:

1. The 'numinous' geophysical and geochemical characteristics peculiar to the Hinglaj region would have strongly prompted the establishment of a religious cult there at an early date.
2. The place may have been formerly a seat of the cult of the non-Indian goddess Nanā/Nanaia.

16 The Cāraṇas worship a long line of medieval and contemporary *sagatīs* (i.e. *śaktis*), deified women who became recognized as full or partial incarnations of Hiṅglāj Devī.

GEOPHYSICAL AND GEOCHEMICAL FEATURES OF THE HINGLAJ AREA IN MAKRAN

Mud volcanoes, whose location in Baluchistan is tied to fault systems, are a very important manifestation of gas in the coastal zone of the Makran desert. Here the extrusion of fluid, gas-charged mud has built crater-like, cone-shaped surface features suggestive of miniature volcanoes. Most of the cones, up to 400 feet in elevation, are found close to sea level, bordering on the Arabian Sea. Submarine mud volcanoes, too, have been identified on the abyssal plain off Makran. Active mud vents have been reported as ephemeral islands that have emerged suddenly in various locations and at various times offshore from the Makran desert in the aftermath of earthquakes. Explosive mud eruptions in the volcanoes of Makran occur only when violent seismic activity is exerting pressure on the methane gas trapped below. The action is similar to the eruption of a geyser, with rumbling noises and much agitation in the crater (Snead 1964: 546–49, 556–57; Delisle 2005: 159–60).

The 190-foot-high Chandrakup (Skt. *candra* 'moon' and *kūpa* 'well'), a mud volcano which most people consider an embodiment of Śiva, is a major object of worship during pilgrimages to Hiṅglāj Devī. After climbing the cone located along the route to the goddess's shrine, pilgrims drop offerings into the mud crater and interpret their destiny in terms of the number and size of the gas bubbles or jets of liquid mud that follow as if the Chandrakup were a kind of 'oracular well' (Schaflechner 2014: 186–92). Moreover, as reported by several British colonial authors, close to the cave shrine at Hinglaj, atop a hill, there is a sacred water pool from whose bottom bubbles of gas, perhaps having the same source as that of the gas vents of the mud volcanoes of Makran, rise to the surface of the water carrying mud in suspension. Pilgrims throw offerings into the water, and if bubbles rise in sufficient quantities and in a short time to the surface, they consider that their sins have been forgiven. The rising of the bubbles of gas is considered a sight (Skt. *darśana*) of goddess Hiṅglāj.[17] It cannot be determined whether the pool in question is fed by sulfur springs; yet, sulfur, deposited from sulfurous springs, is found in Makran south of the large Kandawari mud volcano in the Haro range and in the hills adjacent to the Ormara group of mud volcanoes, where its odor can be detected.[18]

17 Burnes (1834: 33); Carloss (1840: 194); Leech (1843: 474); Minchin$_a$ (1907: 37); Crooke (1913: 716).

18 Minchin$_a$ (1907: 12); Vredenburg (1909: 208); Snead (1964: 555).

Furthermore, abundant submarine gas seeps and vents composed of methane and hydrogen sulfide, the latter producing a strong smell of sulfur, are observed offshore along the Makran coast (von Rad et al. 2000: 10–19).

It is significant to note that sulfurous springs are connected with the other important 'tomb of Bībī Nānī' venerated by the Muslim Baluchs besides the one coinciding with the aniconic cult image of Hiṅgulā at Hinglaj – namely, the one located near the Bolan Pass, which is associated in the legend with the 'tomb' of the Muslim saint Pīr Ghaib (whose name means 'the vanished saint'). The latter, who was supposedly Bībī Nānī's brother, and who is worshiped as Śiva by the Hindus, is said to have miraculously produced a sulfur spring at the spot where he walked into solid rock (inside which his 'concealed' tomb is fancied to lie) on being pursued by an army of *kāfirs* (infidels). His sister Bībī Nānī, locally regarded as 'the guardian-saint of the water supply,' is buried under a road bridge some ten kilometers south of the said spring.[19] The sulfur spring related to the Muslim legend of Bībī Nānī and Pīr Ghaib in the Bolan Pass may provide a parallel to the pool with bubbling of gas associated in cult with the Hindu goddess Hiṅgulā, identified with Bībī Nānī by the Muslim Baluchs.

Gas venting is not a phenomenon confined to the mud volcanoes scattered along the Makran coast and offshore on the continental shelf. Indeed, in the area north of Hinglaj there are old mud ridges (high, broad hills with steep sides formed by thick packages of mud) that have no direct connection with central vents. Such hill ranges are called 'Extrusive Mud Formation' by geologists. This formation occurs over distances of tens of kilometers along fault structures. In past geological eras, huge amounts of gas-charged viscous mud were extruded to the surface, in analogy to circular mud volcanoes. However, even today earthquakes can trigger the eruption of some inactive mud volcanoes along these mud ridges.[20] During the Makran earthquake of 1945 the ignition of large volumes of methane gas that erupted under great pressure near Hinglaj caused a red glow to appear in the sky which could be seen from a great distance and gave rise to stories of volcanic eruptions. The flames of the great fireball leaped thousands of feet high into the sky and were seen burning for days over two mountains north of Hinglaj that are part of the above-mentioned 'Extrusive Mud Formation' (Sondhi 1947: 147, 149). Tales of mud vents on fire are likewise current in the nearby Haro Range (Snead 1964: 557). This

19 Minchin$_b$ (1907: 37–38); Suvorova (2004: 22). See also Rashid (2008).

20 Snead (1964: 546–48); Ambraseys and Melville (1982: 90); Delisle et al. (2002: 94).

natural phenomenon has certainly occurred many times in the Hinglaj region from the prehistoric period, thereby conferring a numinous quality to the area which may have ultimately been conductive to the establishment of a religious center there.

Earthquake-triggered eruptions of mud volcanoes, spontaneous outbursts of flammable gas (generally methane), and the occurrence of sulfur springs, are considered in South Asia to be geophysical and geochemical manifestations of the Goddess. From time immemorial, they have attracted the attention of the people, who have regarded them with mixed fear and reverence. As remarked by White, '[s]uch mineral theophanies are quite a commonplace at *pīṭhas* of the Goddess' (1996: 480n150). In this connection it is worth mentioning that, at the very opposite end of the Indian subcontinent, a Śākta pilgrimage center similar in a sense to that dedicated to Hiṅgulā/Hiṅglāj was established in a tectonically active region where eruptions of mud volcanoes and gas seepages are as prevalent as in Makran, that is, the Chittagong/Arakan hills and the adjacent coastlands straddling Bangladesh and Burma. The shrine in question, dedicated to goddess Bhāvanī, is located on the Chandranath hill near Sitakund in Chittagong district, Bangladesh (Sircar 1973: 47).[21] Numerous gas shows occur in the Sitakund anticline, and near the base of the hill on which the temple stands, a perpetual flame fed by combustible gas bursts out of the rock. In the surrounding hills other temples have been built over gas seepages, one of which is known as the 'burning spring' because it features a jet of inflammable gas coming up through a water spring (O'Malley 1908: 190). This 'burning spring' became highly active soon after the great 1897 Assam earthquake (Oldham 1899: 41). As to mud volcanoes, occurring more frequently in nearby Arakan (Burma), two of them were reported on the Sitakund anticline during the great 1762 earthquake that also caused fountaining of mud throughout this hilly area (Steckler et al. 2008: 372–73).

A conceptually similar cult is the one of the goddess Jvālāmukhī in the Kangra valley of sub-Himalayan Panjab (now part of Himachal Pradesh). The Sanskrit term *jvālāmukhī* ('flame-mouthed') indicates 'any place from which issues subterranean fire or inflammable gas,'[22] including terrestrial gas vents/seeps and, in a more generic sense, the mouths of volcanoes (White 1996: 234). The temple of Jvālāmukhī houses blue flames of natural gas (mostly methane) (Valdiya 2001: 34) issuing from crevices in the rock

21 Although the *Pīṭhanirṇaya* lists this shrine, too, among the fifty-one Śākta *pīṭhas*, Sircar (ibid.: 34) comments it is 'late and unimportant.'

22 MWSED 428 s.v. *jvālā*.

and is, also in this case, regarded as one of the fifty-one Śākta *pīṭhas* (Sircar 1973: 44). There is no presiding image in the temple – what is worshiped as a deity is the main ('eternal') flame itself. Moreover, at this *pīṭha* of the Goddess, too, a (small) water body with bubbling of combustible gas, which is ignited by the temple priests, is actively worshiped. A typologically similar goddess-shrine, sacred to both Hindus and Buddhists, is the one of Jvālā Māī ('Flame Mother') at Muktinath in Nepal, where three 'eternal' blue flames fueled by natural gas burn, respectively, upon the soil, stone and water inside a small Buddhist *gomba* (Snellgrove 1961: 199–201).

Medieval trade routes followed by the Panjabis have caused Hindu places of worship of the *jvālāmukhī* type to be established as far as Azerbaijan and northwest Iraq. Indeed, at Surakhany near Baku in Azerbaijan there once stood a 'fire-temple' run by Hindu priests. The temple was located in the vicinity of petroleum deposits, the flammable gases from which once used to be worshiped by Hindu pilgrims as 'Jvālā-jī' – another name for goddess Jvālāmukhī of Kangra valley (Stewart 1897: 311–18; Williams Jackson 1911: 39–57). At the site of Bābā Gurgur, in the Kirkuk area of Iraq, there are 'eternal fires' fed by petroleum and natural gas seeping through cracks in the rock. Colonel Wilford reports that in his time (late eighteenth century) the site was occasionally visited by Hindu pilgrims who called it a *jvālāmukhī* (1799: 393–94).[23] This is interesting in the light of the hypothesis, discussed in the last section of this chapter, that the shrine of Hiṅglāj Devī in Makran was earlier dedicated to the East Iranian version of goddess Nanā/Nanaia, for Hoffmann (1880: 273) identifies Bābā Gurgur with the site of the Parthian city of Demetrias that is mentioned by Strabo (XVI.1.4) as one having a source of petroleum with (natural?) fires and the sanctuary of Nanaia.[24] The presence in the area of 'self-sustained fires' (Pahl. *ataxš ī a-xwarišnīh*) fed by petroleum and by gas jets (referred to as objects of a cult in Sasanian times; Boyce and Grenet 1991: 48) would have attracted Zoroastrian veneration, and it seems possible that a shrine of goddess Nanaia, whose cult emphasized natural phenomena, was established there in connection with such sacred fires. In course of time, the hypothesized Nanaia cult at Bābā Gurgur near Kirkuk could also have attracted Hindu pilgrims from afar as it might have been the case with the supposed Nanā

23 Wilford spells the name of this site as 'Curcoor' (i.e. Curcur), and one of the alternative transliterations of the Kurdish name Bābā Gurgur (*bābā* 'father', *gur* 'flame') is 'Bābā Kurkur.'

24 The name of the goddess worshiped at Demetrias, for which the manuscripts of Strabo's *Geographika* give the form *Anaia (emendation of *Anea), is generally read by modern scholars as Nanaia: see de Jong (1997: 274–75).

cult at Hinglaj (if indeed there ever was one) before the goddess worshiped there was converted to a Hindu deity.

According to various sources,[25] a plume of fire from a gas vent – an 'undying' flame said to spontaneously emerge from the earth – is maintained at the cave temple of Hiṅglāj Devī down to the present day, yet this feature is not confirmed in most other accounts of the shrine. In any event, Hiṅgulā/Hiṅglāj seems to have been primarily worshiped in the form of burning flame after her cult spread eastwards from Baluchistan-Sindh (Weinberger-Thomas 1999: 156–57; Mishra 2006: 57). For example, the householder Nāths of Rajasthan, a caste of married yogis whose legendary guru is Gorakṣanātha, still today worship Hiṅgulā in the form of a flame arising from a jar of water in a rite held on the eleventh night after a death (Gold 1988: 102–04). Yogis of the Nātha Siddha order also play a very important role in cult practices at the Jvālāmukhī temple in Kangra valley with its holy flames (White 1996: 121; Sharma 2006: 102), as well as in pilgrimages to the Hiṅglāj Devī shrine in Makran with its sacred bubbling of mud in the volcano and the gas seepages,[26] all of these being manifestations of the Goddess as Sacred Fire.[27]

The worship of goddess Hiṅgulā in the form of natural fire is interestingly performed as far as Odisha in the village of Gopalprasad near Talcher (Angul district). The area, part of the large Talcher coalfield, is so rich in coal that natural gas on occasion spontaneously ignites on the surface. During her festival, held every year in the month of Caitra (March–April), the goddess manifests herself in this place in the form of jets of gas which either self-ignite or are ignited by one of her priests, to whom she is believed to appear in dream and indicate the exact spot of her annual manifestation. The flame is thereafter kept burning for days by adding coal to it. The coalfield itself is regarded by Hiṅgulā's devotees as the body of the goddess. Hiṅgulā of Odisha is, therefore, best described as a fire-goddess residing in the subsoil, from which she is believed to spring forth as jets of inflammable gas (Brighenti 2001: 229–30; Mallebrein 2011: 266–67).

25 Erndl (1987: 167); Gold (1988: 103 n. 48); White (1996: 122); Kamphorst (2008: 237).

26 Briggs (1938: 105). Nātha pilgrims en route to Hinglaj worship the bubbles in the Chandrakup crater as 'Bhabaknāth' (cp. Pan. *bhabak* 'glare,' Sin. *bhabhaka* 'flare,' H. *bhabaknā* 'to burst into flame'), supposedly an embodiment of Guru Gorakṣanātha. See Burton (1851: 322–23).

27 In the daily rites of the great temple of Jagannātha at Puri in Odisha, Hiṅgulā is worshiped as the fire on which the *mahāprasāda* (sacred food offerings) is cooked in the temple's kitchen.

WAS THERE EVER A SHRINE DEDICATED TO THE IRANIAN GODDESS NANĀ AT HINGLAJ?

As mentioned in the first section of this chapter, the veneration paid by Muslim groups in Baluchistan to the two grave-shrines of the ill-described female saint 'Bībī Nānī' existing in that country, the one coinciding with the Hindu shrine at Hinglaj and the other located near the Bolan Pass, led scholars since the early nineteenth century to postulate that the deity originally worshiped at those two sites was the ancient eastern Iranian goddess Nanā. This identification, it must be observed, only rests on the similarity of the two names, 'Nānī' and 'Nanā.' The question now is – is there any other evidence for the existence of a shrine of goddess Nanā at Hinglaj in ancient times?

In the opinion of the present writer, too little attention has been paid, in connection with this problem, to a striking feature observed at the site of Hinglaj, namely, the massive crescent-shaped figure near the top of the cliff opposite the cave shrine of Hiṅglāj Devī, apparently a natural formation in the rock wall caused by erosion that is clearly visible as one looks

Figure 2.1. Naturally(?) formed crescent-shaped figure on the rock wall opposite the shrine of Hiṅglāj Devī, Hinglaj, Pakistan, 2010. Courtesy of Jürgen Schaflechner.

upward from the entrance to the shrine (figure 2.1). A moon crescent was the main symbol of goddess Nanā among the eastern Iranians from the Hellenistic to the early medieval period (see below), and the writer's initial hypothesis was that the aforesaid 'crescent-like rock formation' at Hinglaj could be part of the figures of the sun, (crescent) moon and stars 'painted with red ochre on a spot inaccessible to mortals on the opposite mountain [with respect to the cave shrine of Hiṅglāj Devī],' which were first mentioned by Captain Hart (1840: 80, 96), and by many scholars after him.

In the light of more detailed investigations, however, it was ascertained that the figures referred to by Hart are, as a matter of fact, hewn out of the rock (and painted over with red ochre) on the ceiling of a natural recess in the rock wall, accessible from the Hinglaj gorge through a narrow and dangerous ledge, that opens in the cliff *a few metres below* the said crescent-shaped figure. These astral symbols – which, unlike the crescent-shaped figure, are barely visible from the bottom of the gorge – are locally said to have been engraved on that almost inaccessible part of the mountain by Rāma himself when he visited the shrine (Hart 1840: 80, 96; Schaflechner 2014: 222–23).[28] They could have been etched in the rock there at any time from the medieval period onwards either because '[s]un and moon are [...] absolutely pivotal images in the hathayogic system of the Nātha Siddhas, who are the perennial pilgrim's guides to and custodians of Hiṅglāj' (White 1996: 205) or because the sun and moon icons often featured in medieval Rājput memorial stones and steles depicting *satīs* (self-immolating widows) as 'visual references to Hindu marriage vows, which specify that "as long as there is sun and moon this couple will remain together" [and] as the act of *satī* is believed to ensure the married couple's union in heaven for eternity, or "as long as there is sun and moon"' (Belli Bose 2015: 221). In the latter interpretation, and keeping in mind that the worship of Hiṅglāj Devī was prominent among the Rājputs, the sun and moon symbols engraved on the rock at the site of Hinglaj would have acted as 'witnesses' of the self-immolating goddess Satī's eternal union with Śiva.

What, then, about the other, much larger and naturally(?) formed crescent symbol visible high on the rock wall opposite the Hindu shrine at Hinglaj? Could it have had any connection with the postulated worship of the Iranian goddess Nanā at that site in ancient times?

28 I am thankful to Jürgen Schaflechner for sending me the photographs of both the 'crescent-shaped figure' on the top of the cliff opposite the shrine of Hiṅglāj Devī and the sun, moon and star symbols hewn into the rock in the cave-like structure below it.

Nanā (not identical to Inanna/Ištar as previously assumed by many scholars) was an astral goddess worshiped in Mesopotamia and Elam from the mid-third millennium BCE (Potts 2001: 23–24, 28–30). Her cult, though it was not part of the Avestan heritage, was probably imported to Bactria and other eastern satrapies in the Achaemenid period. There is, however, no clear evidence of Nanā from Achaemenid Iran (Ambos 2003: 258; Shenkar 2014: 119).[29]

In the Seleucid period the cult of Nanā/Nanaia is attested in inscriptions from Susa (Ambos 2003: 250), while a seal from Seleucia-on-Tigris possibly indicates she had already been assimilated to the Greek goddess Artemis (Invernizzi 1998: 92–96). The syncretistic goddess portrayed in this seal holds a large moon crescent, the chief emblem of Nanā in the later Iranian world. Although in Mesopotamia she was connected with the moon-god Sin, before the Seleucid period Nanā was not a moon goddess, and it is likely that she owed her lunar character to her identification with Artemis by the Hellenized Iranians. Thus, at Susa, in the Arsacid period, Nanā's astral aspect and function as a city-goddess were transferred to the syncretistic cult of Artemis-Nanaia, supposedly represented on unlabeled Parthian coins issued by Mithradates II around 110 BCE (Azarpay 1976: 537).[30] On Parthian coins of Mithradates II (c. 123–88 BCE) and Orodes II (c. 58–39 BCE) from Susa, only a crescent is represented. It is suggested this could be a symbol of Artemis-Nanaia as the latter deity is depicted together with a moon crescent on some later coins of the Kings of Elymais (Susa), vassals of the Arsacids (Ambos 2003: 251). It is also known that, in the Arsacid period, a sanctuary of Nanā existed in the ancient Parthian capital and royal residence, Nisa in present-day Turkmenistan, and a temple of Artemis-Nanaia must have existed in Susa as frequently recorded in classical sources (Rosenfield 1967: 86; Harmatta et al. 1994: 322). It is, thus, apparent that in Hellenistic and Parthian Iran Nanā/Nanaia was often depicted with a moon crescent, the attribute of a lunar deity.

The earliest evidence for a cult of Nanā in Bactria comes from the coins issued by the local Scythian rulers Sapadbizes (or Sapalbizes), Arseiles and Pulages in the late first century BCE. They show on the reverse a crescent symbol over a lion (the animal attribute of this deity in Central Asia), explicitly identified by the legend as the goddess 'Nanaia' (Ghose 2006: 98).

29 According to N. Sims-Williams, the name of the goddess Nanā was borrowed into the (East Iranian) Bactrian language from the (West Iranian) Old Persian language in the Achaemenid period (2002: 227).

30 Ambos (2003: 233) remarks that from the cuneiform sources no clear evidence can be obtained that Nanā was a moon goddess.

Here Nanā is, therefore, represented aniconically as a lunar crescent, possibly after the Parthian model referred to above. According to Francfort (2012: 87), these coins would be the earliest evidence for the introduction of Nanā into Bactria via the Parthians. One more piece of evidence for the worship of Nanā by Scythian rulers in Bactria is provided by the gold belt from the burial site of Tillya Tepe (first half of the first century CE), whose medallions show the image of a female deity mounted on a lion that is often identified with Nanā (Ghose 2006: 98). As regards the Indo-Scythians or Śakas of Taxila, their king Maues (first half of the first century BCE) issued a coin depicting a goddess with a crescent on top of her head (Mitchiner 1976: 470). This deity could be a Śaka form of Artemis-Nanaia. Finally, on Kuṣāṇa coins and seals of the second to fourth centuries CE Nanā is invariably represented anthropomorphically and wearing a diadem with a crescent. Since the coins are inscribed with the name of the goddess, her identification is, in this case, certain (Shenkar 2014: 119–21). As a dynastic goddess her role was concerned with bestowing the divine legitimation of kingship to the Kuṣāṇa king Kaniṣka I, who, as is recorded in his Rabatak inscription (northern Afghanistan, early second century CE), 'has obtained the kingship from Nanā and from all the gods' (Sims-Williams and Cribb 1995–1996: 78).

To sum up, due to its strong resemblance to the main symbol of goddess Nanā on coins from Susa and Bactria, the mineral formation resembling a moon crescent on the rock at Hinglaj may have been interpreted as a sign of Nanā by Parthian and/or Śaka invaders of Baluchistan around the turn of the current era, thus prompting them to establish a Nanā shrine on the spot.

The image of the goddess Nanā continued to be represented by the eastern Iranians (Bactrians, Chorasmians, Sogdians and Khotanese Śakas) till the seventh–eighth centuries CE, which fact testifies to her popularity as the 'Great Goddess' of Eastern Iran till the advent of Islam in that part of the world (Shenkar 2014: 121, 124–27). That the worship of Nanā, if indeed there ever was a sanctuary dedicated to her at Hinglaj in ancient times, may have survived in Makran till the Arab conquest in the seventh century CE, is therefore not impossible a priori. In Alexander's time, and possibly later, too, the Hinglaj area was included in the territory of the Oreitai (Gr. 'Mountaineers'), taken by some historians as ethnic Iranians, perhaps somewhat culturally mixed with Indians.[31] If the Oreitai were Iranians,

31 Tarn (1948, Vol. II: 15, 250–51, 254); Hamilton (1972: 603–604 n. 6); Eggermont (1975: 63). On the Oreitai see Arrian, *Indika* 21.8, 22.10, 25.2 – all from Alexander's admiral, Nearchus.

the East Iranian cult of Nanā could have been introduced into their country from the north via Seistan and Arachosia in Achaemenid times or, more probably, by Indo-Scythian and especially Indo-Parthian rulers of Baluchistan in the first century BCE–first century CE.

Tanabe, based on the iconography of Nanā on Kaniṣka I's coins, argues this goddess was not only associated with astral phenomena (cf. her moon symbol) and royal power, but also with earth and water (1995: 209–10), while Rosenfield writes her cult 'emphasized natural phenomena' (1967: 88). In view of this, one may suggest that the extraordinary natural phenomena occurring in the Hinglaj area in Makran (see above, pp. 35–39), too, could well have caused groups of Iranian votaries of the 'Great Goddess' Nanā to establish a center of her worship at that place. If this was indeed the case, the shrine at Hinglaj might have continued to be dedicated to Nanā till the Arab conquest of Makran, after which local converts to Islam might have started to frequent it as the supposed 'tomb of Bībī Nānī', thus, 'Islamizing' both the name and veneration of Nanā. A local Muslim tradition recorded by Stein identifies Bībī Nānī as a saintly wife of the Prophet's son-in-law 'Alī, miraculously translated to Hinglaj from some place in West Asia (1943: 202); hence, the first Arab invasion of Makran (644 CE) would have most likely occurred during her supposed lifetime.

The hypothesis outlined above, admittedly speculative, does not, in any event, contradict the textual evidence from Sanskrit sources which, as already discussed (see above pp. 32–34), are silent about a goddess named Hiṅgulā, and all the more so about a shrine dedicated to her in Makran, before the ninth–tenth centuries CE. These textual data (or, rather, the absence thereof) appear fully compatible with the possibility that the Hindu cult of Hiṅgulā may have superseded a pre-existing goddess cult at Hinglaj sometime during the early medieval period. Such a chronological hypothesis would be in line with the expansion of Śāktism in north India as an independent and powerful religious faith in the last four centuries of the first millennium CE (Bhattacharyya 1974: 85–108), particularly after the diffusion and popularization of the *Devīmāhātmya* ('Glorification of the Goddess'; DM in MāP LXXXI–XCIII), a text dating from around the sixth or perhaps fifth century CE (Coburn 1984: 1). The process of Sanskritization of any pre-Hindu cult hypothetically established at the numinous site of Hinglaj from antiquity could have taken place at any time during that period.

ABBREVIATIONS AND REFERENCES

Gr. = Greek H. = Hindi Ku. = Kurku Mun. = Mundari Pahl. = Pahlavi P. = Pali
Pan. = Panjabi Skt. = Sanskrit Sin. = Sindhi

Advani 1970	Advani, K.B. (1970). *Shah Latif.* New Delhi: Sahitya Akademi.
Agrawala 1964	Agrawala, V.S. (ed.) (1964). *Vāmana Purāṇa. A Study.* Varanasi: Prithivi Prakashan.
Ambos 2003	Ambos, C. (2003). 'Nanaja – eine ikonographische Studie zur Darstellung einer altorientalischen Göttin in hellenistisch-parthischer Zeit.' *Zeitschrift für Assyriologie*, 93(2): 231–72.
Ambraseys and Melville 1982	Ambraseys, N.N. and C.P. Melville. (1982). *A History of Persian Earthquakes.* Cambridge: Cambridge University Press.
Azarpay 1976	Azarpay, G. (1976). 'Nanā, the Sumero-Akkadian Goddess of Transoxiana.' *Journal of the American Oriental Society*, 96(4): 536–42.
Bakhru 2001	Bakhru, H.K. (2001). *Indian Spices and Condiments as Natural Healers.* Mumbai: Jaico Publishing House.
Belli Bose 2015	Belli Bose, M. (2015). *Royal Umbrellas of Stone. Memory, Politics, and Public Identity in Rajput Funerary Art.* Leiden and Boston: Brill.
Bhattacharyya 1974	Bhattacharyya, N.N. (1974). *History of the Śākta Religion.* New Delhi: Munshiram Manoharlal.
Bhattacharyya 2002	Bhattacharyya, N.N. (2002). *Tantrābhidhāna. A Tantric Lexicon.* New Delhi: Manohar.
Boyce and Grenet 1991	Boyce, M. and F. Grenet (1991). *A History of Zoroastrianism*, Vol. III. Leiden: E.J. Brill.
Briggs 1938	Briggs, G.W. (1938). *Gorakhnāth and the Kānphaṭa Yogīs.* Calcutta: Y.M.C.A. Publishing House.
Brighenti 2001	Brighenti, F. (2001). *Śakti Cult in Orissa.* New Delhi: D.K. Printworld.
BrVP	*Brahmavaivartapurāṇa.* Sen, R.N. (tr.) (1922). *Brahma-Vaivarta Puranam. Ganesa and Krisna Janma Khandas.* Allahabad: Panini Office.
Burnes 1834	Burnes, A. (1834). *Travels into Bokhara*, Vol. III. London: John Murray.
Burton 1851	Burton, R.F. (1851). *Sindh, and the Races That Inhabit the Valley of the Indus.* London: W.H. Allen.
Carloss 1840	Carloss, T.G. (1840). 'Account of a Journey to Beylah, and Memoir on the Province of Lus.' *Journal of the Asiatic Society of Bengal*, 8(87): 184–202.
Chatterjee 1990	Chatterjee, A. (1990). *Studies in Kauṭilya-Vocabulary.* Delhi: Parimal Publications.
Coburn 1984	Coburn, T.B. (1984). *Devī-Māhātmya. The Crystallization of the Goddess Tradition.* Delhi: Motilal Banarsidass.
Crooke 1913	Crooke, W. (1913). Hinglāj. In J. Hastings (ed.), *Encyclopedia of Religion and Ethics*, Vol. VI, pp. 715–16. Edinburgh: T. & T. Clark.

DBhP — *Devībhāgavatapurāṇa*. Swami Vijnananda (tr.) (1922). *The Srimad Devi Bhagavatam*. Third edition. Allahabad: Panini Office.

de Jong 1997 — de Jong, A. (1997). *Traditions of the Magi. Zoroastrianism in Greek and Latin Literature*. Leiden: Brill.

Dehejia 1986 — Dehejia, V. (1986). *Yoginī Cult and Temples. A Tantric Tradition*. New Delhi: National Museum.

Delisle 2005 — Delisle, G. (2005). Mud Volcanoes of Pakistan: An Overview. In G. Martinelli and B. Panahi (eds.), *Mud Volcanoes, Geodynamics and Seismicity*, pp. 159–69. Dordrecht: Springer.

Delisle et al. 2002 — Delisle, G. U. von Rad, H. Andruleit, C. von Daniels, A. Tabrez, A. Inam. (2002). 'Active Mud Volcanoes On- and Offshore Eastern Makran, Pakistan.' *International Journal of Earth Sciences*, 91(1): 93–110.

Desai 2013 — Desai D. (2013). 'The Goddess Hiṅghalāja of the Yoginī Temple at Khajuraho.' In I. Keul (ed.), *'Yoginī' in South Asia. Interdisciplinary Approaches*, pp. 109–16. New York: Routledge.

Deshpande 1995 — Deshpande, M.N. (1995). 'Some Aspects of Folk Religion in the Konkan and Desh Regions of Maharashtra.' In G.D. Sontheimer (ed.), *Folk Religion and Oral Tradition as a Component in Maharashtran Culture*, pp. 165–80. New Delhi: Manohar.

DM — *Devīmāhātmiya*. Coburn, T.B. (trans.) (1991). *Encountering the Goddess. A Translation of the Devī-Māhātmiya and a Study of Its Interpretation*. State University of New York Press, Albany.

Eck 1982 — Eck, D.L. (1982). *Banaras. City of Light*. New York: Knopf.

Eggermont 1975 — Eggermont, H.L. (1975). *Alexander's Campaigns in Sind and Baluchistan and the Siege of the Brahmin Town of Harmatelia*. Leuven: Leuven University Press.

Erndl 1987 — Erndl, K.M. (1987). *Victory to the Mother. The Goddess Cult of Northwest India*. PhD Dissertation, University of Wisconsin-Madison.

Fausbøll 1891 — Fausbøll, V. (1891). *The Jātaka, Together with Its Commentary*, Vol. v. London: Kegan Paul, Trench, Trübner and Co.

Francfort 2012 — Francfort, H.P. (2012). 'Tillya Tepe and Its Connections with the Eurasian Steppes.' In J. Aruz and E. Valtz Fino (eds.), *Afghanistan. Forging Civilizations along the Silk Road*, pp. 88–101. New York: The Metropolitan Museum of Art.

Gajwani 2000 — Gajwani, S.L. (2000). *A Sufi Galaxy. Sufi Qalandar Hazrat Sai Qutab Ali Shah, His Spiritual Successors, and Select Disciples*. Ulhas Nagar: H.M. Damodar.

Ghose 2006 — Ghose, M. (2006). 'Nana: The "Original" Goddess on the Lion.' *Journal of Inner Asian Art and Archaeology*, 1: 97–113.

Gold 1988 — Gold, A.G. (1988). *Fruitful Journeys. The Ways of Rajasthani Pilgrims*. Berkeley and Los Angeles: University of California Press.

Hamilton 1972 — Hamilton, J.R. (1972). 'Alexander among the Oreitae.' *Historia*, 21(4): 603–08.

Harmatta et al. 1994 — Harmatta, J., B.N. Puri, L. Lelekov, S. Humayun and D.C. Sircar (1994). 'Religions in the Kushan Empire.' In J. Harmatta, B.N. Puri and G.F. Etemadi (eds.), *History of Civilizations of Central Asia*, Vol. II, pp. 313–30. Paris: UNESCO Publishing.

Hart 1840 — Hart, S.V.W. (1840). 'A Pilgrimage to Hinglaj.' *Transactions of the Bombay Geographical Society*, 3: 77–105.

Hoffmann 1880 — Hoffmann, G. (1880). *Auszüge aus syrischen Akten persischer Märtyrer*. Leipzig: F.A. Brockhaus.

Invernizzi 1998 — Invernizzi, A. (1998). 'Osservazioni in margine al problema della religione della Mesopotamia ellenizzata.' In E. Dabrowa (ed.), *Ancient Iran and the Mediterranean World*, pp. 87–99. Crakow: Jagiellonian University Press.

Kamphorst 2008 — Kamphorst, J. (2008). *In Praise of Death. History and Poetry in Medieval Marwar (South Asia)*. PhD Dissertation, Leiden University.

KauAŚ — Kauṭilya: *Arthaśāstra*. Olivelle, O. (trans.) (2013). *King, Governance, and Law in Ancient India. Kauṭilya's Arthaśāstra*. New York: Oxford University Press.

Kinsley 1986 — Kinsley, D. (1986). *Hindu Goddesses. Visions of the Divine Feminine in the Hindu Religious Tradition*. Delhi: Motilal Banarsidass.

Leech 1843 — Leech, R. (1843). 'Brief History of Kalat, Brought down to the Deposition and Death of Mehrab Khan, Braho-ee.' *Journal of the Asiatic Society of Bengal*, 12(138): 473–512.

Lo Bue 1981 — Lo Bue, E. (1981). 'Statuary Metals in Tibet and the Himalayas.' In W.A. Oddy and W. Zwalf (eds.), *Aspects of Tibetan Metallurgy*, pp. 3–67. London: British Museum.

Mallebrein 2011 — Mallebrein, C. (2011). 'Tutelary Deities at Royal Courts in Orissa.' In H. Kulke and G. Berkemer (eds.), *Centres out There? Facets of Subregional Identities*, pp. 255–72. Delhi: Manohar.

MāP — *Mārkaṇḍeyapurāṇa*. Banerjea, K.M. (ed.) (2004 [1862]). *The Marcandeya [sic] Purana in the Original Sanskrit*. Bibliotheca Indica no. 29. New Delhi: Cosmo Publications.

Masson 1834 — Masson, C. (1834). 'Memoir on the Ancient Coins Found at Beghram, in the Kohistan of Kabul.' *Journal of the Asiatic Society of Bengal*, 3(28): 153–78.

Mayrhofer 2001 — Mayrhofer, M. (2001). *Etymologisches Wörterbuch des Altindoarischen*, Vol. III. Heidelberg: Winter.

Mazumdar 1921 — Mazumdar, B.C. (1921). *Typical Selections from Oriya Literature*, Vol. I. Calcutta: University of Calcutta.

MBh — *Mahābhārata*. GRETIL e-text based on Bhandarkar Oriental Research Institute critical edition, Pune, India, 1999, <http://gretil.sub.uni-goettingen.de/gret_utf.htm#MBh> (accessed 19 October 2015).

Minchin$_{a}$ 1907 — Minchin, C.F. (ed.) (1907). *Las Bela*. Baluchistan District Gazetteer Series, Vol. VIII. Allahabad: Pioneer Press.

Minchin$_{b}$ 1907 — Minchin, C.F. (1907). *Sarawan, Kachi and Jhalawan*. Baluchistan District Gazetteer Series, Vol. VI. Bombay: Times Press.

Mishra 2006	Mishra, H.K. (2006). *Goddesses and the Centres of Goddess Worship in Early Rajasthan (7th to 15th Century)*. PhD Dissertation, Jawaharlal Nehru University, New Delhi.
Mitchiner 1976	Mitchiner, M. (1976). *Indo-Greek and Indo-Scythian Coinage*, Vol. 5. London: Hawkins.
MWSED	Monier-Williams, M. (1899). *A Sanskrit-English Dictionary*, new edition. Oxford: Clarendon Press.
Norman 1983	Norman, K.R. (1983). *Pāli Literature*. Wiesbaden: Otto Harrassowitz.
O'Malley 1908	O'Malley, L.S.S. (1908). *Chittagong*. Eastern Bengal District Gazetteer Series, Vol. I. Calcutta: Bengal Secretariat Book Depot.
Oldham 1899	Oldham, R.D. (1899). 'Report on the Great Earthquake of 12th June 1897.' *Memoirs of the Geological Survey of India*, 29: 1–379.
Potts 2001	Potts, D.T. (2001). 'Nana in Bactria.' *Silk Road Art and Archaeology*, 7: 23–35.
Rashid 2008	Rashid, S. (2008). 'The Cult of Bibi Nani.' *The Daily Times* (Islamabad), 25 January 2008, <http://discoveringpakistan.wordpress.com/2013/02/16/the-cult-of-bibi-nani/> (accessed on 25 August 2015).
Rocher 1986	Rocher, L. (1986). *The Purāṇas*. Wiesbaden: Otto Harrassowitz.
Rosenfield 1967	Rosenfield, J.M. (1967). *The Dynastic Art of the Kushans*. Berkeley and Los Angeles: University of California Press.
Saletore 1973	Saletore, R.N. (1973). *Early Indian Economic History*. Bombay: N.M. Tripathi.
Schaflechner 2014	Schaflechner, J. (2014). *Hinglaj Devi. Identity, Change, and Solidification at a Hindu Temple in Pakistan*. PhD Dissertation, Heidelberg University.
Scharf 2003	Scharf, P. (2003). *Rāmopākhyāna. The Story of Rāma in the Mahābhārata*. London: Routledge Curzon.
Schimmel 1975	Schimmel, A. (1975). *The Mystical Dimensions of Islam*. Chapel Hill: North Carolina University Press.
Sharma 2006	Sharma, M. (2006). 'Shaktism in Himachal.' In J.S. Grewal (ed.), *Religious Movements and Institutions in Medieval India*, pp. 95–109. New Delhi: Oxford University Press.
Shenkar 2014	Shenkar, M. (2014). *Intangible Spirits and Graven Images. The Iconography of Deities in the Pre-Islamic Iranian World*. Leiden: Brill.
Sims-Williams 2002	Sims-Williams, N. (2002). 'Ancient Afghanistan and Its Invaders: Linguistic Evidence from the Bactrian Documents and Inscriptions.' In N. Sims-Williams (ed.), *Indo-Iranian Languages and Peoples*, pp. 225–42. Oxford: Oxford University Press.
Sims-Williams and Cribb 1995–1996	Sims-Williams, N. and J. Cribb. (1995–1996). 'A New Bactrian Inscription of Kanishka the Great.' *Silk Road Art and Archaeology*, 4: 75–142.
Sircar 1960	Sircar, D.C. (1960). *Studies in the Geography of Ancient and Medieval India*. Delhi: Motilal Banarsidass.
Sircar 1973	Sircar, D.C. (1973). *The Śākta Pīṭhas*, 2nd ed. Delhi: Motilal Banarsidass.

Snead 1964 — Snead, R.E. (1964). 'Active Mud Volcanoes of Baluchistan, West Pakistan.' *Geographical Review*, 54(4): 546–60.

Snellgrove 1961 — Snellgrove, D. (1961). *Himalayan Pilgrimage*. Oxford: Bruno Cassirer.

Sondhi 1947 — Sondhi, V.P. (1947). 'The Makran Earthquake, 28th November 1945: The Birth of New Islands.' *Indian Minerals*, 1(3): 147–54.

Steckler et al. 2008 — Steckler, M.S., S.H. Akhter and L. Seeber (2008). 'Collision of the Ganges-Brahmaputra Delta with the Burma Arc: Implications for Earthquake Hazard.' *Earth and Planetary Science Letters*, 273(3–4): 367–78.

Stein 1943 — Stein, A. (1943). 'On Alexander's Route into Gedrosia: An Archaeological Tour in Las Bela.' *Geographical Journal*, 102(5–6): 193–227.

Stewart 1897 — Stewart, C.E. (1897). 'Account of the Hindu Fire-Temple at Baku, in the Trans-Caucasus Province of Russia.' *Journal of the Royal Asiatic Society of Great Britain and Ireland*, 29(2): 311–18.

Suvorova 2004 — Suvorova, A. (2004). *Muslim Saints of South Asia. The Eleventh to Fifteenth Centuries*, trans. M. Osama Faruqi. London: Routledge Curzon.

Tambs-Lyche 2004 — Tambs-Lyche, H. (2004). *The Good Country. Individual, Situation, and Society in Saurashtra*. Delhi: Manohar.

Tanabe 1995 — Tanabe, K. (1995). 'Earliest Aspect of Kaniṣka I's Religious Ideology: A Numismatic Approach.' In A. Invernizzi (ed.), *The Land of the Gryphons. Papers on Central Asian Archaeology in Antiquity*, pp. 203–15. Florence: Le Lettere.

Tarn 1948 — Tarn, W.W. (1948). *Alexander the Great*. Cambridge: Cambridge University Press.

Valdiya 2001 — Valdiya, K.S. (2001). *Himālaya. Emergence and Evolution*. Hyderabad: Universities Press.

VāmP — *Vāmanapurāṇa, Adhyāyas* 1–69. GRETIL e-text based on: Gupta, A.S. (ed.) (1967). *Vāmana-Purāṇa*. Varanasi: All India Kashiraj Trust, <http://gretil.sub.uni-goettingen.de/gretil/1_sanskr/3_purana/vamp__u.htm> (accessed 8 September 2015).

von Rad et al. 2000 — von Rad, U., U. Berner, G. Delisle, H. Doose-Rolinski, N. Fechner, P. Linke, A. Lückge, H. A. Roeser, R. Schmaljohann and 20 more. (2000). 'Gas and Fluid Venting at the Makran Accretionary Wedge off Pakistan.' *Geo-Marine Letters*, 20: 10–19.

Vredenburg 1909 — Vredenburg, E. (1909). 'Report on the Geology of Sarawan, Jhalawan, Mekran and the State of Las Bela.' *Records of the Geological Survey of India*, 38(3): 189–215.

Weinberger-Thomas 1999 — Weinberger-Thomas, C. (1999). *Ashes of Immortality. Widow-Burning in India*, trans. J. Mehlman and D.G. White. Chicago: University of Chicago Press.

White 1996 — White, D.G. (1996). *The Alchemical Body. Siddha Traditions in Medieval India*. Chicago and London: University of Chicago Press.

Wilford 1799	Wilford, F. (1799). 'A Dissertation on Semiramis, the Origin of Mecca &c. from the Hindu Sacred Books.' *Asiatic Researches*, 4: 376–400.
Williams Jackson 1911	Williams Jackson, A.V. (1911). *From Constantinople to the Home of Omar Khayyam*. New York: Macmillan.

Chapter 3

From Iron to Sapphire: Indian Myths and Rituals about Saturn, the Implacable Lord of Celestial Spheres

MONIA MARCHETTO AND MANUEL MARTIN HOEFER[1]

The figure of Saturn is a prominent one in Indian tradition. Under names like Śani, Śanaiścara, Manda and Kroḍa, he is the most revered of all planets and his worship is spread across the whole of the Indian subcontinent.[2] A profusion of references are found in both *śruti* and *smṛti* texts down to Agamas, astronomic sources, devotional works and ritual manuals describing the praxis for his propitiation. Most Indians are aware of his powerful influence over human life and adopt different measures to counter his destructive effects. Here follow some general considerations about Saturn, which will be propaedeutic to a more insightful analysis of the worship of Śani.

According to the Indian tradition, Śani is one of the nine planets (*grahas*).[3] The Sanskrit root √*gṛh* means 'grasping,' 'seizing,' 'laying hold

1 Monia Marchetto is Visiting Lecturer of Hindi Language and Literature at Ca' Foscari University of Venice and Scientific Coordinator and Research Fellow at VAIS (Venetian Academy of Indian Studies) (Italy).
Manuel Martin Hoefer is Research Fellow at VAIS (Venetian Academy of Indian Studies) (Italy).

2 The mathematician, astronomer and geographer Claudius Ptolemy (second century CE) states in Tetrabiblios (II.3) that the Indian subcontinent is under the influence of Capricorn, one of the two zodiac signs over which Saturn rules.

3 These are: Sūrya (Sun), Candra (Moon), Maṅgala (Mars), Budha (Mercury), Bṛhaspati (Jupiter), Śukra (Venus), Śani (Saturn), Rāhu (Caput Draconis) and Ketu (Cauda Draconis). Rāhu and Ketu are the ascending and descending lunar

of something' (VI: 243–44).[4] *Grahas* are not inanimate bodies but sentient beings; to be precise, they are considered deities with specific qualities and characteristics, and some measure of free will (Rao 1995: 14). According to Śāstric literature, *grahas* are endowed with 'consciousness'; in this sense they are said to be alive. In BPHS II.7 the sage Pārāśara explains to his pupil Maitreya why planets are alive: the supreme deity Viṣṇu manifested himself as the nine *grahas* to bestow on living beings the fruit of their actions (karma). The god assumed the auspicious form of the *grahas* to destroy demons, to sustain the strength of men and to establish *dharma*.[5] At the conclusion of their earthly mission, the *avatāras* will merge into their respective *grahas* again.[6]

The science which studies the *grahas*, Jyotiṣa, is ancillary to the Vedas (Vedāṅga) and deals mainly with the measure of time, astronomy, astrology and fields such as architecture and medicine. The main purpose of Jyotiṣa is to determine the correct and most auspicious time for performing Vedic sacrifices. Jyotiṣa has three main parts: (1) *siddhanta* deals with the origin, final destination and measurements of the cosmos, with mathematical calculations concerning the position of planets, prediction of the eclipses and so on; (2) *hora* investigates the influence of celestial bodies over the lives of individuals; (3) *saṃhitā* deals with miscellaneous topics, such as the effects of heavenly bodies on weather, architecture, pharmacopeia and other matters.

Different kinds of beings bear the name *graha*: planets (*grahas*), invisible heavenly bodies (*upagrahas*) and malevolent beings that need to be pacified to get rid of their oppressive grip (*balagrahas*).[7] Let us now deal with the nine planets, the *navagrahas*.

---

nodes. Indian tradition classifies them too under the category 'planet' and defines them as *chāyā graha*, shadow planets. These two points of intersection between the Moon orbit and the ecliptic gain a 'body' during lunar and solar eclipses, of which they represent a sort of negative image. For exhaustive explanations about *grahas*, see Rao 1995.

4 The etymological meaning of *graha* is different from that of 'planet', a word that comes from Gr. ἀστέρες πλανῆται (*asteres planētai*: wandering stars).

5 Rāma is the manifestation (*avatāra*) of the Sun, Kṛṣṇa of the Moon, Narasiṃha of Mars, Buddha of Mercury, Vāmana of Jupiter, Paraśurāma of Venus, Kūrma of Saturn, Varāha of Rāhu, Mīna of Ketu.

6 According to BPHS II.13, living spirits (*jivātman*) move from planets and take birth as human and other beings, live their lives according to their destiny and at the end merge in the *grahas*.

7 *Balagrahas* are not heavenly bodies. They are defined as *grahas* because they 'catch' their victim inflicting sufferings (Skt. *pīḍā*), hence their alternative

The leader (*grahapati*) of the planets is Sūrya, the Sun. The Sun represents at a cosmological level the transcendent Light, which is the Supreme Principle according to many Indian traditions. The other planets are particular functions of Sūrya himself.[8] Another important concept is that of Kāla Puruṣa, the Primordial Man, the manifested universe as Personification of Time. From this point of view, the twelve zodiac signs starting from Aries are, respectively, his head, face, arms, heart, stomach, hip, space below the navel, private parts, thighs, knees, ankles and feet (BPHS iv.3–4). The Sun is the spirit of the Kāla Puruṣa, the Moon his mind, Mars his strength, Mercury his speech, Jupiter the essence of all knowledge and happiness, Venus is desire and Saturn is suffering (*duḥkha*) (JP ii.1).

Śani and the other planets are considered deities and they are invoked in hymns (*stotra*) specifically assigned to them and their tutelary deities. In the case of Śani, these are Prājāpati (*adhidevatā*) and Yama (*pratyadhidevatā*) (see below). According to Jyotiṣa, the nine planets are defined by many features, the most important of which are caste, gender,[9] metal, color, humor (*doṣa*),[10] quality (*guṇa*), the stage of human life, the region of space and the deity to which they correspond (table 3.1).

As for Śani, besides the details listed above, we are informed that his form is effulgent and his character intense. He belongs to the lineage of the *ṛṣi* Kāśyapa. He is the son of Sūrya and his second wife Chāyā, hence his alternative names Sauri and Chāyātmaja. Since *chāyā* means shadow, it is probably for this reason that Śani's complexion is dark (black or blue) and one of his names is Asita (Skt. black; dark colored). He wears blue garments, and therefore is known as Nīlāmbara. As he is identified with old age, he is described as crooked and lame, hence his names Vakra (curved; bent) and Paṅgu (crippled [in the legs]). His qualities are inactivity, ignorance, darkness (*tamas*). He is associated with cremation grounds, barren land,[11] marshes, dirty places and prisons. He rules over servants and malevolent

name *grahapīḍā*. Although *balagrahas* can be easily confounded with planets (cf. Guenzi 2008: 191–97; Mani 1989: 297–98), an accurate examination of classical texts permits one to ascertain their different domains (Rao 1995: 40–41).

8 For this reason seven are the horses which drive the mythological chariot of the Sun (Rāhu and Ketu are not counted among them).

9 In Jyotiṣa planets can be masculine, feminine or hermaphrodite, whereas in Indian iconography they are all masculine.

10 According to Āyurveda the *tridoṣa* are: wind (*vāta*), bile (*pitta*) and phlegm (*śleṣman* or *kapha*).

11 Not by chance Śani is believed to inhabit Saurashtra (Kathiawar peninsula), an arid region in the western state of Gujarat in India.

Table 3.1. The features of the nine planets, based on popular almanacs (*pañcaṅga*) and on Rao (1995: 33–38, 105).

	Sūrya	**Candra**	**Maṅgala**	**Budha**	**Bṛhaspati**	**Śukra**	**Śani**	**Rāhu**	**Ketu**
Gender	M	F	M	H	M	F	H	M	M
Age	middle age	middle age	young age	childhood	old age	middle age	decrepitude	old age	old age
Color	red	white	red	green	yellow	mixed	dark	dark	dark
Caste	kṣatriya	vaiśya	kṣatriya	śūdra	brāhmaṇa	brāhmaṇa	śūdra (or no caste)	outcaste	outcaste
Humor	bile	phlegm	bile	all three humors	all three humors	phlegm	wind	wind	wind
Quality	*sattva*	*sattva*	*tamas*	*rajas*	*sattva*	*rajas*	*tamas*	*tamas*	*tamas*
Gem	ruby	pearl	coral	emerald	yellow sapphire	diamond	blue sapphire	hessonite	cat's eye
Metal	copper	silver	copper	mercury	gold	silver	iron	lead	lead
Deity	Sūrya	Ambu	Skanda	Viṣṇu	Indra	Devī	Brahmā	deity of eclipse	deity of eclipse
Planet	Sun	Moon	Mars	Mercury	Jupiter	Venus	Saturn	Caput Draconis	Cauda Draconis

old women.[12] He is valorous in nature (*vīra*) but also cruel. He is malefic among other malefic planets (Rāhu and Ketu) and for this reasons Indians are historically terrorized by him. He rides a vulture (*gr̥dhravāhana*)[13] and in his four arms he carries a dart, a bow, a lance and a spear (Rao 1995: 106). *Śrītattvanidhi* (an iconographic treatise of the nineteenth century) describes Śani's chariot as yoked to eight vultures of dark blue color; his icons are made of black iron and installed on a platform the form of which is that of an arrow facing west (ibid.: 108). In many temples he is represented riding a buffalo and holding in his two hands an iron club and a *yoga daṇḍa* (a staff used as an aid to concentration during repetition of ritual formulae).[14] At a popular level, Śani is often depicted mounted on a crow.[15]

The figure of Śani is complex and performs different functions. From the point of view of Jyotiṣa, he is the most important for predictive astrology (*phala jyotiṣa*). It is thus of the utmost importance to ascertain whether his cruel nature, disposition and outlook may have some positive effects over human beings and their destiny.[16] Another concept is helpful for an understanding of the prominence of Śani. In consonance with traditional cosmological views, our world, the planets, the lunar mansions, the zodiac signs, the circle of Ursa Major (Sapta R̥ṣi) and the orbit of the Polar Star (Dhruva Maṇḍala) are contained in the Kāla Cakra, the Wheel of Time, which is another name for the universe. Let us give a brief description of it.

At the center of Jambudvīpa (Rose-Apple Island), where present humanity lives, lies the mythical mountain Meru. Its peak reaches the Dhruva Maṇḍala, the celestial North Pole whose shape is that of a small

12 At a popular level, Śani is said to be associated to iron objects, living beings of dark color, cruel deeds; he is the cause behind the fear of thieves, diseases and addiction to gambling.

13 The vulture is famous for its strength, velocity, cruelty, egotism and miserly nature.

14 These *mūrti*s appear to be of recent origin, however.

15 The crow is supposed to be a messenger of Yama and plays an important part during *piṇḍadāna*, a funeral ritual related to the deceased becoming an ancestor (*pitr̥*). On the tenth day, *piṇḍa*s are offered to the dead; if there is a delay on the part of the crow in partaking of the food, it is understood that the deceased has some unfulfilled wish. After taking care of the desire of the departed, *piṇḍa*s are offered again to the crows. See Filippi (1996) and Zeiler (2013).

16 This debate was very much alive over the centuries in Europe. The judgment about the effects of Saturn changed notably according to historical periods and cultural milieus; it went from very adverse opinions regarding him as responsible for all human miseries to his rehabilitation as being responsible for the spiritual growth of an individual (Klibansky et al. 1952).

ring. Śiśumāra, the starry form of the god Viṣṇu holding his Wheel of Time, is shaped like a crocodile, although in some astrological renditions he has the form of a scorpion. At his tail end is Dhruva, the Polar Star that rotates at pace with the movement of the tail and makes all the other stars and planets rotate too, though at different speeds. The orbit preceding Dhruva Maṇḍala is that of Ursa Major; the next is that of the zodiac signs. Then comes the orbit of the lunar mansions and after it the Graha Cakra, the sphere where the nine planets revolve.[17] Due to their slow path, the sidereal spheres appear to move evenly. In fact, Ursa Major, the zodiac signs and lunar mansions revolve at different speeds: they reflect the slow movements of the starry vault such as the precession of equinox (related to Ursa Major), nutation of the axes mundi and obliqueness of zodiac signs.

The Graha Cakra is naturally included in the Kāla Cakra, but the two spheres differ in the direction of their revolution. The Kāla Cakra moves around Meru and keeps it on its right side while the Graha Cakra shares the general movement of the Kāla Cakra but at the same time it moves in the reverse direction keeping Meru on the left. (We should imagine the sidereal orbits moving faster than the planets revolving in their own orbits.) Alberuni cites the astronomer Brahmagupta on a concept of fundamental importance for the comprehension of ancient geocentric cosmology:[18]

> The wind makes all the fixed stars and the planets revolve towards the west in one and the same revolution; but the planets move also in a slow pace towards the east like a dust atom moving on a potter's wheel in a direction opposite to that in which the wheel is revolving. (Sachau 1992: 280)

From the point of view of human beings, Śani's orbit is the foremost distant sphere between the planetary orbits of the Graha Cakra.[19] The planets move eastward inside the Graha Cakra while beyond it the orbits of the other spheres pace westward. This fact has led us to consider that the spheres beyond Śani pace uniformly with the Primo Mobile of the Ptolemaic system which according to Indian mythology is the movement of Śiśumāra. From an astronomic point of view, the planet Śani is the limit

17 For the entire cosmological structure, see Rao (1995: 207–08). Among the various texts, KūP I.41.1–27 gives a detailed description of Graha Cakra.

18 Echoes of this thought resound in Dante's Divina Commedia (*Paradiso* XXII.106–11) and in Plato's Timaeus (XXXVI).

19 Śani's circumambulation across the twelve zodiac signs is very slow. He takes 29.5 years to complete an orbit around the Sun. His epithets Śanaiścara ('Quietly moving') and Manda ('Slow') highlight this astronomical fact.

of the Graha Cakra, that is to say of the sphere where the planets move influencing the elements (*tattva*) and the qualities (*guṇa*) of our environment and of ourselves. At the same time, Śani is also a means of access to higher realms of existence if we consider the human realm as the starting point of our investigation. In relation to this, the name Saptāṃśu, the seventh ray, which is an epithet of Śani, bears interesting implications: the seventh ray is what brings one out of the realm over which Śani dominates (see below).[20]

The interest about Śani is still very much alive in contemporary India; almost every inhabitant of the subcontinent is aware of the deity's mythological background and his worship is observable in numerous shrines. In general, Indians believe that his influence over human life is dominant. Various methods for increasing his positive effects have been developed or, on the other hand, for controlling his negative impact. Remedies differ according to geographical area and caste milieu. Śani is present in every human being's life and the extent of his influence can be evaluated in Jyotiṣa through three basic facts:

1. the orbiting of Śani around the zodiac lasts twenty-nine years and a half;
2. the planet affects human beings when entering the twelfth sign starting from the natal position of the Moon in the horoscope of a person;
3. the effects of Śani's presence last seven and a half years, when the planet leaves the second sign from the natal Moon sign (H. *sāṛhe sātī*).

During an average lifespan, an individual will have to face three revolutions of Śani. The beginning of the third round is of particular significance, as in this period one may face death or, conversely, achieve high status, fame, peace of mind and so on. The outcome varies because Śani bestows upon each person positive or negative experiences on the basis of the merit (*puṇya*) or demerit (*pāpa*) accumulated in previous lives.[21] (In case

20 The seven rays of the Sun are called *grahayonayaḥ*, matrixes of planets. Svaraka, the seventh ray, 'nurtures' Śanaiścara (KūP I.43.1–10).

21 Jyotiṣaśāstras affirm that if Śani's position in the natal chart is higher (*ucca sthāna*) then the effects are positive; if conversely the position is lower (*nīca sthāna*), the results are poverty, misery, suffering. Śani is called also Nīdhiman, the judge, because he inflicts suffering only on wrongdoers. For this reason, it is said that even during *sāṛhe sātī* (periods of affliction caused by the transit of Śani) a pious and devoted person will not suffer.

of individuals who enjoy longevity, Śani will grasp the person in a fourth round that will surely bring death.) In general, when speaking about the planet's negative effects, Indians refer to his pernicious glance (*dṛṣṭi*). This is explained in a Purāṇic myth, where Śani while engaged in severe austerities refused to satisfy his wife's passionate desires. In fact, he did not even look at her. Enraged by this, she cursed him, saying that he would destroy everything he looked at henceforth. To avoid this, therefore, he keeps his head bent (Rao 1995: 107).[22]

Śani, who is said to be the strongest among maleficent *graha*s, causes destitution and suffering, but also teaches endurance and effort. When he is hostile, he can reduce a king to a pauper. Conversely, when he is favorable, he bestows vigor, perseverance, health and talent more than any other planet. In this case, he can make a person a recognized and compassionate leader. It is said that even when he inflicts terrible suffering, in reality he is removing ignorance; he 'purifies' a person thus preventing wrong deeds, and makes one understand the principles in one's life. He may cause the greatest troubles, or on the contrary bestow the greatest blessings.[23]

In general, ordinary persons are advised by *jyotiṣī*s to take remedial measures (*upāyas*) when Śani's grip becomes coercive. These prescriptions consist in performing rituals, reciting specific *mantra*s and adopting certain rules in everyday life. Plenty of popular booklets in all Indian languages describe them in detail. People perform such *upāya*s in temples, at home or near particular trees (see below). Ritual specialists are not needed although expert brahmins are usually hired.[24]

Śani's icons (*mūrti*s) are worshiped widely in India. His images should be of black iron (*lauha*) and the ornaments of blue sapphire (*indranīla*) (Rao 1995: 108). Both men and women can worship Śani,[25] and the use of iron

22 Many popular stories are related to Śani's *dṛṣṭi*. One of the most famous regards Gaṇeśa or Gaṇapati, whose human head was burned after his mother, Pārvatī, caused Śani to look upon Gaṇapati.

23 According to a popular narrative, Daśaratha, king of Ayodhyā and Rāma's father, went to Śani's abode to beseech mercy for his subjects when he learnt that a period of terrible suffering was about to come. The hymn he recites is known as *Daśarathakṛta Śani stotram*. Popular booklets contain various hymns (*stotram*) dedicated to Śani and extracted from major Purāṇas like *Skanda*, *Padma* and *Brahmāṇḍa*.

24 For the interaction between *jyotiṣī*s and *pujārī*s, see Guenzi (2005). Interesting research about one of the most widespread caste of *pujārī*s (the *ḍakauts*) in Śani temples in northwest India has been carried out by Bellamy (2014).

25 In Shingnapur women cannot touch Śani's icon. In general, a person cannot venerate Śani if there has been a birth or a death recently in his house.

and black items is fundamental. In several temples, the *mūrti* is an anthropomorphic black sculpture situated in a subsidiary shrine together with those of all other planets. Occasionally the icon is isolated in a dedicated portion of a temple or in a consecrated building (sometimes just under a canopy). In south India, painted icons of the god are carried in procession in colorful carts along villages and cities during established festivities (Swami Sivapriyananda 1990: 74–75). A celebrated icon, that of Shingnapur in Maharashtra, is constituted of a big black rock installed on an open-air platform.[26] In many places, Śani's icons are black sculptures depicting the god riding a black buffalo (see below).[27]

Observing Śani's vow (*vrata*) is effective only if done on a Saturday (Skt. Śanivāra), especially during the path of the Sun towards the south (*dakṣiṇāyana*) and mainly during the month of Śrāvaṇa.[28] The ritual should begin at sunset, as the planet is the Lord of the West. The devotee should eat only prescribed food,[29] perform ablutions and make offerings to the ancestors. Wearing black or dark colored cloths, while uttering specific

26 The temple of Śani at Shingnapur is one of the most, if not actually the most, famous in India. Devotees consider the icon to be self-manifested, i.e. without human intervention. Inhabitants of the place believe so strongly in Śani's protection against thieves that doors have no locks (which usually are made of iron). Many buildings do not even have doors. Other places of pilgrimage are spread all over India, from Himachal Pradesh to Uttar Pradesh, elsewhere in Maharashtra, Madhya Pradesh, Karnataka, Tamil Nadu and so on. At Devipattinam, in Tamil Nadu, the icon of Śani is an informal stone pillar immersed in the water; along with other pillars representing the remaining planets it is worshiped in a shrine situated close to the seashore. Many lists of temples dedicated to Śani can be found on specialized websites along with descriptions of the rituals offered. The proliferation of interest in Śanideva is connected with the growing aspirations of Indians to quickly improve their wealth and wellbeing by various means.

27 Another element in common with Yama is the dog. In the case of Śani, the devotee can feed and look after black dogs – especially if old and sick – to pacify the god. For the same purpose, a person can also feed crows and black cows. The cow is worshiped by tying sacred thread over her horns, and offering her incense, a burning lamp and sweets.

28 The *vrata* should continue for thirty-three Saturdays or at least for seven. The same number occurs in Śani's *yantra*, where thirty-three is the sum of the nine numbers engraved on the plate. Śani *vrata* is said to pacify Rāhu and Ketu too. The new moon day of the month of Jyeṣṭa (May–June) of the Hindu calendar is celebrated as Śani Jayantī, the 'birth anniversary' of the god.

29 During Śani Pūjā, the devotee should abstain from consuming meat, fish and alcohol.

formulae (*mantras*), s/he performs *pūjā*. Water is first offered to the feet of the god; then it is poured into the consecrated shrine. More water is drunk for purification purposes and then offered for the bath of god. After that, one should present black cloths, camphor, incense, dark colored fragrant essential oil, mustard oil,[30] dark colored flowers or *bilva* leaves[31] and so on. Further offerings are: lamps (especially fueled by mustard oil), black rice, black sesame and other edible items of dark color. The offerings can vary considerably from caste to caste, and based on family customs. Usually a coconut is offered to gods; the nut is broken with an abrupt stroke and its water poured on the icon. In place of it, in Shingnapur, a devotee can offer an iron horseshoe that had belonged to a black horse. At the end of the worship, a reward (*dakṣiṇā*) is offered to the *pujārī*. The devotee can also give special offerings to brahmins. Among these are iron objects, black sesame, black cloths, footwear, oil of dark color, black gram (H. *uṛad*; *Vigna mungo* (L.) Hepper), barley, a musk product called *kastūrī*, coal and black cows. One can offer even gold, or blue sapphire. The sources of the *mantras* that the devotee recites are Vedic, Purāṇic or Tantric.[32] At some shrines, an iron ring is offered to Śani inside a lamp fueled by mustard oil; after that the ring is worn following certain regulations (for the first time on a Saturday, during a new moon day and so on). Rings forged by a blacksmith out of horseshoes are particularly appreciated.[33]

As already pointed out, the icon of Śani can be inside a temple or in one's own house. Apart from these places, a *pīpala* tree[34] can be the abode of Śani. If the tree is already consecrated to the god, it is usually surrounded by a

30 During certain rituals, Śani's icon in Shingnapur is literally flooded with mustard oil.

31 *Bilva* is a variety of *Aegle marmelos* (L.) Corrêa, commonly known as Bengal quince, wood-apple and so on. It is considered sacred as it associated with Śiva.

32 Śani's *mantra* should be recited 23,000 or 92,000 times. When one-tenth of the repetition has been done, the devotee should perform a fire sacrifice following prescribed ritual rules.

33 Iron seems to play an important function in appeasement and possession rituals evoking Śani (Zeiler 2015: 80n9). At Śani Devālayam in Deonar, Mumbai, resides an oracle of Śani. On Saturdays the chief priest of the temple is possessed by the god. He then sits on a chair entirely covered with long protruding iron nails and listens to the devotees' troubles. On the basis of the nature of afflictions (deeds of past lives, black magic, evil eye, deceased's spirit and so on) he suggests appropriate remedies (*yajña*, *pūjā*, donation, abstinence, etc.). See HVPh 2011.

34 *Pīpala* or *aśvattha* (*Ficus religiosa* L.) is a large semi-evergreen tree considered holy in some Asian countries.

platform upon which is an icon of Śani, often along with that of other gods and/or with *nāga kalas*, steles representing cobras. The mode of worship is similar to that discussed above. The presence of the snake is very interesting because it bears witness to the link between Śani and Yama. One prescription for propitiating Śani is feeding snakes with milk, believed to be their favorite food, and protecting them. In India the serpent is mostly associated with the world of the dead. Serpents easily get furious but they are worth propitiating as their blessings are beneficial for human beings, as are those of ancestors satisfied by their worship.[35] The benevolence of the forefathers further guarantees prosperous progeny. Snakes too, through their relation with waters, are symbols of fertility and represent the future generations.

If a *pīpala* tree is to be consecrated as an abode of Śani, the devotee will build an altar at the base of the tree; on this a twenty-four petaled lotus will be outlined with black rice grains. Then an iron icon of the planet will be enshrined for worship. Finally, the devotee will fix a seven-skeined thread seven times around the tree and worship it too in the appropriate manner.[36]

Other forms of *pūjā* include the worship of a *yantra*. This diagram is usually engraved on a copper plate. Worship, which in this case is an individual practice, is done along with the repetition of a prescribed amount of *mantras*.[37]

Another technique for propitiating Śani is associated with the blue sapphire (*indranīla*).[38] The use of this precious stone is controversial. Its association with Śani is undisputable and commonly accepted in many milieus. Expert *jyotiṣīs* suggest wearing it under certain conditions. Āyurvedic compendia recommend the use of the powder obtained from grinding and purifying the gemstone. The combination of this powder with other ingredients is believed to heal certain diseases, to improve physical vigor, to counter poisoning and to work as an aphrodisiac, a heart tonic, et cetera.[39]

35 For an exhaustive analysis of snakes in Indian culture, see Vogel 1926.

36 For a detailed account about the relation between the *pīpala* and Śani, see Haberman 2013: 106–31.

37 On *yantras* as objects of worship see Zeiler (2015).

38 Other names of this gemstone are *śouriratnam*, *nīlāśmā* and *tr̥ṇagrāhī*.

39 Before being administered, sapphire has to be purified through a long process (Sudarshan 1999: 91–93) in phases (*śodhana*, *māraṇa*, *ratnadruti*) which emphasize the idea that gemstones are endowed with life (*māraṇa* means 'killing of a living being'). Iron too (along with many other minerals) is purified with similar processes and used as a medicine for curing different diseases (Sudarshan

In general blue sapphire is reputed a powerful gem. It is used by *yogi*s and by those who are afflicted by Śani. Whatever the reason, wearing it always produces positive effects, although judicious *jyotiṣī*s advise that it be used with caution.[40]

According to traditional sciences, precious stones are classified in two groups: the *navaratna*s, the nine (most important) gemstones, and the *uparatna*s, minor gemstones. The blue sapphire is included in the first group along with ruby (*māṇikya*), pearl (*mukta*), coral (*pravāla*), emerald (*marakata*), yellow sapphire (*puṣparāga*), diamond (*vajra*), hessonite (*gomeda*) and cat's eye (*vidūra*).[41] A blue sapphire should be of a uniform color, heavy, unctuous, rounded, smooth and lustrous (Sudarshan 1999: 99). This stone, which is well known for its smoky purple/blue coloration,[42] displays diverse shades of blue depending on different trace combinations of titanium and iron, which are the coloring agents. Highly evaluated gems are neither very dark nor pale. The most treasured ones are of a pure 'cornflower-blue' or 'Kashmir blue', a soft and vivid blue. A pure blue sapphire gemstone is transparent and glittering, apparently lively; its qualities are in complete opposition with that of iron which is opaque (it is like the 'crash site' of light), gloomy, apparently inert. Iron and blue sapphire seem to represent two complementary aspects of Śani: the entanglement of the spirit in the complex development of one's existence and the process of releasing it from that entanglement.[43]

1999: 103, 114–18). In Rasāyana (alchemy), the blue sapphire is praised for bestowing longevity. On alchemy and metallic medicines in traditional medicine, see Dash 1986.

40 Some maintain that blue sapphire should be used only when Śani is very positive in one's natal chart.

41 The group of precious stones of secondary importance includes glass, conch, et cetera (Sudarshan 1999: 91).

42 Various texts mention gems and their qualities. In KauAŚ II.15.5 the treasure of a king is discussed, and we learn that sapphire 'may have blue streaks, the color of Kalāya [*Lathyrus sativus* L.] flower, a deep blue color, the radiance of a rose apple, or the luster of a dark cloud,' or be a 'delighter' or a 'flowing-middle.'

43 From the point of view of mineralogy, iron is not found in pure form in nature but as an ore, mixed with other minerals from which it has to be extracted. The common word for iron, *lauha*, was a general term under which different metals were classified in Indian treatises: gold, silver, copper, tin, lead and so on. Sapphire, conversely, is a pure gem; it needs only to be polished to get rid of the impurities that cover it. India has large deposits of iron ore and is one of the most important producers of iron in the world. Iron occurs mainly in the form of oxides: hematite and magnetite. Indian treatises classified iron through its physical properties, e.g. strength, ductility, fracture and magnetism (which are

A blue sapphire should be as pure as possible. In fact, not all stones can be used for propitiating a 'malevolent' planet, owing to their having some defects. Hence proper examination is thus indispensable.[44] We learn in BS LXXX.1 that: 'Since a jewel endowed with good characteristics ensures good luck to a king, and one with bad ones, disaster, connoisseurs ought to examine their fortune depending on jewels.'

The blue sapphire is usually worn on a ring or an amulet and it is set in such a way that it directly touches the skin. Amulets can be of different shapes. Some types have a blue sapphire set on a plate made of iron or of a mixture of iron and silver; the plate is shaped like a sword and on the back of it the Śani Yantra is engraved. One who wears such item should worship the *yantra* on Saturdays for a certain time and then make donations to brahmins as per Śani Pūjā. The metal plate for the talisman can also be made of one part gold, two parts silver, three parts lead, one part copper and five parts iron (Johari 1996: 215).[45]

In agreement with ritual prescriptions, an enraged Śani can be pacified also through the worship of Hanumān,[46] Bhairava and Kālī. Devotees worship at Hanumān temples on Saturdays to be delivered from Śani's grip. *Pujārīs* often perform Śani's *pūjā* in front of Hanumān's icon.[47]

---

in agreement with contemporary classifications). India is not as dominant in the world regarding sources of sapphire, although the gemstones from Kashmir are highly valued for their purity. India is a large market for gemstones of various kinds, which may be imported from other countries. In Indian mythology, gemstones are described as originating from titans, or great sages. BS LXXX.3 states: 'Gems, they say, were born of the bones of the demon Bala; while others state that they were born of sage Dadhīcī; yet others declare that the wonderful variety of gems is caused by the characteristic qualities of the earth.' Dadhīcī offered his bones out of which the gods made the *vajrāyudha*. With this weapon Indra slayed the mighty Vṛta, who was holding the primordial waters captive.

44 A traditional test of the authenticity of a blue sapphire is to ascertain whether it imparts its color to milk when suspended in it (Sudarshan 1999: 100). The defects generally found in precious stones are loss of transparency or presence of a stain, discordant color, (lined or bubbled) spots and cracks.

45 We cite this book in spite of its lack of academic perspective because the reference to Śani Yantra in it is accurate.

46 Oral narratives relate that Rāvaṇa, the *asura* king of Laṅka, had seized and imprisoned all the planets and heavenly bodies to prevent the passing of time. Hanumān, on his way to Sītā, set Śani free and earned his gratitude.

47 See the video recorded in Pandey 2014. This short clip bears witness to the vitality and resourcefulness of Indian culture. The *pūjā* is performed on behalf of an Indian who lives abroad, in England, and it has been recorded as a proof of his accomplishment.

Further, most Indians recite the *Hanumān Cālīsā* and the *Sundarakāṇḍa* from Gosvāmī Tulasīdāsa's *Rāmacaritmānasa* (sixteenth century), which narrates the story of Hanumān conveying a message from Rāma to his beloved wife Sītā after the latter was abducted by Rāvaṇa. Reporting Rāma's fervent words, Hanumān instills in Sītā's heart a strong devotion and an indomitable strength. She feels relieved from her agony and gathers determination to resist Rāvaṇa till her period of suffering ends.[48] This explains why on Saturdays devotees flock to temples dedicated to Hanumān, the divine messenger who instills firmness in the hearts of distressed human beings and inspires them to overcome obstacles.

Śani is worshiped within Śākta and Śaiva traditions too. Many similarities can be found between Śani and manifestations of Durgā (notably Kālī) and Śiva. For instance, all these deities have wrathful aspects (their icons have fierce protruding eyes and so on), are associated with time, which inexorably leads everything to its fulfillment and exhaustion, and their *pūjās* are *tamasik* in nature (although Śani does not require animal sacrifice). Devotees of Śani are known to recite *Devīmāhātmya* (MāP LXXXI–XCIII) whereas the black and fierce dog-riding Bhairava, in his form of Kāla Bhairava, the supreme guardian of Banaras and – according to some – of India, may be fully identified with Śani.[49] When the negative influence of Śani is detected, one may resolve to offer a *pūjā* to Mṛtyuñjaya, a hypostasis of Śiva as vanquisher of death.

One of the most common and loved texts is *Śani Māhātmya*, a part of *Śani Vrata*. This includes a well-known folk narrative and is a celebrated remedy against Śani's affliction. It is told in temples and written in many booklets in all languages. There are different versions of it, but the basic plot has been standardized. The main character of the story is Vikramāditya, the wise king of Ujjain. Once he fell under the negative influence of Śani Deva because he did not recognize the *graha*'s supremacy over other planets.

48 BPHS II.5 establishes a correspondence between Rāma and the Sun. Indian *paṇḍits* are inclined to interpret the episode of Sītā's abduction as the leaving soul (*jivātman*) which is driven away from its source, the spiritual Sun (*ātman*), and kept under the spell of the delusive power of ignorance that informs the cosmos. From this point of view, some correlate Rāvaṇa with Śani and his misleading function (i.e. confounding those who are not able to distinguish the real from the unreal).

49 Right in front of the entrance of the Kāśī Viśvanātha Mandir, the main temple of Banaras, there is a shrine with an icon venerated as Śani, even though at a closer look – under the garments – the *mūrti* appears to be a sculpture of Kāla Bhairava.

He thus lost his sovereignty and was reduced to the level of a beggar in a land distant from his own. He had to work as a servant and as a result of false accusations, he was punished and his hands and feet cut off. An oilseed crusher (*telī*) took pity on him and gave him work in his mill.[50] At the very end, Vikramāditya's atonement for his negative karma produced new qualities in him, among which prevailed pity and compassion towards mankind. Vikramāditya acknowledged Śani's greatness; the god was pleased, restored his kingdom and made him a rightful monarch.[51]

Further to that, Śani appears to be prominent in some ascetic Śaiva traditions. In BPHS LXXXI.1–3, it is stated that there are *yogas* (particular combinations of planets) that favor wandering renunciation (*pravrajya*), i.e. asceticism. This happens if one's house in the natal chart contains four strong planetary presences. If among these Śani is the most powerful, the person will become a *nāgā sādhu*, a member of one of the three subgroups of the Daśanāmī Śaiva monastic order founded by Ādi Śaṅkarācarya (780–822 CE).

More generally, while ordinary persons should take remedial measures against Śani's *sāṛhe sātī*, or afflicting transits, it is not so for ascetics or practitioners of spiritual disciplines (*sādhakas*). *Sādhakas* are not afraid of the period during which Śani's influence is expected to be negative. On the contrary, they welcome it. In fact this implies the exhaustion of previous karmic effects and a consequent greater freedom from the grip of the cycle of death and rebirth which prevents one from final deliverance (Swami Sivapriyananda 1990: 75).[52] The following passages – *pallavi*, *anupallavi* and *caraṇam* – from Sanskrit lyrics composed between the seventeenth and nineteenth centuries and regularly sung today in south Indian temples, further illustrate the two different functions of Śani just stated:

> I always meditate upon the slow-moving Śani, the son of Sūrya and the courageous one.

50 The figure of the *telī* is meaningful. The king had to turn the millstone for many years (the millstone represents the passage of time), and the offering of oils (mainly mustard and sesame) is fundamental in Śani Pūjā.

51 See also Svoboda (1997). Other methods of propitiating Śani include the recitation of his 108 names (*Śani Aṣṭottaraśatanāmāvali*) or his 1008 names (*Śrī Śanaiscara Sahasranāma Stotram*), and devotional hymns like *Śanaiścarastotram* (excerpted from BrāP), *Śanaiścarastavarāja* (excerpted from BhavP) or *Śani Cālīsā*.

52 This function is clearer considering Śani from the point of view of alchemy (*rasāyana vidyā*) where he is symbolized by iron, the most solid matter, and then transformed into gold, the most precious metal.

> Who causes fear in people plunged in the ocean of worldly existence and is the harbinger of calamitous events. Who grants uniquely auspicious rewards for devotees favored by Śiva's benign glances.
> Who with a body of dark lustre like collyrium, brother of Yama, riding on his vehicle the Crow, decorated with blue dress and a blue flower wreath, with ornaments embedded with blue precious stones [...] yielding all desires, the fire capable of splitting the time-wheel (Kāla Cakra) [...].[53]
> (NGK: 83–86)

Śani is also associated with Prajāpati and Yama, the former being the primordial deity (*adhidevatā*) of the planet, the latter the tutelary deity (*pratyadhidevatā*).[54] The function of Prajāpati is to rule over an entire cycle of existence; according to the doctrine of Yoga, he is a reflection of Īśvara, the Principle of manifestation of the universe with which any *sādhaka* aims to be reunited. Yama, conversely, is the lord of death.[55] He rules over the posthumous destiny of each and every person by judging merit and demerit. In the form of Dharmarājā (the king of *dharma,* the supreme law that rules the universe), Yama distributes pains and rewards. While Śani punishes or rewards human beings during their life according to *prarabdha karma* (the actions of all previous lives) so that they may intervene and rectify their doings, Yama does so after death. Yet at that point, in the abode of Yama, one can only bear with the fruits of one's previous life – be it merit or demerit – without any chance to intervene and improve upon them.[56]

Yama is also half-brother to Śani. According to some Purāṇic myths (Mani 1989: 182, 680, 683), Sūrya had three sons by Chāyā, the maidservant

53 While for the profane, the god is just one of the instruments for actualizing the effects of one's own karmic bonds, for the *sādhaka* Śani can become the 'fire capable of splitting the time-wheel' and the means of jumping out of the cosmic existence.

54 See Rao (1995: 153–60) for an exhaustive investigation of Vedic hymns related to the *adhidevatā* and *pratyadhidevatā* of Śani.

55 Yama is frightening, his complexion is dark blue, his face is terrible to look upon, his eyes are red, he rides a buffalo (symbol of ignorance and death) and holds an iron rod and a noose. Using these instruments his messengers pull out the souls from human beings at the time of their death and send them to Yama's realm. For a complete exposition of doctrines and rituals related to Yama and death, see Filippi (1996).

56 References to Śani are scattered across Itihāsa and Purāṇas. Many passages illustrate the effects of Śani's influence on a human being – usually the protagonist – who gains the favor of the god through atonement for release from karmic bondage which Śani himself had previously caused.

of his wife Saṁjñā: Manu, Tapatī and Śani. From his legitimate wife, the Sun had already begotten three other sons: Manu, Yama and Yamī. Saṁjñā herself had created Chāyā because she could not bear anymore her spouse's brightness and heat. She left her maid to attend her husband, who was unaware of this replacement, and went to perform austerities in the forest. One day Chāyā cursed Yama, so revealing that she loved her own sons more than Saṁjñā's. Upon discovering the deceit, Sūrya sent Chāyā away while Saṁjñā's father Viśvakarma reduced the heat and effulgence of the Sun by an eighth, so that his daughter could live again with her husband.[57]

In conclusion, Śani appears to be feared for his cruel, hideous, unyielding, *tamasik* nature. Yet, even though he is often associated with the heavy burden of nemesis, at the same time he can be instrumental in the palingenesis of the individual: one can be pushed by Śani to abandon the transient world and develop the desire for realization of the self (*ātman*).

ABBREVIATIONS AND REFERENCES

Gr. = Greek H. = Hindi Skt. = Sanskrit

Bellamy 2014 — Bellamy, C. (2014). 'The age of Śani in Modern Delhi.' *Nidān. International Journal for the Study of Hinduism*, 26: 22–41.

BhavP — *Bhaviṣyapurāṇa*. (1917). Mumbai: Veṅkaṭeśvara Press.

BPHS — Pārāśara: *Bṛhatpārāśarahorāśāstra*. Sharma, G.C. (ed.) (1995). *Bṛhatpārāśara-horāśāstra*. New Delhi: Sagar Publications.

BrāP — *Brāhmaṇḍapurāṇa*. Tagare, G.V. (tr.) (1984). *Brāhmaṇḍa Purāṇa*. Translated and annotated. Delhi: Motilal Banarsidass.

BS — Varāhamihira: *Bṛhatsaṃhitā*. Bhat, M.R. (ed. and trans.) (1993). *Varāhamihira's Bṛhat Saṃhitā*. Delhi: Motilal Banarsidass.

Dash 1986 — Dash, B. (1986). *Alchemy and Metallic Medicines in Ayurveda*. Delhi: Concept Publishing Company.

Divina Commedia — Dante Alighieri: *La Divina Commedia*. Petrocchi, G. (ed.) (1966–1967). *La Commedia secondo l'antica vulgata*. Milano: Mondadori. Text available at http://world.std.com/~wij/dante/ (accessed 28 September 2015).

57 In many texts, Sūrya is identified with *ātman* (the self). We can thus interpret the birth of Śani from Sūrya and his entrance into the world as the effect of a progressive clouding of the self by means of the numerous 'covers' of the manifested world. The effect of Śani appears similar to the veil of Māyā: all human beings are under his spell and do not recognize their real identity, the *ātman*. *Phala jyotiṣa* confirms this concept and says that only Śani, Rāhu and Ketu are able to deprive the Sun of its strength. In the case of all the other planets, it is the Sun that takes away their power.

Filippi 1996	Filippi, G.G. (1996). *Mṛtyu. Concept of Death in Indian Traditions*. New Delhi: D.K. Printworld.
Guenzi 2005	Guenzi, C. (2005). 'L'influence (*prabhāva*) de la planète et la colère (*prakopa*) du dieu Śani (Saturne) entre astrologie et pratique de culte à Bénarès.' *Bulletin d'études indiennes*, 23: 391–444.
Guenzi 2008	Guenzi, C. (2008). 'Planètes, remèdes et cosmologies. La thérapeutique astrologique à Bénarès.' In Županov, I.G. and C. Guenzi (eds.), *Divins Remèdes. Médicine et Religion en Asie du Sud*, pp. 191–218. Paris, Éditions de l'École Des hautes Études en Sciences Sociales, Collection Puruṣārtha.
Haberman 2013	Haberman, D.L. (2013). *People Trees. Worship of Trees in Northern India*. Oxford: Oxford University Press.
HVPh	Hindu Vedic Philosophy (Holy Pilgrimage – Theertha Kshetra) (2011). 'Holy Pilgrimage – Nava Graha Temples in Tamilnadu State – 3', <http://hinduphilosophyholypilgrimage.blogspot.it/2013/04/holy-pilgrimage-nava-graha-temples-in_6296.html> (accessed 25 August 2015).
Johari 1996	Johari, H. (1996). *The Healing Power of Gemstones in Tantra, Ayurveda and Astrology*. Vermont: Inner Traditions-Bear & Company.
JP	Vaidyanātha Dīkṣita: *Jātakapārijātaḥ*. Dīkṣita, V. (1978?). *Jātaka Pārijātaḥ*, with an English Translation and Copious Explanatory Notes and Examples by Subramanya Sastri. Vol. 1. New Delhi: Ranjan Publications.
KauAŚ	Kauṭilya: *Arthaśāstra*. Olivelle, P. (trans.) (2013). *King, Governance, and Law in Ancient India. Kauṭilya's* Arthaśāstra. New York: Oxford University Press.
Klibansky et al. 1952	Klibansky, R., E. Panofsky and F. Saxl. (1952). *Saturn and Melancholy. Studies in the History of Natural Philosophy, Religion and Art*. London: Thomas Nelson & Sons Ltd.
KūP	*Kūrmapurāṇa*. Tagare, G.V. (tr.) (1981). *Kūrma Purāṇa*. Translated and annotated. Delhi: Motilal Banarsidass.
Mani 1989	Mani, V. (1989). *Purāṇic Encyclopaedia. A Comprehensive Dictionary with Special References to Epic and Purāṇic Literature*. Delhi: Motilal Banarsidass.
MāP	*Mārkaṇḍeyapurāṇa*. Banerjea, K.M. (ed.) (2004 [1862]). *The Marcandeya [sic] Purana in the Original Sanskrit*. Bibliotheca Indica no. 29. New Delhi: Cosmo Publications.
NGK	Mutthuswamy Dikshitar: *Navagraha Krtis*. Translation by T. Mccomb (n.d.), http://www.sadagopan.org/SundaraSimham/SS100%20-%20Navagraha%20mantrams.pdf (accessed 25 August 2015).
Pandey 2014	Pandey, Pundit B.D. (2014). 'Shani Sade Sati Puja for Shilpa, Shani Puja, Shani Shanti Puja.' *AapkiKundli*, https://www.youtube.com/watch?v=Fczd3q8zoQA (accessed 28 August 2015).
Rao 1995	Rao, R. (1995). *Navagraha-kosha*. Vol. 1. Bangalore: Kalpatharu Research Academy.

Sachau 1992	Sachau, E.C. (ed.) (1992). *Alberuni's India. An Account of the Religion, Philosophy, Literature, Geography, Chronology, Astronomy, Customs, Laws and Astrology of India about AD 1030.* Delhi: Munshiram Manoharlal.
Sivapriyananda 1990	Swami Sivapriyananda. (1990), *Astrology and Religion in Indian Art.* New Delhi: Abhinav Publications.
Sudarshan 1999	Sudarshan, S.R. (1999). *Rasa-dhātu-kośa (Metallic and Mineral Drugs in Āyurveda).* Bangalore: Kalpatharu Research Academy.
Svoboda 1997	Svoboda, R. (1997). *The Greatness of Saturn. A Therapeutic Myth.* Delhi: Sadhana Publications.
Tetrabiblos	Claudius Ptolemy: *Tetrabiblos.* Claudio Tolomeo. (1989). *Le previsioni astrologiche (Tetrabiblos)*, a cura di Simonetta Feraboli. Milano: Fondazione Lorenzo Valla, Arnoldo Mondadori Editore.
Timaeus	Plato: *Timaeus.* Platone. (1953). *Repubblica, Timeo, Crizia*, a cura di Francesco Adorno. Torino: UTET.
VI	Macdonell, A.A. and A.B. Keith. (1995). *Vedic Index of Names and Subjects.* Delhi, Motilal Banarsidass.
Vogel 1926	Vogel, J.P. (1926). *Indian Serpent-lore. The Nāgas in Hindu Legend and Art.* London: Probsthain.
Zeiler 2013	Zeiler, X. (2013). 'Dark Shades of Power: The Crow in Hindu and Tantra Religious Traditions.' In F.M. Ferrari, and T.W. Dähnhardt (eds.), *Charming Beauties and Frightful Beasts. Non-Human Animals in South Asia Myth, Ritual and Folklore*, pp. 199–216. Sheffield/Bristol: Equinox.
Zeiler 2015	Zeiler, X. (2015). 'Yantras as Objects of Worship in Hindu and Tantric Traditions: Materiality, Aesthetics and Practice.' In K.A. Jacobsen, M. Aktor and K. Myrvold (eds). *Objects of Worship in South Asian Religions: Forms, Practices and Meanings*, pp. 67–84. London: Routledge.

Section Two

Science and Health

Chapter 4

Mineral Healing: Gemstone Remedies in Astrological and Medical Traditions

ANTHONY CERULLI AND CATERINA GUENZI[1]

From pre-modern times to the present day, in written treatises and oral traditions, gemstones have played varied and significant roles in treating physical, emotional and social ailments in South Asia. Known in Sanskrit and a number of north Indian vernaculars as *ratna* and *maṇi*, precious stones (which we refer to in what follows synonymously as gems and gemstones) have a long history as powerful remedies. In the Sanskrit literature of the classical knowledge systems of Jyotiṣa ('astral science') and Āyurveda ('knowledge for longevity'), as well as the allied field of Rasaśāstra ('alchemy'), for example, *ratna* and *maṇi* are described as valid therapies to prevent and remedy an array of personal miseries, difficulties and diseases. In this chapter, we reflect on some of the chief characteristics of gemstones as healing remedies in India, drawing on Sanskrit literature from the astral sciences, classical Āyurveda, and Ratnaśāstra ('the science of gemstones'). We also occasionally complement the literature with contemporary observations of gemstone healing among astrologers and their patients in the north Indian city of Banaras.[2]

1 Anthony Cerulli is Associate Professor of Asian Languages and Cultures at the University of Wisconsin at Madison (USA).
Caterina Guenzi is Associate Professor at the École des Hautes Études en Sciences Sociales (EHESS) and member of the Centre for South Asian Studies (CNRS/EHESS), Paris (France).

2 The fieldwork in Banaras that is referenced in this chapter was conducted by C. Guenzi at different times between 1999 and 2008.

Among our primary considerations are the following points: applications of gemstones in the healing process; ritual preparations of gemstones for healing usages; associations of certain gemstones with the planets; potential good and bad consequences of using gemstones to heal or prevent misfortune and illness; the significance of visibility when using gemstone remedies to heal; and the relationship of a gemstone's preciousness – including costliness – to its therapeutic capacities. These qualities, we suggest, differentiate the use of precious stones from other healing remedies, such as devotional practices and tantric amulets containing sacred formulas or diagrams.

HEALING FROM THE OUTSIDE IN: WEARING, INGESTING AND PREPARING GEMS

Whereas astrologers base the prescription of gems on a person's birth chart and calculation of the planetary conjunctions (*yoga*) and periods (*daśā*) impacting a person's life, Āyurvedic physicians typically direct a patient to obtain gemstones, or the powders or oxides derived from gems, according to an assessment of a patient's humoral makeup. Āyurveda holds that the properties of certain gems have the capacity to affect positively the body's three humors (*tridoṣa*): wind, bile and phlegm (*vāta*, *pitta* and *kapha* or *śleṣman*). The activity, balance and location of the humors in the body determine a person's wellbeing or ill health.

A curative prescription of gemstones may take a number of forms. In therapeutic use in astrological and medical contexts, gemstones, gems pulverized into powders, and liquids that have been infused with gem essence are often prescribed either to be worn or ingested. The 'wearing of gemstones' (*ratnadhāraṇa*) is a basic practice of 'gemstone therapy' (*ratnacikitsā*). Such a prescription might require a patient to wear a precious stone for a period of time as short as a few days or as long as a few years. In some cases, a patient might be asked to don a gem somewhere on his or her body permanently. As we discuss below, a variety of benefits await the person who wears precious stones. The following aphorism in the *Carakasaṃhitā* (c. 100 BCE–100 CE), one of the principal texts of Āyurveda's classical Sanskrit literature, sums up the general value of gemstones for healing:

> Wearing gems and ornaments bestows prosperity, auspiciousness, long life, radiance, and impedes deterioration. Delightful and desirable, it [i.e., wearing gems] brings vitality.[3]

The wearing of gems is also recommended in more specific and pressing circumstances. So, for example, the *Carakasaṃhitā* suggests that wearing diamonds, emeralds, rubies, lapis lazuli, pearls, a collection of other semi-precious minerals and an herbal talisman protect one from snake poisons.[4]

Apart from the wearing of gems, in the medical context an Āyurvedic doctor might also require a patient to consume gemstone powder (*cūrṇa* or *piṣṭa*) or ashes (*bhasma*) in a mixture of milk and honey. A patient might be asked to imbibe water or alcohol in which soaking precious stones have ionized the liquids over a period of time (usually at least twelve hours). The rationale for ingesting gemstone powder or ashes or drinking liquids ionized by precious stones is based on a belief that gems are made up of the very same minerals that comprise the human body. Whenever gems are to be consumed internally as remedies, the essence of a gemstone must be extracted for medicinal use. Particular preparations to extract the healing capacities of gemstones include *svedana-yantra*, an 'apparatus for steam fomentation' (Dash 1986: 215) and *kṣāra-drāva*, 'a liquid preparation with alkalis and salts' (Ray 1991: 141).

Gemstones used prophylactically against planetary influence are usually set in golden or silver rings to be worn on the hands. An early-medieval work on alchemy (*rasaśāstra*), medicine and gemology, the *Rasaratnasamuccaya*, explains that when wearing gems in rings (*mudrādhṛtam*), if the gems are positioned according to their corresponding planets, alliances with those planets form.[5]

In contemporary practice observed in Banaras, to ensure a patient obtains the correct stone(s) for his or her ailment, a practitioner will write a prescription indicating the kind of gemstone(s) needed and the type of metal in which it (they) should be set; the number of carats desired for optimal effect; propitious times (*muhūrta*) to buy and to wear the stone(s); locations on the body the stone(s) should be worn; and any other relevant details. Gems set in rings are said to absorb and filter planetary rays

3 CaS, Sūtrasthāna, v.97: *dhanyaṃ maṅgalyamāyuṣyaṃ śrīmadvyayasanasūdanam/ harṣaṇaṃ kāmyamojasyaṃ ratnabharaṇadhāraṇam*. See also CaS, Sūtrasthāna, v.110, 'wearing gems' (*ratnadhāraṇe*) and vi.31, 'decorated with pearls and gemstones' (*muktāmaṇivibhūṣitaḥ*).

4 CaS, Cikitsāsthāna xxiii.252–53.

5 RS iv.7.

and transmit them to the body in a balanced dose. Jewelers selling gemstones are therefore asked to cut the stones in a conic shape to facilitate the absorption of planetary rays: according to local explanations, while the upper surface serves as a plateau for capturing the rays, the point of the cone that touches the skin transmits the appropriate amount of planetary rays. Although rings are generally the favorite vessel for brandishing the stones because of their consistent exposure to sunlight, gemstones used for healing may in some cases be worn on necklaces, bands placed around the arm, or belts wrapped around the waist. Instances in which precious stones might not be worn openly on the body often involve children, who may hurt themselves if the stones are sharp and easily handled or swallowed, and women and teenagers who might not be willing to display their 'planetary problems.'[6] Contrariwise, it is frequently the case that men, and especially 'householders' (*gṛhasthas*) from middle and upper classes – such as businessmen, civil servants, politicians, academics, actors, doctors, lawyers and so on – proudly display their expensive gemstones that, on their fingers, are seen as symbols of their professional and familial responsibilities.

OF GEMS AND PLANETS: *NAVARATNA* AND *NAVAGRAHA*

In the astral and medical sciences, precious stones – including, in some cases, semi-precious stones and minerals – that are used as remedies and their affiliated planets and celestial bodies generally align according to table 4.1.

In astral science (*jyotiḥśāstra*), *ratnaparīkṣā*, the formal 'examination of precious stones,' appears as a subject beginning around the sixth century CE when Varāhamihira composed the *Bṛhatsaṃhitā*. An authoritative treatise on divination, the *Bṛhatsaṃhitā* has four chapters (LXXIX–LXXXII) devoted to the inspection of precious stones, in particular, diamonds, pearls, rubies and emeralds. Varāhamihira explains that the physical characteristics of precious stones reveal whether they are auspicious or inauspicious (*śubha* or *aśubha*) and whether they portend desired or undesired (*iṣṭa* or *aniṣṭa*) events, such as wealth or loss, pleasure or pain, health or disease, fertility or sterility, and the like. The destiny of the ever-important pillar of classical

6 According to some astrologers in Banaras, when a gemstone is worn under clothing its power should be compensated by increasing the stone's number of carats. Thus, for a nine-carat stone worn as a ring, an astrologer should prescribe a twelve-carat stone to be worn around the waist.

Table 4.1. Planets (*grahas*) and associated gems.

Planet	Gemstone
1. Sun (Sūrya)	ruby (*padmarāga, māṇikya*)
2. Moon (Candra)	pearl (*mauktika, muktā*)
3. Venus (Śukra)	diamond (*vajra, hīraka*)
4. Mars (Maṅgala)	coral (*pravāla*)
5. Mercury (Budha)	emerald (*marakata, gārutmata, tārkṣya*)
6. Jupiter (Bṛhaspati)	topaz (*pītāśmā, puṣparāga, puṣyarāga*)
7. Saturn (Śani)	blue sapphire (*nīla, indranīla*)
8. Rāhu	zircon (*gomeda*)
9. Ketu	cat's eye, beryl or lapis lazuli (*vaiḍūrya*)[7]

Indian society, the king, Varāhamihira says, is 'dependent on gems' (*ratnāśrita*).[8] Even more, Wojtilla has observed that the term *ratna* itself is 'an attribute of *cakravartin*,' the universal ruler, who in ancient India was often described as *ratnin* (1973: 215), a person 'possessing precious things.' This descriptive plays with the idea that the king is at once wealthy and powerful because of the treasures he possesses as well as healthy and successful because of the powers and influence his precious trove affords him.

During Varāhamihira's time, however, the connection between gemstones and planets was not yet established. In the *Bṛhatsaṃhitā*, different types of stones – catalogued according to size, shape and color – are linked to Vedic deities such as Indra, Yama, Viṣṇu, Varuṇa, Agni and Vāyu.[9] But planetary deities are not mentioned. The absence of such a connection is also seen in Varāhamihira's *Bṛhajjātaka*, where planets (*graha*) are linked to many sets of cosmological elements, including colors, directions, qualities, body parts, seasons and metals. But no mention is made of gemstones.[10] What's more, Varāhamihira mentions twenty-two gemstones (three of which are sub-varieties), and he does not seem to be aware of the concept of 'nine gems' (*navaratna*) that becomes important later. We can therefore say that prior to the sixth century an astrologer was supposed to be well-versed in gemology, since precious stones were seen as important wonders

7 About the ambiguity concerning the gems designated as *vaiḍūrya* and, more generally, about the difficulties in identifying which stone corresponds to a gemological term, see Finot (in AM: pp. XLV–XLVII); Sarma (in RaPa: pp. 67–68); Winder (1990); and Biswas (1994).

8 BS LXXIX.1.

9 BS LXXIX.8–10.

10 See, for example, BJ II.

endowed with natural powers. Yet the use of gems as prophylactics against planetary influences according to a person's horoscope had not yet been established before that time.[11] Absent in the *Sārāvalī* of Kalyāṇavarman (800 CE) as well, the nine planetary gems are only mentioned in later horoscopic treatises, such as the *Jātakapārijāta* of Vaidyanātha – a south Indian text that is largely based on the *Bṛhajjātaka* and the *Sārāvalī* (and composed probably at some point between the twelfth and the mid-fifteenth centuries),[12] the *Phaladīpika* of Mantreśvara, and the *Muhūrtacintāmaṇī* (1600 CE) – a treatise on 'auspicious moments' (*muhūrta*).[13]

This raises some important questions. How, where and when did the connection between 'nine gems' – *navaratna* – and 'nine planets' – *navagraha* – develop? To answer this series of imbricated questions, we propose the following three aspects require attention: First, the origin of the concept *navaratna* and its uses must be established. When did nine become a standard number for gemstones, a question pointed out by Sircar (1972), and for what purposes?[14] Next, the meaning of *navaratna* must be elaborated. Did this concept designate a fixed set of gemstones, or could these specific stones vary from one context to the next? Finally, what is the relationship between the idea of *navaratna* and astrological practices? Did this idea originate in astrological contexts in order to 'fit' the number of planetary deities? Or, are the astrological applications of *navaratna* a later development? These questions lead to a vast area for philological research into the astrological and gemological literature, indeed crossing other fields besides, including ritual studies, medical history and religious studies.[15] As we are commissioned in this chapter merely to sketch out the basic uses and understandings of gemological healing, these bigger questions

11 In the *Bṛhatsaṃhitā*, the auspiciousness of gemstones is nevertheless 'context-sensitive': according to their colors, the varieties of gemstones are said to be more or less auspicious for different social classes (*varṇa*). See, for example, the passage about diamonds at BS LXXIX.11.

12 See Pingree (1981: 89–90).

13 In horoscopic literature, the nine gemstones are enumerated in the second chapter, dealing with the characteristics and qualities of the planets, *grahasvarūpaguṇa* (JP II.21; PhDi II.29; MC IV.9–11).

14 See Sircar (1972).

15 In this regard, we would like to thank Christèle Barois and Dominic Goodall for their generous suggestions – through correspondence and informal exchanges – pointing out different sources (manuscripts, inscriptions, objects) and ritual domains (planets' worship, placing of consecration deposit in temples, installation of images or *liṅgas*, preparation of jewel-water, etc.) from where the history of the *navaratnas* could be unveiled.

are well beyond the scope of our work. That said, we shall try to identify just a smattering of the landmarks in this massive and underexplored sea of research.

In gemological literature as well, the idea of *navaratna* appears somewhat late. Lacking in ancient lapidaries, such as Buddhabhaṭṭa's *Ratnaparīkṣā*, the concept of *navaratna* is attested in the *Agastimata*, a later south Indian lapidary whose period of composition is still uncertain.[16] The *Agastimata* seems to be the oldest lapidary known to us that presents a system of correspondences between nine precious stones – ruby, pearl, coral, emerald, topaz, diamond, sapphire, zircon and cat's eye – and a group of the nine planetary deities in a fashion that became standardized later on and that is followed even today in most regions of India (see table 4.1). It should be noted, however, that the connection between the gemstones and the nine planetary deities in the *Agastimata* is mentioned only in the last two verses of the treatise.[17] Furthermore, these verses refer to gemstones that are not described in the treatise and subvert the conventional order of gemstones used in most lapidaries (including the *Agastimata*), which is: diamond, pearl, ruby, sapphire, emerald and crystal. The association between planets and gems established in the last two verses of the *Agastimata* may therefore be an addition or an innovation that the author(s), or later redactors, interpolated into the standard content of ancient gemological literature.

Ṭhakkura Pherū, a Svetāmbara Jain scholar serving at the sultan's court in Delhi at the beginning of the fourteenth century, states in his *Rayaṇaparikkhā* that gemstones have an apotropaic power and that they protect those who wear them against planetary influence.[18] This statement is not surprising, since it comes from an author familiar with all branches of the astral sciences. Planetary considerations are well integrated into the *Rayaṇaparikkhā*, for example, appearing early in the introduction and linked to the origin myth of gemstones found in most lapidaries.[19] According to the *Rayaṇaparikkhā* version of the myth, twelve gemstones were formed from Bali's body and nine were seized by the planets (*grahas*, lit. 'seizers'). Each gem was picked up by the planet that matched it in color. Besides this mythical etiology, planetary deities are also evoked

16 The *Agastimata*'s terminus post quem is the sixth century (Finot in AM: p.xi) and its terminus ante quem is the beginning of the fourteenth century, when Ṭhakkura Pherū composed his *Rayaṇaparikkhā*, a Prakrit text on gemology that contains passages closely following *Agastimata* (Sarma in RaPa: p.15).

17 AM vi. 343–44.

18 RaPa 0.3.16.

19 RaPa 0.2.9–11.

in other passages of the *Rayaṇaparikkhā* that deal with the prophylactic qualities of sapphire and topaz.

The myth of the origin of gemstones in the *Garuḍapurāṇa* is perhaps the most well-known version of the story, where numerous precious stones (*ratnāni*) and semi-precious stones (*uparatnāni*) spring from the body of the mighty demon Bala. The story states that, although he had previously conquered Indra and driven the gods out of heaven, Bala had magnanimously agreed to support any sacrifices (*yajñas*) the gods would perform in the future. In an effort to regain their seats of privilege in heaven, the gods planned an elaborate sacrifice and, remembering Bala's earlier agreement to assist a future sacrifice, they approached Bala and jestingly asked him to offer his body as the sacrificial oblation. Bound by his earlier promise, Bala acquiesced. The gods in fact meant business, and they seized the opportunity to kill Bala. Dismembering him for the sacrifice, each piece of Bala's body became a 'seed of a gemstone' (*ratnabīja*), and as the Yakṣas, Siddhas, and others raced to collect the precious gem-seeds, a cohort of gods swooped in on their *vimānas* to snatch them up. A mid-air scuffle ensured, and some of the gem-seeds fell to the ground.[20] They fell into oceans and rivers, onto the Himalayan mountains, and throughout India's dense wildernesses. Wherever Bala's body parts landed, in those locations mines of precious stones formed. Gems that grew in the earth in these locations, the *Garuḍapurāṇa* explains, were endowed with healing powers of all sorts, both prophylactic and remedial, and they contained powers to cure physical, emotional and spiritual illnesses. From Bala's body diamonds, pearls, rubies, emeralds, sapphires, cat's eyes, topazes and other precious and semi-precious stones arose. His bones materialized as diamonds. From his teeth came pearls, and from his blood rubies appeared. His bile turned into emeralds. His eyes generated sapphires and his skin made topaz. The great cries Bala bellowed just before he died turned into cat's eyes, while his semen produced zircon, and his intestines became coral. The fat from his sacrificed body dripped to earth, creating quartz, and the remnants of his form turned into agate.[21]

20 GP LXVIII.4–6: *tasya sattvavaśuddhasya viśuddhena ca karmaṇā / kāyasyāvayavāḥ sarve ratnabījatvamāyayuḥ // 4 // devānāmatha yakṣāṇāṃ siddhānāṃ pavanāśinām / ratnabījamayaṃ grāhaḥ mumahānabhavattadā // 5 // teṣāṃ tu patatāṃ vegādvimānena vihāyasā / yadyatyapāta ratnānāṃ bījaṃ kvacana kiñcana // 6 //*

21 The details of the 'Bala sacrifice' that created precious and semi-precious stones, including details about the inspection of *ratnāni* to ascertain their power and value to health and longevity, are spread across chapters 68–80 of GP.

The *navaratna* and their association with the *navagraha* are attested in the *Rasaratnasamuccaya* as well. Attributed to someone named Vāgbhaṭa, the text is a medieval alchemical and iatro-chemical compendium. No consensus has been reached regarding the precise date of the work. Murthy (1991: 156) places it as early as the ninth century, for example, whereas Biswas (1987: 29) considers the work a product of the thirteenth century. In the introduction to his edition of the text, Satpute (RS xi) locates it even later in time, between the fourteenth and fifteenth centuries. Meulenbeld has argued that the text was produced after the fifteenth century and most likely in the sixteenth century. His position is the most persuasive in the extant secondary literature, grounded as it is on extensive references in the *Rasaratnasamuccaya* in coeval and earlier works. Meulenbeld suggests that scholars who defend an earlier date for the text tend to sacrifice philological evidence, such as intertextual references that suggest an obvious awareness of other texts and authors which points 'to a period posterior to the first half of the fifteenth century,' in an effort to associate the work's author, Vāgbhaṭa, son of Siṃhagupta, with the author(s) of the same name and pedigree who ostensibly wrote one (or both) of the Āyurvedic classics, the *Aṣṭāṅgahṛdayasaṃhitā* and *Aṣṭāṅgasaṃgraha* (Meulenbeld 2000, iia: 670). For his part, Satpute firmly states that the Vāgbhaṭa of the *Rasaratnasamuccaya* is not the same person as the author(s) of the medical works (RS xiv).

Not an original work per se, the *Rasaratnasamuccaya* is a compilation of earlier Rasaśāstric treatises from whose authors Vāgbhaṭa claims to have borrowed freely.[22] While the text is an important benchmark in Indian alchemical literature for its sophisticated cataloging and standardizing of earlier and often scattered works on *rasaśāstra*, for our purposes its fourth chapter, titled *ratnāni*, is especially illuminating. Of the *navaratna*, the text declares diamond, ruby, sapphire, emerald and topaz to be the five superior healing gems known as the *pañcamahāratna*.[23] The text also presents the same correspondences between the nine gemstones and nine planets found elsewhere in Sanskrit literature.[24] Particularly significant is the *Rasaratnasamuccaya*'s articulation of the positive and negative effects and

22 RS i.2–10. Meulenbeld provides an exhaustive list of the works and authors to which the author of the RS refers (2000, iia: 667–68).

23 RS iv.5: *padmarāgendranīlākhyau tathā marakatottamaḥ / puṣparāga savajrākhyaḥ pañcaratnavarāḥ smṛtāḥ*.

24 RS i.6: *māṇikyamuktāphalavidrumāṇi tārkṣyañca puṣpaṃ bhidurañca nīlam / gomedakañcātha vidūrakañca krameṇa ratnāni navagrahāṇām*.

uses of both precious and semi-precious gems in the treatment of illness and misfortune, which we discuss in the next section.

Introduced later in astrological, gemological and alchemical literature, the idea of *navaratna* may have originated in ritual contexts. Take, for example, the domestic rituals described in texts like the *Mayamata* (circa ninth–twelfth centuries) and *Kāśyapaśilpa* (circa eleventh–twelfth centuries), where nine gemstones are placed in the so-called consecration deposit (*garbhanyāsa*) to ensure the prosperity of people residing in a building.[25] The list of gems given in the *Mayamata* is almost identical to the list of gems assigned for prophylactic use against planetary influence, but there is no discussion of planetary deities.[26] As described in both the *Mayamata* and the *Kāśyapaśilpa*, a ruby should be placed in the centermost compartment of the consecration deposit, with eight other compartments that correspond to the eight cardinal directions surrounding it. As the precious stone of the Sun, the ruby is also usually placed in the center of *navaratna* rings prescribed against planetary influence.[27] Although it is difficult to imagine that the perfect matching between the set of nine gemstones used for the consecration deposit and the nine planetary gemstones is a matter of coincidence, further research needs to be done to understand how knowledge was transmitted from one field of expertise to the other.

BENEFICIAL AND DANGEROUS EFFECTS

When using gemstones as a remedy against planetary influences, astrologers and their clients refer to a complex set of principles and rules that aim to identify the properties of a stone and its effects (*phala*) on the person who will wear it. These principles may be based on both Sanskrit gemological literature, works of Rasaśāstra like the *Rasaratnasamuccaya*, and

25 On the *Mayamata* see Dagens (in M: pp. 142–43 and 200–01); on the *Kāśyapaśilpa* see Ślączka (2007: 99–101).

26 The only difference is the mention of sapphire (*mahānīla*) instead of zircon (*gomeda*). The list given in the *Kāśyapaśilpa* presents more important variations, since it includes items such as a crystal (*sphaṭika*), a conch (*śaṅkha*) and a sun-stone (*sūryakānta*) instead of coral, emerald and zircon (Ślączka 2007: 263). Nevertheless, from the point of view of the directions corresponding to the gems, the *Kāśyapaśilpa* is closer to astrological traditions: the rule of placing ruby in the center, diamond in the East and pearl in the Southeast correspond to the instructions for the *navagraha* ring, as described in MC IV.9.

27 See e.g. MC IV.9.

local lore. They are also sometimes described in popular and devotional literature (in vernacular languages) regarding the propitiation of planetary deities.

In Banaras, astrologers prescribe gemstones as a remedy (H. *upāy*) either to increase (*baṛhānā*) positive planetary influence (*grahaprabhāv*) or to reduce (*ghaṭānā*) negative planetary effects.[28] A person may thus wear a stone to strengthen the planet that is in a favorable position in her horoscope, or she may use a gemstone as an antidote against a malefic planet (*krūr graha*, *pāp graha*). In both cases, however, precious stones are supposed to modify planetary influence, and astrologers as well as clients typically agree that because of their power, gemstones can be dangerous (*khatarnāk*). That is why before purchasing a gemstone from a jeweler, clients ask astrologers to check its quality. Practitioners have to check not only whether or not the stone is authentic (*sahī*, *aslī*), but they also must determine whether or not it presents any faults (*doṣ*). Clients are asked to watch carefully the 'behavior' of the stone they wear, since any change is regarded as an omen (*śakun*) that may indicate a change in the stone's efficacy.

In the *Rasaratnasamuccaya*, before he launches into a detailed explanation of the beneficial and dangerous effects of both precious and semi-precious stones, their applications and their properties, Vāgbhaṭa states in no uncertain terms:

> It's said that wherever gemstones are used in the preparation of quicksilver (*rasa[karman]*), rejuvenation therapy (*rasāyana*), as gifts, for wearing, or for worshipping the gods, always use well preserved, well produced, and splendid gemstones.[29]

The exercise undertaken to determine whether or not gemstones are defective and their potential negative effects is described in Sanskrit treatises dealing with *ratnaparīkṣā*, 'examination of precious stones.' In *ratnaparīkṣa* literature a distinction is made between the 'quality' (*guṇa*) and 'fault' (*doṣa*) of gemstones, as well as a distinction between 'auspicious' (*śubha*) and 'inauspicious' (*aśubha*) gemstones. While jewels endowed with good characteristics are said to bring health, good luck, wealth, glory and sons, gemstones bearing defective characteristics may lead to pain, sorrows, disasters, poison, poverty and death.

28 For a discussion about these two ways of dealing with planetary influence, see Jindal (1997: 11).

29 RS I.8: *rase rasāyane dāne dhāraṇe devatārcane / surakṣyāṇi sujātīni ratnānyuktāni siddhaye.*

According to Varāhamihira's *Bṛhatsaṃhitā*, precious stones that are glossy (*snigdha*), illuminated with rays (*prabhānulepin*), transparent (*svaccha*), shining (*arciṣmat*), heavy (*guru*), well-shaped (*susaṃsthāna*), gleaming from inside (*antaḥprabhā*) and intensely colored (*atirāga*) are seen as endowed with good qualities.[30] Those that are turbid (*kaluṣa*), with a dull luster (*mandadyuti*), scratched (*lekhākīrṇa*), mixed with mineral substances (*sadhātu*), broken (*khaṇḍa*), badly perforated (*durviddha*), not beautiful (*na manojña*) or mixed with gravel (*saśarkara*) are seen as defective.[31] Similarly, astrologers in Banaras say that stones lacking brilliance (*camak*), or containing a spot (*dāg*), a mark (*chīṃṭ*), a filament (*reśā*) or a 'net' (*jāl*) are defective and may bring undesired effects. Commenting on the ubiquity of defects in precious stones, the *Rasaratnasamuccaya* states that in all gemstones sharpness, asymmetry, spottiness, scratches and water bubbles are the five most common defects.[32] Nevertheless, in some cases these faults may also be seen as a sign of the authenticity of the stone, since synthetic stones have no irregularities, and on occasion astrologers may ask a client to wear a stone even though it's defective.

Medicinally, in chapter four of *Rasaratnasamuccaya* a series of verses (or *ślokas*) outlines the particular effects of the *navaratna* when they are ingested in the form of ashes. Some of the text's account of the specific curative effects of these gems includes the following. Ingestion of the ash of diamond (*vajra*) gives the body strength by regulating the body's three humors. Diamond ash cures impotence and increases the volume of a man's semen. It is effective in strengthening a weak heart and removes lung diseases. When taken in conjunction with other medications, diamond is said to intensify the curative powers of those drugs.[33] The ashes of ruby (*māṇikya*) are beneficial for digestion, increasing semen and sexual vigor (*vṛṣya*); it is especially effective against diseases caused by the wind (*vāta*) and phlegm (*kapha*) humors.[34] Ingestion of the ashes of emerald (*marakata*) is effective against fever, vomiting, indigestion and anemia.[35] The ashes of topaz (*puṣparāga*) target poisoning and diseases caused by the wind and phlegm humors; topaz ash alleviates desiccation of the body and removes

30 BS LXXXI.3.

31 BS LXXXI.3.

32 RS IV.33: *grāsastrāsaśca binduśca rekha ca jalagarbhatā / sarvaratneṣvamī pañca doṣāḥ sādhāraṇā matāḥ*.

33 RS IV.32.

34 RS IV.12.

35 RS IV.22.

impurities in the blood.[36] The ash of blue sapphire (*nīla*) works especially well for breathing (*śvāsa*) problems and increases sexual productivity; sapphire ash targets ailments caused by all three humors.[37] The ingestion of zircon (*gomeda*) ash rectifies diseases originating from the phlegm humor, anemia and skin diseases; zircon is also said to increase intelligence (*buddhiprabodhanam*).[38] The ash of cat's eye (*vaiḍūrya*) targets blood impurities and indigestion; it is an overall bodily cleansing substance that promotes longevity.[39] Ashes of pearl (*muktā*) are effective at improving digestion, skin complexion and strength; pearl ash has expansive healing capacities that reach illnesses caused by all three humors, heart diseases, urinary disorders, severe cough and brittle bones.[40] The ash from coral (*pravāla*) is used to treat severe desiccation, blood disorders, cough and eye diseases.[41]

During astrological counseling in Banaras, several criteria are taken into consideration to make sure that a particular stone and any accompanying stones chosen to support it suit the person who will wear them. An inauspicious gemstone is thought to produce effects opposite to the outcomes desired. If, for example, a pearl (the Moon's gemstone) is worn as protection against a cold and headache – disorders that are usually attributed to the influence of the Moon – an inauspicious or low quality pearl may increase the risk of getting such ailments (according to the *Rasaratnasamuccaya*, a low quality pearl displays a rough surface, darkened or reddish color, is riddled with holes and is not perfectly round).[42] Some gemstones are considered to be more dangerous than others, and their powers need to be managed more carefully both by the professional prescribing it and the patient wearing it. This is the situation with coral (Mars), sapphire (Saturn), and diamond (Venus). Dangers associated with these stones are linked to the perceived malefic powers of Mars and Saturn, and the potentially precarious association of Venus's gem (diamond) with unbridled sexuality.

When gemstones are used in a healing context, they are generally observed for a few days before they are activated and worn. Although positive effects of precious stones are said to manifest very slowly, in some cases taking months to appear, negative effects are often said to

36 RS IV.25.
37 RS IV.53.
38 RS IV.56.
39 RS IV.59.
40 RS IV.14.
41 RS IV.19.
42 RS IV.16.

be immediately noticeable. Before committing themselves to wearing a stone for an extended period, people often test the stone's likelihood to produce positive or negative effects by placing it underneath their pillow for a night. After a full night's sleep, by evaluating their dreams from the night before they can determine whether or not the stone will be helpful or harmful in the future. In other cases, such as inauspicious coral and sapphires, which are thought to increase the risks of traveling (typically road accidents), people wear these stones only at home initially; in a relatively controlled domestic environment, the potential hazards of wearing coral and sapphire are augured by small domestic injuries and out of the ordinary family disputes. If such incidents are observed to occur during the testing period, the stones are discarded.

In most cases of gemstone therapies, astrologers also consider the kinds of thoughts a person has while wearing a stone for the first few days to be an important signal about the stone's potential efficacy or risk. In addition to their effects on physical wellbeing, as we saw in remarks from the *Rasaratnasamuccaya* about the efficacy of zircon, gemstones are used for their strong influence on the human mind (*buddhi*) and for bolstering psychic strength (*manobala*). After wearing an inauspicious gemstone, a person's thoughts might get misled or muddled. As one client of an astrologer in Banaras reported about the effects of a bad sapphire, 'a sapphire that does not suit you means that you will take a snake for a rope and a rope for a snake.' In the case of diamond, a person may have impure, sexual and uncontrollable thoughts that may lead to indecency (H. *caritr hīnatā*, lit. 'lack of character').

Bad effects of gemstones do not affect only the person who wears the stone; they can also harm the wearer's family and people closely associated with him or her. This principle is stated in the classical literature, and astrologers and patients in contemporary India also frequently note it. So, for instance, the *Bṛhatsaṃhitā* says that a diamond endowed with inauspicious signs (*aniṣṭalaksaṇa*) brings the destruction of a person's kinsmen (*svajana*, BS LXXIX:18). Similarly, the *Navagraha Upāsanā*, a devotional pamphlet about the worship of the nine planetary deities, declares that:

> a ruby without glitter (*camak*), [signals] a brother's suffering (*dukh*); a ruby with the color of milk, [signals] the death of cattle (*paśunāś*); a bicolored ruby, [signals] the suffering (*dukh*) of the wearer's father or the wearer him or herself; a ruby with a spot (*chīṃṭ*) or a filament (*jāl*), [augurs] disputes [...] (Shastri n.d.: 18).

In Banaras today people often attribute their kinsmen's problems to the negative effects of gemstones. A professor in the History Department at Kashi Vidyapith elaborated this belief by explaining that the day he started wearing a forty-carat gemstone, his wife lost a sari she had brought to the dry cleaners, and he experienced unusual disputes with his colleagues. These incidents were for him visible signs of a gemstone that was too powerful and thus inappropriate for therapeutic use. Similarly, it is common to see parents wearing gemstones in the hopes of securing success or health for their children, as well as wives wearing coral in order to protect their husbands from Mars's damaging influence.

To prevent people from wearing defective, 'wrong' (H. *galat*) or 'inappropriate' (*anucit*) gemstones, astrologers not only have to check the physical properties of the stone, but they also must ascertain the compatibility between the gemstone and the zodiacal signs (*rāśi*, lunar sign, or *lagna*, ascendant sign) of the person seeking help. This correlation is vital, since, according to a person's planetary affiliations, gemstones are not equally compatible with all horoscopic signs. Astrologers should also determine if a person is simultaneously wearing gemstones that correspond to planets that are inimical to one another: because Sun and Saturn are enemies, for instance, one should not wear ruby and sapphire at the same time.[43]

A stone's 'dosage' or power (the English term is used in Hindi), which corresponds to the number of carats to be worn, is also an important characteristic that may cause unfavorable effects. While a gemstone that is too light may be ineffective, one that is too heavy may produce undesired effects.[44] To be auspicious, a gemstone should also be bought, manufactured and worn at the proper 'moment' (*muhūrta*). For some astrologers, an auspicious moment is defined according to the different sections of the twenty-seven lunar mansions (*nakṣatras*): in the case of a ruby, for instance, buying of the stone should happen during the first division of *maghā nakṣatra*, its manufacturing in the second, its wearing in the third.[45]

43 *Navaratna* rings are an exception, since these rings are supposed to balance all planetary influences. Nevertheless, astrologers very rarely prescribe *navaratna* rings, which most people buy and use independently from astrological consultations.

44 According to a jeweler working in a very popular shop in the center of Banaras and having experience from other cities, astrologers in Banaras rarely prescribe planetary gemstones under seven carats, while in other cities, lighter gemstones are commonly prescribed.

45 For a complete list of the *nakṣatras* corresponding to the different gemstones, see Joshi (n.d.: 193–204) and Jha (2006).

To further ensure the achievement of the desired outcomes of a gemstone remedy, most astrologers indicate a favorable day (H. *abhīṣṭa dīn*) of the week when the gemstone should be worn for the first time (Monday for pearl, Tuesday for coral, and so on), and the appropriate finger on which it should be worn.

Before wearing a gemstone, a consecration ritual commonly performed for divine images in temples called *prāṇa pratiṣṭhā* should be performed. This ritual is meant to invoke the planetary deity associated with the stone. In Banaras, the *prāṇa pratiṣṭā* may be performed by the astrologer who prescribes the stone or by the person who will wear it. While submerging the finger on which a gemstone is worn in milk mixed with water from the Ganges (Ganga) River – or in a *pañcāmṛta* mixture of 'the five nectars': Ganges water, milk, curd, honey and clarified butter – the person seeking aid from the stone should recite to him- or herself the *bīja mantra* ('seed-*mantra*') that corresponds to the planetary deity affiliated with the stone. While wearing topaz for its connection to Jupiter (Bṛhaspati), for example, one should pronounce *auṃ hrīṃ klīṃ hūṃ bṛhaspataye namaḥ*. The materials used in the ritual preparation of gemstones are thought to act like antennas or radar by 'picking up' or capturing the radiation and waves emitted from the stone. This metaphor is common in contemporary gemological literature: 'If you wear it at the right moment and perform the ritual, then the gemstone will become active (*sakriya*) and will spread cosmic radiations (*ākāśīya vikiraṇa*) just like an antenna (*enṭinā*).' (Joshi n.d.: 195) Another account corroborates this sentiment: 'According to my own research, I found out that gemstones work just like *radars (rāḍār)*. Just like a radar that, after receiving the signal, makes visible an aircraft moving in the sky, in the same way precious stones establish a contact (*sampark*) with planets moving in the solar system.' (Jaiswal 2006: 137)

Described both as icons inhabited by deities and mechanical devices, gemstones should always be carefully examined during their therapeutic activity. Any change observed or perceived is seen as a sign of altered efficacy. Thus, clients regularly ask astrologers the meaning of a gemstone that moves (H. *hiltā hai*), becomes scuffed (*ghistā jā rahā hai*), breaks (*ṭūṭ jatā hai*) or is lost (*kho gayā*). A crack, for instance, is generally understood as evidence that a stone has protected a person from an accident and that it should be replaced as quickly as possible since its power has been used up. The loss of a gemstone is presumed to be a bad omen, indicating that the stone's associated planetary influence is so negative that all means of protection are gone. It may even happen that the smell of a gemstone is interpreted as a symptom. Thus a businessman from Mumbai complained to an astrologer about the bad smell of his zircon (*gomeda*, Rāhu's gemstone):

> Client: You prescribed me a zircon but it stinks so badly that I was not able to wear it, even for a week. It stinks only on me, not on other people. On me, it stinks so badly that people refuse to sit next to me. And I am not even able to eat with this hand...
>
> Astrologer: Do you smell badly when you sweat?
>
> Client: Not at all, I can wear the same socks for a month and it smells good...
>
> Astrologer: Zircon is a gemstone that protects you from external obstacles (*uparibādhā*), from sorcery (*ṭonā-ṭoṭkā*) and from black magic (*kālā-jādu*). It is time to change it.

In the astrologer's view, foul smell is an indication that the gemstone has absorbed a negative external influence.

CONCLUSION: DOES 'PRECIOUS' MEAN COSTLY?

We conclude this chapter with some remarks about the costliness of precious stones. Price is an important aspect throughout most gemological literature, going back to Kautilya's *Arthaśāstra*. Although the price of gemstones is generally thought to be proportional to the valuableness of a stone, and therefore a mark of its quality, astrologers nowadays are well aware that many clients cannot afford to pay several thousand rupees to procure the most precious stones. This is a problem stated in classical treatises as well. The author of the *Muhūrtacintāmaṇi* and its commentators, for instance, seem fully aware of the fact that gemological remedies are not affordable for everyone. They thus describe different options so that people can choose according to their capacity (*yathāśakti*). One (quite expensive) option described is the *navaratna* ring (MC 4.9):

> In order to please the planets, one should make a gold ring with a round shape, divided into nine parts (*navadhā*). The nine gems should be set in the nine divisions starting in the eastern region.

Table 4.2. The placement of the nine planetary gemstones according to the spatial directions (MC 4:9).

Northwest (*vāyavya*) Cat's eye for Ketu	North (*uttara*) Topaz for Jupiter	Northeast (*īśān*) Emerald for Mercury
West (*paścima*) Blue sapphire for Saturn	Center Ruby for Sun	East (*pūrva*) Diamond for Venus
Southwest (*nairṛtya*) Zircon for Rāhu	South (*dakṣiṇa*) Coral for Mars	Southeast (*agni*) Pearl for Moon

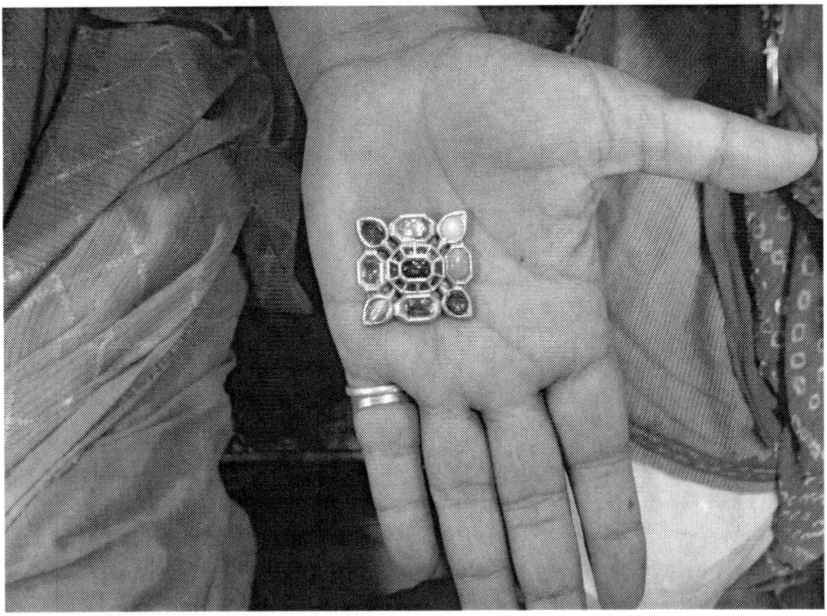

Figure 4.1. *Navaratna* jewel. Photograph by C. Guenzi.

Another option advances the idea that planetary stones should be worn in individual gold rings, according to the conventional associations between *navagraha* and *navaratna*.[46] A third and cheaper option is available to people who cannot afford the first two.[47] As clearly stated in the Hindi *pītāmbarā* commentary of the *Muhūrtacintāmaṇi*, whenever it is not possible to wear more expensive stones (*adhik mūlyavān ratna*), one should wear less expensive stones (*alpa mūlyavān ratna*). Thus, for Venus and Moon, a

46 MC IV.10.
47 MC IV.11.

silver ring (*raupya*) can be worn instead of diamond and pearl; for Jupiter, a pearl (*muktā*) may substitute topaz; for Saturn, an iron ring (*lauha*) will replace sapphire, and so on.

In astrological practices observed in Banaras, the principle of *yathāśākti*, of paying according to one's capacity, is vital in determining the efficacy of a gemstone. It is commonly said that one should always pay for a gem, even if it is just a nominal fee (*nāmmātra mūlya*): 'You should never wear a stone for free, a stolen gem or a gem lost (by someone else) on the road.' (Joshi n.d.: 194). In order to accommodate their clients' financial means, astrologers very often prescribe *uparatna*, 'semi-precious stones,' as substitutes for the truly 'precious' (*kīmatī, mūlyavān*) gems. Therefore, instead of ruby, a gemstone worth several thousand rupees, they might prescribe cheaper substitutes like *sūryakānta* (sun-stone), *saugandhika* (the 'perfumed' one), *tāmṛā* (a reddish-blue artificial stone), or star ruby (*maisūrī māṇikya*), each of which can be purchased for a few hundred rupees. Usually these stones are prescribed for shorter periods than the most precious ones because their therapeutic powers are considered limited. They typically should not be worn more than a few months or years, whereas the power of authentic precious stones is not supposed to diminish if the stone remains intact. The genuine precious stone may in fact last several years or even an entire lifetime. In the words of one practitioner, while an authentic stone is as powerful as sunlight, an artificial or semi-precious stone is just like an electric torch: its batteries should be renewed regularly.

ABBREVIATIONS AND REFERENCES

H. = Hindi Skt. = Sanskrit

AM — *Agastimata*. Finot, L. (ed. and trans.) (2013). *Les lapidaires indiens*. Reprint of 1896 edition. *Agastimata* text and French translation, pp. 77–139. Paris: Hachette livre, BNF.

Biswas 1987 — Biswas, A.K. (1987). '*Rasa-ratna-samuccaya* and Mineral Processing State-of-Art in the 13th Century A.D. India.' IJHS, 22(1): 29–46.

Biswas 1994 — Biswas, A.K. (1994). '*Vaidūrya, Marakata* and Other Beryl Family Gem Minerals: Etymology and Traditions in Ancient India.' IJHS, 29(2): 139–54.

BJ — Varāhamihira: *Bṛhajjātaka*. Jha, S. (ed.) (1974). *Bṛhajjātaka, with Bhaṭṭotpala's Sanskrit Commentary*. Varanasi: Thakur Prasad.

BS — Varāhamihira: *Bṛhatsaṃhitā*. Dvivedi, K.C. (ed.). (1997). *Bṛhatsaṃhitā*. Varanasi: Sampurnanand Sanskrit Vishwavidyalay (Sarasvatībhavana-Granthamālā, 97).

CaS — Agniveśa: *Carakasaṃhitā*. Trikamji Āchārya, J. (ed.) (1992). *Carakasaṃhitā of Agniveśa, with the Āyurveda-Dīpikā Commentary of Cakrapāṇidatta*. 5th edition. New Delhi: Munshiram Manoharlal.

Dash 1986 — Dash, B. (1986). *Alchemy and Metallic Medicines in Āyurveda*. New Delhi: Concept Publishing Company.

GP — *Garuḍapurāṇa*. Pandey, S.R. (ed.) (1986). *Garuḍapurāṇa of Kṛṣṇadvaipāyana Vyāsa*. Vidyabhawan Prachyavidya Granthamala 3. Varanasi: Chowkhamba Vidyabhawan.

IJHS — *Indian Journal of History of Science*

Jaiswal 2006 — Jaiswal, A.K. (2006). *Rudrākṣa aur Ratna*. Delhi: Nai Sadi Book House.

Jha 2006 — Jha, D. (2006). 'Jyotiṣ Śāstrānusār Rogõ kā Nidān.' *Jyotiṣa Vaijñānikī*, 4: 30–36.

Jindal 1997 — Jindal, M.R. (1997). 'Ratnõ kā Prayog.' *Triskandha Jyotiṣam*, 15: 11–14.

Joshi n.d. — Joshi, K.A. (n.d.). *Sampūrṇa ratna vijñān*. Delhi: Manoj Pocket Books.

JP — Vaidyanātha: *Jātakapārijāta*. Chaudhary, K. (ed.) (1953). *Jātakapārijāta of Vaidyanātha*. Varanasi: Chowkhambha Sanskrit Sansthan (Kashi Sanskrit Series 10).

M — *Mayamata*. Dagens, B. (ed. and trans.) (1970). *Mayamata. Traité sanskrit d'architecture*. Pondicherry: Institut Français d'Indologie (Publications de l'Institut français d'indologie, 40).

MC — Rāma: *Muhūrtacintāmaṇi*. Joshi, K. (ed. and comm.) (1992). *Muhūrtacintāmaṇi of Rāma*. Varanasi: Motilal Banarsidass.

Meulenbeld 2000 — Meulenbeld, G.J. (2000). *A History of Indian Medical Literature*, Volume IIA. Groningen: Egbert Forsten.

Murthy 1991 — Murthy, S.R.N. (1991). 'Role of Gems in Indian Medicine.' *Ancient Science of Life*, 10(3): 156–64.

PhDī — Mantreśvara: *Phaladīpikā*. Ojha, G.K. (ed.) (2001). *Phaladīpikā of Mantreśvara*. Varanasi: Motilal Banarsidass.

Pingree 1981 — Pingree, D.E. (1981). *Jyotiḥśāstra. Astral and Mathematical Literature*. Wiesbaden: Otto Harrassowitz.

RaPa — Ṭhakkura Pherū: *Rayaṇaparikkhā*. Sarma, S.R. (tr.) (1984). *Ṭhakkura Pherū's Rayaṇaparikkhā. A Medieval Prakrit Text on Gemmology*. Aligarh: Viveka Publications.

Ray 1991 — Ray, M. (1991). 'Minerals and Gems in Indian Alchemy.' IJHS, 26(2): 133–54.

RS — Vāgbhaṭa: *Rasaratnasamuccaya*. Satpute, A.D. (ed. and trans.) (2003). *Rasaratnasamuchchaya* [sic], *Chapters 1–11 (Sanskrit with English Translation)*. The Chaukhamba Ayurvijnan Studies. Delhi: Chaukhamba Sanskrit Pratishthan.

Shastri n.d. — Shastri, J. (n.d.). *Navagraha Upāsanā*. Delhi: Dehati Pustak Bhandar.

Sircar 1972	Sircar, D.C. (1972). 'The Number of Ratnas.' In Hazra, R.C. and S.C. Banerji (eds.), *S.K. De Memorial Volume*, pp. 75–81. Calcutta: Firma K.L. Mukhopadhyay.
Ślączka 2007	Ślączka, A.A. (2007). *Temple Consecration Rituals in Ancient India. Text and Archaeology*. Leiden: Brill.
Winder 1990	Winder, M. (1990). 'Vaidurya.' *Bulletin of Tibetology*, 26(1–3): 27–37.
Wojtilla 1973	Wojtilla, G. (1973). 'Indian Precious Stones in the Ancient East and West.' *Acta Orientalia Academiae Scientiarum Hungaricae*, 27(2): 211–24.

Chapter 5

Mercury Tonics (*Rasāyana*) in Sanskrit Medical Literature

DAGMAR WUJASTYK[1]

The use of mercury has a long history in Indian medicine with clear evidence of mercurial treatments found in Sanskrit medical treatises from about the seventh century. Mercury medicines were understood to be particularly potent remedies and as such were often used in the treatment of serious diseases that were difficult to cure. Mercury preparations were also used in rejuvenation therapy (*rasāyana*) – one of the eight classical areas of Indian medicine. *Rasāyana* sections in Sanskrit medical works describe the preparation and application of tonics for restoring health, stopping and even reversing the aging process, generally improving physical appearance and significantly increasing lifespan.

While rejuvenation therapies are already described in the oldest Sanskrit medical works known to us, the first mention of mercury as a *rasāyana* ingredient occurs in a seventh century treatise called *Aṣṭāṅgahṛdayasaṃhitā* ('Heart of medicine'). From about the eleventh century, mercury became a common ingredient in *rasāyana* therapy with works such as Cakrapāṇidatta's *Cakradatta* (eleventh century) and Vaṅgasena's

1 Dagmar Wujastyk is Principal Investigator of the AyurYog project (ayuryog.org) at the Institute for South Asian, Tibetan and Buddhist Studies at the University of Vienna (Austria).
The author wishes to acknowledge the generous support of the Verein zur Förderung des wissenschaftlichen Nachwuchses der Universität Zürich. The final version of this paper was completed with the support of the European Research Council (ERC) Starting Grant Ayuryog 639363.

Cikitsāsārasaṃgraha (eleventh/twelfth century) recording a number of recipes for mercurial tonics.

This chapter will provide a brief survey of the uses of mercury preparations in *rasāyana* therapy, discussing their position within rejuvenation therapy in regard to other *rasāyana* formulations and exploring their link with Indian alchemical traditions.

THE PRACTICE OF *RASĀYANA*

In the context of classical Indian medicine or Āyurveda, the Sanskrit term '*rasāyana*' is used to describe a sub-discipline of medicine; the therapies that together constitute this sub-discipline; and finally, the medical formulations used in these therapies. Broadly speaking, *rasāyana* therapies and medicines are meant to prolong life, restore youthfulness and promote physical health and mental acuity, and the formulations could therefore perhaps be generally termed 'tonics.' However, the various medical authors present a wide variety of approaches to the subject and in some works, *rasāyana* formulations are presented as medicines for specific diseases rather than for general health, rejuvenation or longevity. Somewhat confusingly, the term *rasāyana* is also used to refer to the Indian alchemical tradition, which postdates Āyurveda by several centuries at least in its written form. In alchemical literature, some substances and formulations are also called *rasāyanas*. Here again, *rasāyana* can denote quite different functions.

Both in medical and in alchemical works, a wide range of substances are used for *rasāyana* purposes: plant materials, animal products, minerals and metals. In this chapter, I shall focus on metals, and in particular, on mercury as constituents of *rasāyana* formulations. I will provide an overview of the uses of mercury in *rasāyana* therapy, discussing the substance's position in regard to other *rasāyana* materials and exploring how or if the use of mercury in Āyurvedic *rasāyana* links with the *rasāyana* of the Indian alchemical tradition.

MERCURY: THE BEST OF ALL SUBSTANCES

A typical alchemical work will often begin with the praise of mercury, extolling its virtues as an elixir of immortality and as a powerful medicine, and presenting it as an object and subject of worship. Consider, for example, this proverb from the *Rasahṛdayatantra* (tenth/eleventh centuries), one of the earliest alchemical works known to us at present:[2]

> After it has been solidified, it cures disease, having been bound, it gives liberation;
> having been thoroughly killed, it gives immortality. What is more compassionate than mercury?[3]

A further passage emphasizes the unique potential of mercury to act as a rejuvenant as well as to make its user immortal:

> Just as all beings are inevitably absorbed in the supreme spirit, the king of essences (i.e. mercury) alone bestows immortality and youth to the body.[4]

And the superiority of mercury to other substances is explained as follows:

> No potion of roots or metals that is itself unstable, can be burnt, rotted or dried up is capable of producing its (i.e. the body's) stability. Woods and herbs are absorbed in lead, lead in tin, and tin in turn in copper, copper in silver, silver in gold and gold in mercury.[5]

Mercury's apparent ability to absorb other substances is interpreted as a sign of its being at the top of a power hierarchy, both because of being the last, itself unabsorbable substance, and because in alchemical procedures, it would have been given other substances and metals to absorb, and therefore is then considered to have assimilated the essence or power of these other substances.

2 See White (1996: 146) on the date of the *Rasahṛdayatantra*.

3 The passage in the RH I.3 reads: *mūrchitvā harati rujaṃ bandhanam anubhuya muktido bhavati* | *amarīkaroti sumṛtaḥ ko 'nyaḥ karuṇākaraḥ sūtāt* || A parallel passage in RS (I.34) reads: *mūrchitvā harati rujaṃ bandhanam anubhūya muktido bhavati* | *amarīkaroti hi mṛtaḥ ko 'nyaḥ karuṇākaraḥ sūtāt* || See White's discussion of the passage in various alchemical works and parallels in the *Haṭhayogapradīpika* (1996: 274–75 and 495n55).

4 RH I.13: *paramātmanīva niyataṃ layo yatra sarvasattvānām* | *eko'sau rasarājaḥ śarīram ajarāmaraṃ kurute* || See also the parallel passage in RS I.43, which substitutes '*satatam*' for '*niyatam*'.

5 *tatsthairyaṃ na samarthaṃ rasāyanaṃ kimapi mūlalohādi* | *svayamasthirasvabhāvaṃ dāhyaṃ kledyaṃ ca śoṣyaṃ ca* ||11|| *kāṣṭhauṣadhyo nāge nāgaṃ vaṅge vaṅgamapi līyate śulve* | *śulvaṃ tāre tāraṃ kanake kanakaṃ ca līyate sūte* ||12|| See also RS I.40–41 for a parallel passage.

However, in its natural, liquid state, mercury is unstable, can be burnt, or dried up, though not rotted. And to counter these or other unwelcome properties, alchemists developed a series of eighteen procedures that would make mercury stable. The terms 'solidified,' 'bound' and 'killed' we saw in the quotation from the *Rasahṛdayatantra* above are technical terms that refer to some of these stages in the processing of mercury.

The eighteen procedures or *saṃskāras* the alchemists devised were thought to remove any faults in mercury, in particular its instability or volatility, contamination with other matter, its heaviness and its reaction to fire, leaving an end product that was then considered suitable for use in transmutational alchemy.[6] The alchemists would thus use this product for what they called *dehavāda*, the 'piercing of the body,' which was meant to confer immortality, but also to let the user achieve spiritual liberation in a living body. For medical purposes, eight procedures were deemed sufficient by the alchemists, which should include alchemical *rasāyana* formulations against diseases.[7]

It is clear from the above quotations that Indian alchemists regarded mercury as a superior substance that was uniquely suitable for the purposes of rejuvenation and the prolongation of life. This sentiment is echoed in later medical literature, which often takes its cue from the alchemical works in regard to the use of mercury and other metals. For example, the *Bhāvaprakāśa*, an important medical compendium from the sixteenth century,[8] states the following in its entry on mercury in the glossary section:

> Mercury has all the six flavours and is unctuous; it mitigates all three humours and is a *rasāyana*;
> it assimilates to itself and it is a powerful aphrodisiac; it always strengthens eyesight;
> it is declared the destroyer of all diseases and especially removes all skin diseases.
>
> After it has been solidified, it cures disease, having been bound, it lets one move in the sky;
> having been killed, it gives youth. What is more compassionate than mercury?

6 For a detailed description and analysis of these procedures, see White (1996: 265–69).

7 It should be noted that medical works typically do not quite follow the alchemical lists of *saṃskāras* in their descriptions of the processing of mercury, neither in number nor in sequence. See Wujastyk (2013: 20) on the eight medical *saṃskāras*.

8 See Meulenbeld (1999–2002, IIA: 246) on the date of *Bhāvaprakāśa*.

Be it an incurable disease or one for which there is no treatment, mercury removes the diseases of men, elephants and horses.[9]

The *Bhāvaprakāśa*'s entry on mercury combines both alchemical and medical elements. The middle part of this statement is familiar from the above quotation from the *Rasahṛdayatantra*. The difference is that Bhāvamiśra, the author of the *Bhāvaprakāśa*, promises the ability to walk in the sky, i.e. fly, rather than spiritual liberation, and youth, rather than immortality. This does not, however, necessarily mark a contrast between alchemical soteriological aims and more mundane medical ones: We also encounter the idea that mercury counters disease and affords the consumer the ability to fly in other versions of this proverb in alchemical literature. For example, the *Rasārṇava*, another early and very influential alchemical work (c. twelfth century),[10] compares mercury and breath, stating that both will drive away disease and afford the power of flight.[11]

The first part of Bhāvamiśra's entry on mercury is positioned squarely within a medical framework with its reference to the six flavors, the unctuosity of mercury and the humors. It is one of the central Āyurvedic concepts that the body contains three primal substances called *doṣa* (often translated as 'humor' in analogy to the humors of ancient Greek medicine). These substances are understood to fundamentally sustain the functioning of the body, but also represent potential sources for the arising of disease.[12] Āyurvedic authors write of three (sometimes four) humors: wind (*vāta*), bile (*pitta*) and phlegm (*kapha*), each of which is associated with particular properties.[13] These characteristics are related to the humors'

9 BhāPr. *Pūrvakhaṇḍa Nighaṇṭubhāga Dhātvādivarga* 91–92, 94–95. See also Wujastyk (2013: 16). The passage reads: *pāradaḥ ṣaḍrasaḥ snigdhas tridoṣaghno rasāyanaḥ* ||91.2|| *yogavāhī mahāvṛṣyaḥ sadā dṛṣṭibalapradaḥ* | *sarvāmayaharaḥ prokto viśeṣāt sarvakuṣṭhanut* ||92|| *mūrchitvā harati rujaṃ bandhanam anubhūya khe gatiṃ kurute* | *ajarīkaroti hi mṛtaḥ ko 'nyaḥ karuṇākaraḥ sūtāt* ||94||

10 See Meulenbeld (1999–2002, IIA: 684–85) on the presumed date of the *Rasārṇava*.

11 However, according to the *Rasārṇava*'s version, mercury and breath would give the practitioner neither immortality nor youth, but would revive only themselves after being killed. Ras I.19: *mūrchito harati vyādhiṃ mṛto jīvayati svayam* | *baddhaḥ khecaratāṃ kuryāt raso vāyuś ca bhairavi* ||

12 See, for example, CaS.Vi (I.5) for a definition of the *doṣas* and their functions. See also Scharfe (1999) on the evolution of this concept.

13 SuS.Sū (XXI.28) counts blood as a fourth humor. The more or less standard translation of the terms *vāta*, *pitta* and *kapha* as wind, bile and phlegm, respectively, is again historically connected with the perceived correspondence of these Āyurvedic concepts with Greek ones. The terms 'wind,' 'bile' and 'phlegm'

relative composition in regard to the five elements (wind, ether, earth, fire and water) which are understood to make up the fabric of the material world. The elements have particular properties: For example, water is liquid, unctuous, cold and soft; fire is hot, sharp, light and rough.[14] Disease may arise when there is an imbalance in the proportional quantity of the humors, i.e. a pathological predominance of one or two of the humors; if one or several of the humors spread outside their normal pathways; or if one or several of them are hindered from flowing through their normal pathways. Treatment aims at counteracting a humor's abnormal growth and thus avoiding its overflow from the area that is supposed to contain it. This is done by countering the affected humor(s) with substances that have opposite characteristics to them.

Āyurvedic works present a corresponding system of classifying medicinal substances according to certain categories, typically flavor (*rasa*), potency (*vīrya*), transformed tastes (*vipāka*) and special power (*prabhāva*).[15] Some sources add a further category to the other four, namely quality (*guṇa*). There are six flavors (sweet, salty, sour, bitter, astringent and pungent), three transformed tastes (sweet, sour and pungent), about twenty qualities (pairs of opposites such as light, heavy, unctuous, dry, sharp, smooth, etc.), two potencies (hot and cold), and one special power, all of which are thought to have specific effects on the humors in the body.[16] Just as with the humors, the specific qualities of any substance are related to its relative composition in regard to the five elements. The flavors of substances are formed by the elements they are made of and they therefore share the elements' properties. The *Carakasaṃhitā*, an encyclopedic compilation of medical knowledge that roughly dates to the early centuries of the Common Era,[17] explains that sweetness is produced by a predominance

should here be understood as technical terms with very specific associations. For a modern reader, these terms may be misleading, 'bile' bringing to mind the secretion of the liver, and 'phlegm' a liquid secreted by the mucous membranes rather than the concepts of humoral theory. It is difficult to adequately translate such technical terms since no equivalent concept exists today.

14 See CaS.Sū xxvi.11.

15 See Wujastyk (2003:198) on these categories.

16 This is a simplified synopsis: The different Āyurvedic sources vary in their principles of substance categorization and there is also rather more detail to the classification of substances.

17 Caraka, the compiler of the *Carakasaṃhitā*, probably lived in the first century CE. He presented his work as a revised version of an ancient work on medicine (the *Agniveśatantra*, according to a later redactor). In turn, the version of the *Carakasaṃhitā* that has been transmitted to us is a further redacted version

of the element water in a substance, saltiness by a predominance of water and fire, pungency by wind and fire, bitterness by wind and ether.[18] Once a substance's flavor is identified, it can be used to effect or counter the growth of a humor. For example, pungent, bitter and astringent substances combined together are thought to increase the humor wind, or conversely, a combination of sweet, sour and salty substances is thought to reduce wind.[19] The categorization of medicinal substances according to their flavors is fairly complicated, since one substance often combines in itself more than one flavor. Indeed, the *Carakasaṃhitā* counts sixty-three possible combinations of the six canonical flavors.[20] So, the flavors of drugs do not always tally with their actions upon symptoms that are understood to reflect problems with specific humors. The effects of the drugs are then explained in terms of their flavors having been transformed through digestion (an originally bitter plant having become pungent after digestion, for example) and/or with their potency, i.e. generating heat or cold. Heat, for example, is supposed to reduce the humor wind while cold would further increase it. Finally, the last category, 'special power,' is used to explain specific actions of substances that cannot be attributed to the other qualities. For example, if two plants have the same flavor, transformed taste, qualities and potency, but one acts as a purgative while the other doesn't, purging would be the first plant's special power.

Bhāvamiśra's brief characterization of mercury as having all the six flavors and being unctuous generally defines mercury as a beneficial substance: Having all six flavors points to a balanced distribution of all six elements which would mean that an intake of mercury should not create an imbalance of humors in the body. In a sense, this would make it a neutral substance. It is, however, difficult to see how an even distribution of elements would help counter an already extant imbalance of the humors. Nevertheless, according to Bhāvamiśra, mercury mitigates all three humors, and thus is a panacea, a 'destroyer of all diseases.'

---

that was edited and added to by a scholar called Dṛḍhabala in about the fifth century. For a detailed discussion of the issues surrounding the dating of the *Carakasaṃhitā*, see Meulenbeld (1999–2002, IA: 105–15). Also see the article by Maas (2010) on the complicated transmission history of the *Carakasaṃhitā* after Dṛḍhabala's redaction.

18 See CaS.Sū XXVI.40.

19 See CaS.Vi I.7.

20 See CaS.Sū XXVI.14–22. On the combinatorics of tastes and humors in Āyurveda, see Wujastyk (2000).

RASĀYANA IN ĀYURVEDIC LITERATURE

Given that mercury is specifically defined as a rejuvenating agent in the glossary section, it comes as a surprise, then, that none of the recipes in the *Bhāvaprakāśa*'s chapter on *rasāyana* contain mercury.[21] This is particularly strange, since the *Bhāvaprakāśa* contains many other kinds of recipes with mercury, as well as a large section on how to prepare mercury for medical use.[22] The very short *rasāyana* chapter (only fifteen verses) lists nine recipes, all of which contain plant materials and are meant to rejuvenate, prolong life or enhance intelligence. Only one recipe contains a metal, namely iron. This formulation, however, is attributed with bestowing immortality upon its consumer. The other recipes follow older ones from the early medical treatises to which we will now turn.

First descriptions of *rasāyana* recipes and treatment methods appear in Sanskrit medical literature about two thousand years ago. In the *Carakasaṃhitā*, *rasāyana* is defined as one of eight branches of medicine.[23] In its large chapter on *rasāyana*, the *Carakasaṃhitā* presents the subject as follows:

> Through *rasāyana*, a man gains longevity, memory, mental vigor, health, youthful strength, a great radiance, complexion and voice, an extremely strong body and keen senses, mastery of speech, respect and beauty. *Rasāyana* is a means for attaining the best bodily elements, beginning with *rasa*.[24]

21 The *rasāyana* section is the second of two chapters of the last part of the *Bhāvaprakāśa*, the *uttarakhaṇḍa*.

22 See Wujastyk (2013: 22–31) on the use of mercury in the *Bhāvaprakāśa*.

23 The eight branches of medicine as listed in the *Carakasaṃhitā* are: (1) *kāyacikitsā* (general medicine), (2) *śālākya* (the [surgical] treatment of body parts above the shoulders), (3) *śalyāpahartṛka* (the removal of foreign bodies – surgery), (4) *viṣagaravairodhikapraśamana* (toxicology), (5) *bhūtavidyā* (the treatment of possession by various supernatural beings), (6) *komārabhṛtyaka* (pediatrics), (7) *rasāyana* (restorative therapies) and (8) *vājīkaraṇa* (aphrodisiacs).

24 CaS.Ci I.1.7–8: *dīrghamāyuḥ smṛtiṃ medhāmārogyaṃ taruṇaṃ vayaḥ| prabhāvarṇasvaraudāryaṃ dehendriyabalaṃ param ||7|| vāksiddhiṃ praṇatiṃ kāntiṃ labhate nā rasāyanāt| lābhopāyo hi śastānāṃ rasādīnāṃ rasāyanam ||8||* For a discussion of this passage, see Wujastyk (2015). In this instance, *rasa* refers to food that has undergone the first process of digestion.

Another early Sanskrit medical work, the *Suśrutasaṃhita* (c. third century CE[25]) also contains a *rasāyana* section, describing treatments against afflictions, for improving intelligence and for longevity, and for preventing natural afflictions or vulnerabilities, i.e. thirst, hunger and tiredness.[26]

Despite a general consensus in aims and methods, a comparison of the *rasāyana* recipes in the *Carakasaṃhitā* and the *Suśrutasaṃhitā* shows them to differ greatly from each other with not a single recipe shared between them, though with some overlap in single ingredients. However, a notable shared element of these works' presentations of *rasāyana* is that the majority of their recipes are based on herbal ingredients that are mixed with animal products (milk, butter, ghee and honey). Some formulations also include metals: The *Suśrutasaṃhitā* describes several recipes with gold as an ingredient.[27] These recipes widen the above-mentioned range of *rasāyana* applications in that they are supposed to help the person using them to overcome misfortune, to avert the threat of death, to subdue the king or to confer power and charm.

The *Carakasaṃhitā* uses a wider range of metals for its *rasāyana* medicines: gold, copper, silver and iron.[28] These are typically added in powdered form, rendering the formulation more potent or more suitable for particular patients through their addition. Iron takes a prominent place among the metals in Caraka's *rasāyana* recipes and is added to several of them. It is also used as a main ingredient in the '*lauhādirasāyana*' ('iron, etc. elixir'). This is an important recipe, because it is the first description of a method of processing metals. It foreshadows what later becomes a prominent feature in Indian medicine, namely that metallic, mineral and poisonous herbal ingredients are processed before they are used in formulations. The idea is that the substances are made both safer and more potent through the various processing methods.

The relevant passage in CaS (Ci I.3.15–23) describes how long and thin strips of iron are heated until red-hot and then dipped in a mixture of the three myrobalans, cow's urine and alkali prepared from various plant materials. The resulting powder is mixed with honey and the juice of the

25 For a survey of the proposed dates of the *Suśrutasaṃhita*, see Meulenbeld (1999–2002: IA, 342–52).

26 The *Suśrutasaṃhitā*'s *rasāyana* section is found in chapters XXVII–XXX of the *Cikitsāsthāna* (the section on therapeutics).

27 See SuS.Ci XXVIII.14–24.

28 Gold, silver, copper and iron are mentioned in Ci I.1.58. CI.3.46–47 mentions gold and 'all metals' (*sarvalauha*), which probably means copper, silver and iron, but may also include lead and tin.

Indian gooseberry and stored for a year in a pitcher smeared with ghee. This formulation is described as a particularly potent mixture with a host of wonderful effects on the user's health, looks and mental powers.

Caraka notes that the same method of preparation can also be applied to other metals, and particularly mentions corresponding gold and silver *rasāyanas* as providing longevity and alleviating disease.

Another notable *rasāyana* substance that deserves mention as it is connected to metals and used in multiple *rasāyana* formulations in the *Carakasaṃhitā* (though not in the *Suśrutasaṃhitā*) is shilajit (Skt. *śilājatu*), or rock resin. The Āyurvedic authors differentiate between four kinds of shilajit, each understood to be an exudate of a different metal, namely gold, silver, copper and iron, respectively. Caraka considers the iron kind the best, especially for the purposes of *rasāyana*, and describes it as having a beneficial effect on all three humors.[29]

It should be noted that neither the *Carakasaṃhitā* nor the *Suśrutasaṃhitā* use mercury in their *rasāyana* recipes. This is not surprising, given that there is little evidence for the use of mercury in medicine before the seventh century. There are possible exceptions to this rule: A verse in the *Carakasaṃhitā* in a section on skin disease states that *rasa*, which cures all diseases, should be used by persons afflicted with skin disease. According to Dutt (1922: 27) the commentators interpret *rasa* as mercury.[30] However, as the term *rasa* signifies many other things besides mercury, such as the sap or juice of plants or fruits, et cetera, one cannot decide with any certainty that *rasa* indeed means mercury here, though it also cannot be entirely discounted, especially since sulfur and pyrites are mentioned in the same sentence. In the *Suśrutasaṃhitā*, we find a curious reference to mercury in a chapter on poisons, in which Suśruta claims that playing various musical instruments smeared with anti-poison will cure food poisoning in animals and humans. One of the ingredients of Suśruta's anti-poison paste is *sutāra*, which is a term for mercury found in later medical literature.[31] Finally, also in SuS (Ci XXV.38–42), there is a recipe for an ointment against freckles, wrinkles, eruptions in the face, skin diseases in general and cracked feet, which lists *pārada* as an ingredient.

29 See CaS.Ci I.3.48–65 on *śilājatu*, especially 55–61 for a description of the four different kinds of *śilājatu* and their characteristics.

30 The passage in question is CaS.Ci VII.71.

31 See SuS.Ka III.13–15 and Wujastyk (2003: 78–79) on this section in the *Suśrutasaṃhitā*.

MERCURY IN *RASĀYANA* RECIPES: VĀGBHAṬA'S WORKS

It is certain that medical compounds containing mercury were used by the time the works ascribed to Vāgbhaṭa, the *Aṣṭāṅgahṛdayasaṃhitā* and the *Aṣṭāṅgasaṃgraha*, were written (c. early seventh century CE).[32] Both works prescribe a medicinal paste for the treatment of an eye disease that contains *rasendra*, an unequivocal term for mercury.[33] The *Aṣṭāṅgahṛdayasaṃhitā* also has a recipe for a topical cream against freckles, where the term for mercury is *pārada*.[34] Both works further prescribe *pārada* as a *rasāyana*.[35] Vāgbhaṭa's mention of mercury as a *rasāyana* reads as follows:

> The depleted tissues of the body of one who eats shilajit, honey, false black pepper, ghee, iron, chebulic myrobalan, mercury (*pārada*), and pyrites are replenished within fifteen nights like the moon.[36]

This recipe (or perhaps rather list) occurs within a large group of short recipes most of which are purely herbal while a few contain iron, other metals or shilajit. This recipe stands out, not only for its mention of mercury, but also because it is not quite clear whether the named ingredients should be taken together in one formulation, or whether this is simply a list of restorative substances that might be taken on their own or as components of other formulations.[37]

32 The given date is a rough estimate. The authorship, or perhaps more accurately, compilership of these two works, and consequently also their relative chronology is a matter of some discussion, which has been summarized by Meulenbeld. See his analysis on the identity of Vāgbhaṭa (1999–2002, IA: 597–602), on the dating of the *Aṣṭāṅgasaṃgraha* (ibid.: 613–31) and on the dating of the *Aṣṭāṅgahṛdayasaṃhitā* (ibid.: 631–35).

33 AHS.Utt XIII.36 and ASaṃ.Utt IL.392.

34 AHS.Utt XXXII.31.

35 See AHS.Utt XXXIX.161. Eliade (1969: 279) unfortunately introduced the mistaken idea that mercury was mentioned as a *rasāyana* substance in the Bower Manuscript. This false assertion (which Eliade attributed to Hoernle himself) is then repeated later by other scholars (for example, by Leslie 2009: 45). In the passage in question, the word Eliade understood as mercury is '*rasa*' and Hoernle (1893: 107n123) explicitly noted in his edition and translation of the text that this does not denote mercury. He translated it as 'juice of sugar-cane.'

36 See AHS.Utt XXXIX.161 and ASaṃ.Utt XLIX.392: *śilājatukṣaudraviḍaṅgasarpirlohābhayāpāradatāpyabhakṣaḥ* | *āpūryate durbaladehadhātus tripañcarātreṇa yathā śaśāṅkaḥ* || The same passage occurs in alchemical literature, in Ras XVIII.14 and RS XXVI.12.

37 The surrounding recipes give instructions on the formulation and processing of pills and pastes and the admixture of herbal, metal and mineral ingredients with carrier substances such as ghee or honey.

The reference to the waxing moon gives a time frame for their use, but also presents an image of their effects: The body is filled with the essential substances and the nutrition it needs, making it bigger and stronger, thickened and buffered from within. The suggestion that 'depleted tissues will be replenished' refers to the Āyurvedic concept of the transubstantiation of body tissues. The Sanskrit word for what I have called body tissues is *dhātu*, literally 'constituent part' or 'element.' These are defined as essential parts of the body that evolve in succession from each other. The body, which consists of the five great elements of wind, ether, fire, earth and water, is nourished and constantly built up through the intake of food and drink which, as we have seen, also consist of combinations of the five elements.

In the *Aṣṭāṅgahṛdayasaṃhitā*, the list of *dhātus* begins with chyle (i.e. food that has undergone the first process of digestion). Chyle forms blood, blood forms flesh, flesh forms fat, fat becomes bone, bone makes marrow and marrow develops into semen.[38] And semen is understood to be not only the end product of the succession, but also the most refined and important substance, something akin to life essence. This list of body tissues (or chain of events) is also what Caraka refers to in his definition of *rasāyana*, which concludes with '*rasāyana* is a means for attaining the best bodily elements, beginning with *rasa*.'

Thus we might infer that certain medicinal substances were thought of as providing similar or even superior nourishment to regular food. For a thorough *rasāyana* treatment, the therapy begins with cleansing procedures. The body is rid of waste products, opening the channels in the body for a better absorption of the medicines. Then treatment proper begins with the intake of the *rasāyana*, repeated daily in the morning. In one passage, Caraka notes that the *rasāyana* should be digested before partaking of a light meal of rice with milk and ghee.[39] The suggestion is that normal food intake is strongly reduced, the *rasāyana* functioning as a substitute. The *rasāyana* is now the starting point in a chain of transformation in which the tissues of the body begin to be replaced as one constituent transforms into the next. As it provides the basic building blocks from which the physical parts of the body are put together, one could say that the *rasāyana* becomes the body. The process is the same with regular food, but the difference is that a *rasāyana* provides more suitable materials and

38 See AHS.Sū XI. However, also see Maas (2008: 135–46) for an overview of *dhātu* lists in other medical and non-medical works, in which he shows that other lists have both different beginnings, additional elements and different sequences.

39 See, for example, CaS.Ci I.1.52–53 and 58.

therefore improves the very fabric of the body. Better raw or processed materials that are appropriately mixed together and/or suitably processed should accordingly equal a better body.[40]

MERCURY: JUST ANOTHER INGREDIENT?

It should be noted that Vāgbhaṭa does not promise any special effects for mercury and also does not present it as better, or more potent than any of the other substances in his recipe (or list). It seems that at the time he was compiling his works, mercury was not yet regarded as a particularly noteworthy substance in the Āyurvedic tradition. Mercury was only slowly accepted into Āyurvedic pharmacopoeia: It does not occur as an ingredient in either the *Mādhavanidāna* (c. eighth century), or the *Kāśyapasaṃhitā* (c. seventh century) and it only features in a single recipe in the *Siddhayoga* by Vr̥nda (c. ninth or tenth century).[41] The recipe in question is an ointment against lice made of datura (*dhattūra*) and mercury (*rasendra*).[42]

Then, the eleventh century *Cakradatta* (also called *Cikitsāsaṃgraha*) by Cakrapāṇidatta gives about nine recipes for mercurial medicines, all for internal use.[43] Two of the recipes that contain mercury are *rasāyana* formulations. In both recipes, copper is the main ingredient. The first is called the 'six-part copper compound' (*ṣaḍaṅgatāmrayoga*), the six parts being copper, mercury, mica, long pepper (*pippalī*), false black pepper (*viḍaṅga*) and black pepper (*marica*). A closer look reveals the recipe to be more complicated, however, as the copper, mercury and mica all need to be processed first. In the case of copper, this means rubbing thin leaves of copper with sulfur or salts. It is then heated and immersed in sour gruel mixed with *nirguṇḍī* (*Vitex negundo* L.) paste. The sediment is collected and rubbed

40 The effects of *rasāyanas* on cognitive function and the senses are less easily explained according to this model. However, one should keep in mind that the elements present in foodstuffs or medicines also affect the humors of the body, and these are understood to have a direct influence on the mind and the sense organs and faculties, et cetera. On the categories of mind, sense organs, sense faculties, intellect and being and their relation to the five great elements, see CaS.Sū VIII. On the use of food to maintain positive health and to control the sense faculties, see in particular CaS.Sū VIII.20. Also see Roşu (1978: 157–214).

41 On the dates of the *Mādhavanidāna*, the *Kāśyapasaṃhitā* and the *Siddhayoga*, see Meulenbeld (1999–2002, IIA: 70–72, 39–41 and 81–82).

42 See Meulenbeld (1999–2002, IIA: 80) on mercury in the *Siddhayoga*.

43 For the date of the *Cakradatta*, see Meulenbeld (1999–2002, IIA: 93).

with half its weight in sulfur. This mixture is then roasted in the pit cooking method (*puṭapāka*). That is, a pit is filled with dried cow pats. A sealed vessel containing the mixture is placed on top and covered with an equal amount of cow pats which are then set on fire.

The processing of mica is slightly less complicated: Mica is mixed with the pulp of *hilamocī* (*Enhydra fluctuans* Lour.) root, placed in a sealed earthen saucer and heated intensely. It is then macerated with sour gruel until it loses its luster. In turn, mercury is processed by rubbing it with processed copper. It is then made into a bolus and steam-heated with rain water mixed with *nirguṇḍī* paste.

Finally, the processed powders of copper, mercury and mica are mixed with the three kinds of pepper. The resulting preparation is a *rasāyana* that can be used for various abdominal disorders and wasting disease.[44]

The second copper *rasāyana* is called 'seven-part copper compound' (*saptāṅgatāmrayoga*), though it is difficult to discern which seven parts are meant exactly. In this recipe, leaves of Nepalese copper are covered on all sides with sulfur and placed in a sealed earthen vessel and then roasted in a sand bath, i.e. a. vessel filled with sand and sealed with mud. The sublimated copper is collected.

In the next step, sulfur is melted in an iron pan. The processed copper is added and then mixed with a paste of processed mercury and sour gruel. Eight drops of ghee are added to the mix, which is then taken off the fire, poured into a stone mortar and macerated with *alambuṣā* (water mimosa?) juice. The mixture is then put back on the fire and stirred until it is dry. In a further step, the dry mixture is again mixed with *alambuṣā* juice and formed into a round mass. This is covered with a cloth smeared with a paste of black pepper, long pepper and dried ginger and then placed into a pouch. It is then heated in clarified butter until the contents have become hard. This is taken together with ground black pepper, long pepper and dried ginger and the powder of the three myrobalans. This medicine is a *rasāyana* that can be used for abdominal complaints.[45]

While mercury is certainly an integral part of these recipes, it is not the main ingredient and Cakrapāṇidatta offers no special comment on its characteristics. Indeed, if there is one substance that receives special attention in Cakrapāṇidatta's *rasāyana* chapter, it is iron. Eighty-four of 202 verses in the chapter are dedicated to the subject of how to process iron and how to the use it in *rasāyana* formulations. The simple procedure described in the

44 This is described in CD LXVI.129–35.

45 This is described in CD LXVI.136–51.

Carakasaṃhitā for processing iron is now exchanged for a complicated multiple-stage process that produces an end product called *amṛtasāralauha*, 'immortality-essence-iron.' Iron is mixed with various herbal and mineral substances, and heated and roasted in different processes called *māraṇa* ('killing'), *puṭana* ('enveloping') and *sthālīpāka* ('roasted in a vessel'). These preparation methods are similar to the alchemical *saṃskāras* for mercury. While no reason is given in the *Carakasaṃhitā* for the necessity of processing iron before use, Cakrapāṇidatta explains that the herbs he lists for use within the various stages of processing will remove the defects (*doṣa*) of iron.

Another eleventh century work, Vaṅgasena's *Cikitsāsārasaṃgraha* (also called the *Vaṅgasenasaṃhitā*) contains more mercurial *rasāyana* recipes in its long and somewhat miscellaneous *rasāyana* chapter.[46] While this work is in many ways similar to Cakrapāṇidatta's treatise,[47] its *rasāyana* section differs substantially. There are some fifteen recipes that mention mercury as an ingredient. In most of these recipes, mercury is one of many ingredients and seems secondary to another main substance, such as copper (there are copper compound recipes very similar to Cakrapāṇidatta's ones), iron and mica. A few recipes also contain tin and lead. While iron is the most prominent ingredient in the *Cakradatta*'s recipes, the material that receives the most attention in the *Cikitsāsārasaṃgraha* is mica, which is described as a very potent medical and rejuvenative substance. The indications for the different *rasāyanas* that contain mercury vary from combating abdominal complaints to increasing sexual stamina to rejuvenating and prolonging life span. Several recipes mention *kajjalī*, that is, a mixture of mercury and sulfur. For example, in a very simple recipe called *parpaṭarasāyana* (thin-layered *rasāyana*) mercury, sulfur and copper powder are mixed and roasted in a closed vessel. The resulting compound is taken with honey and said to alleviate all diseases.[48] In several recipes, a number of herbal ingredients are macerated with mercury and other metals (probably in

46 See Meulenbeld (1999–2002, IIA: 227–28) on the date of the *Cikitsāsārasaṃgraha*. The *rasāyana* chapter also contains information on the *marmans*, the vital points in the body, types of diseases categorized as being caused by wind, bile or phlegm, and a therapy using a head pouch called *śirovasti*.

47 Both works are modelled to some extent on the structure of the *Mādhavanidāna* and also borrow from the *Siddhayoga*. See Meulenbeld (1999–2002, IIA: 228) on the relationship between Cakrapāṇidatta's and Vaṅgasena's works. Meulenbeld regards it as likely that Vaṅgasena borrowed from Cakrapāṇidatta.

48 CSS Rasāyanādhikāra 115.

processed form, though this is not completely clear from the text) and the resulting mass is then rolled into pills.[49]

A work from the thirteenth or fourteenth century, the *Śārṅgadhara-saṃhitā*, represents a watershed moment in the history of the use of mercury in Indian medicine, as it presents an elaborately formulated system of processing and using mercury only alluded to in the older medical works. Its long chapter on mercury contains one rather complicated recipe for the purification of mercury (*rasaśodhana*), two recipes for extracting mercury from cinnabar (*daradaśodhana*), four recipes for giving mercury a 'mouth' to 'devour' other metals, four recipes for the 'killing', i.e. turning into ash of mercury, and nearly fifty recipes for medicines prepared from the above products.[50]

The recipes for mercury medicines that follow are diverse in production methods, ways in which they are applied and diseases they are meant to treat. A common denominator of all recipes is the occurrence of sulfur as one of the ingredients. Mercury is ingested mixed with honey or ghee, as a beverage or in the form of pills. It is also applied as an eye ointment, smeared into the nose, rubbed into a small incision in the skin or used topically on areas of the skin affected by skin disease.[51] Diseases or conditions to be treated with the various medicines span from fevers (understood as a disease category, not as the symptom of heightened body temperature) to digestive complaints (diarrhea, constipation, indigestion, colics, etc.), wasting and skin diseases.

Two of the mercurial recipes describe the making of aphrodisiac (*vājīkaraṇa*) tonics. The first potion, called *madanakāmadeva,* 'the god of passion and love,' is meant to enhance the consumer's strength, improve his appearance and prolong his life, as well as functioning as an aphrodisiac. A further recipe is specifically called a *rasāyana* (*loharasāyana*) and is meant to bestow longevity and youthfulness, but also has aphrodisiac properties.

Vājīkaraṇa is another of the eight formal divisions of medicine first introduced by Caraka. It is often coupled with *rasāyana* as the two subjects cover some of the same ground. That is, aphrodisiacs are often also supposed to give youthful strength and improve the general appearance of the consumer, while recipes for *rasāyanas* may list enhanced sexual stamina as one of the expected results of treatment. In many of the medical works,

49 See, for example, CSS Rasāyanādhikāra 93–100, 194–97 and 233–37 for recipes detailing how pills are made in this fashion.

50 ŚS II.12.

51 See ŚS II.12.135, 136, 121–26 and 190–93, respectively.

the *rasāyana* chapter follows the *vājīkaraṇa* one or vice versa. In others, they are combined, or one subsumes the other. In the early medical works, there are very big differences between *rasāyana* and *vājīkaraṇa* formulations and they are easily distinguished, most *vājīkaraṇa* recipes being for foods and ointments for topical use. However, this distinction becomes somewhat blurred in the later medical works, especially as mercury and other metals and minerals become introduced in *vājīkaraṇa* recipes, as well as other *rasāyana* staples, such as the myrobalans, and shilajit.[52] Certainly, the recipes for *vājīkaraṇa* and *rāsayana* potions in the *Śārṅgadharasaṃhitā* do not seem very different in nature from each other.

The recipe for the rejuvenating aphrodisiac *madanakāmadeva* details that (the ashes of?) silver, diamonds, gold, copper, mercury, sulfur and iron are macerated in the juice of aloe vera. The mixture is filled into a glass bottle that is sealed with rock-salt. The bottle is covered with mud plaster and then placed in a receptacle filled with salt. This, in turn, is set on a fire and gently heated for a day. The mixture is then macerated several times with the juices and milky saps of a number of plants. Then, powdered plant materials are added as well as quite a lot of sugar (the weight of sugar is equal to the weight of the rest of the materials). The consumer of this presumably rather costly preparation is promised beauty, strength and vitality as well as the ability to enjoy sexual intercourse with many young women without loss of semen, i.e. without loss of vitality.[53]

For the *loharasāyana* recipe, processed (the technical term used is '*śuddha*,' 'purified') mercury is mixed with purified sulfur to make the mercury-sulfur compound called *kajjalī*. Filings of magnetic iron are added and the mixture is then macerated for three days with aloe vera juice. This is done in sunlight so that 'smoke' (*dhūma*) – presumably the evaporating mercury – rises from the mixture. The mass is then placed in a copper vessel and covered with paddy husk for three days, after which the mixture is again macerated with a host of herbal substances. This is dried in the

52 The *Cakradatta*'s *vājīkaraṇa* section contains a number of recipes with the Indian gooseberry (*āmalaka*) and marking nut (*bhallātaka*). Its mercurial recipe is not similar to *rasāyana* formulations: In this recipe (57.51) an Indian beech (*karañja*) seed is filled with mercury (*parāda*) and covered with gold. It is then kept in the mouth during intercourse and supposed to let the man retain his semen. The *Cikitsāsārasaṃgraha*'s *vājīkaraṇa* section (LXXVIII.66) gives a recipe for a mixture of pyrites, honey, mercury, iron, chebulic myrobalan, shilajit and false black pepper. A similar recipe is also found in the *Bhāvaprakāśa*'s *vājīkaraṇa* section in Uttarakhaṇḍa I.20.

53 See ŚS II.12.259–66.

shade, then ground, and stored. The mixture is administered in small doses with honey and clarified butter and is said to be useful in a number of indications: poor appetite, dyspnea, cough, urinary disorders, gripes, piles and others. It is also supposed to give the consumer strength and good color, sexual vigor and longevity as well as curing any disease.[54]

As already mentioned, the *Bhāvaprakāśa*'s *rasāyana* section contains no recipes with mercury. Similarly, in the *Yogaratnākara*, a later work from about the early eighteenth century,[55] we find no mercurial recipe in the *rasāyana* section, though the author does mention that *gandhakarasāyana*, a preparation based on sulfur, can be taken together with the ash of mercury.[56] But the author lists several mercurials in the *vājīkaraṇa* section (chapter LXXIV) and a few of these are said to have rejuvenating properties.[57] Indeed, the *madanakāmadeva* recipe from the *Śārṅgadharasaṃhitā* is found there.[58]

In a slightly younger work, the *Bhaiṣajyaratnāvalī* (eighteenth or nineteenth century[59]), we find that mercury plays a prominent role in the *rasāyana* section. Ten of its circa forty recipes contain mercury. Notably, although the author Govindadāsa uses a number of recipes from the *Cakradatta*,[60] he does not include the latter's signature recipe, the iron-based *amṛtasāralauha*. Nor does he include Cakrapāṇidatta's (and Vaṅgasena's) mercury-containing copper *rasāyanas*. In fact, most of the recipes that contain mercury and other metals seem to be derived from a particular alchemical work, the *Rasendrasārasaṃgraha*.[61] Meulenbeld (1999–2002, IIA: 727) describes this work by Gopālakṛṣṇa as a treatise on alchemy in the

54 See ŚS II.12.275–89.

55 See Meulenbeld (1999–2002, IIA: 352) on the date of the *Yogaratnākara*.

56 See YR LXXVI.28–31 for the recipe for *gandhakarasāyana* and verse 33 for the statement regarding the addition of mercury ash and gold.

57 For example, the *kāśmīrāvaleha* formulation in verses 50–57 is meant to provide longevity, health strength and radiance, while the *kāmāgnisandīpana* recipe in verses 78–94 is meant to confer strength and eradicate wrinkles and grey hair.

58 See YR LXXIV.112–18.

59 The part of the *Bhaiṣajyaratnāvalī* ascribed to Govindadāsa dates to the eighteenth century, but other parts (chapters II, IV and LXXVI–CVI) were added in the nineteenth century by Brahmaśaṃkara Miśra. The *rasāyana* section (chapter LXXIII) is thought to belong to the older stratum of text. See Meulenbeld (1999–2002, IIA: 336) on the date and the authorship of the *Bhaiṣajyaratnāvalī*.

60 For example, Cakrapāṇidatta's recipe for *śivaguḍikā*, pills containing shilajit (CD LXVI.172–93 and BR LXXIII.151–75).

61 This is a rough count, and includes advice about sniffing water in the morning, et cetera as a recipe.

service of medicine and notes that it contains a large number of mercurial compounds against diseases. The date of this work is uncertain, with suggestions ranging from the fifteenth century to the eighteenth.[62]

CONCLUDING REFLECTIONS

This overview is incomplete as it has not taken into account the many Āyurvedic works written between the sixteenth and the eighteenth centuries, and has also not surveyed the *rasāyana* recipes of alchemical literature. A more thorough examination of the entangled histories of Āyurvedic and alchemical (and *haṭhayoga*) *rasāyana* practices will be the focus of a new research project that will hopefully shed further light on the topic.[63]

However, from the examined works it should have become clear that mercury, while used in a number of Āyurvedic *rasāyana* formulations from the seventh century onwards, did not play a particularly prominent role for this part of medicine. The exception to this rule is the *Bhaiṣajyaratnāvalī*, which dedicates a fairly big part of its *rasāyana* section to mercurials, many of which are derived from alchemical literature. However, even in this work, there is no particular sense that mercury is given pride of place as an anti-aging, life-prolonging substance in the *rasāyana* section itself.[64] Rather, just as in the other Āyurvedic works, the sense is that mercury, while an integral part of the recipes, is just one of many suitable *rasāyana* ingredients within them. A useful and powerful, but not prominent medical substance.

This stands in strong contrast to the attitudes of the alchemists as reflected in the quotation from the *Rasahṛdayatantra* given at the beginning

62 See Meulenbeld (1999–2002, IIA: 730).

63 This is the ERC starting grant project 'Medicine, immortality, moksha. Entangled histories of medicine, alchemy and yoga in South Asia.' See ayuryog.org.

64 In BR II.6, one finds the statement: 'When its faults have been removed, mercury repels death and fevers; purified, it is truly a nectar of immortality, mercury with its faults is poison.' – *doṣahīno yadā sūtas tadā mṛtyujvarāpahaḥ | śuddho 'yam amṛtaḥ sākṣād doṣayukto raso viṣam* || As the preceding verses (3–5) explain, the faults or blemishes of mercury include its contamination with lead, tin and dirt, being flammable, unstable, poisonous and being unable to withstand heat. See also Wujastyk (2013: 24 and 31–32) on the faults of mercury in the *Bhāvaprakāśa* and the *Bhaiṣajyaratnāvalī*, respectively. It should again be noted, however, that chapter two probably belongs to the newer part of the *Bhaiṣajyaratnāvalī*, so that the definition of mercury as a nectar of immortality may not have reflected the original author's views.

of this essay. However, an examination of the actual *rasāyana* chapters of the alchemical works, and the recipes associated with them, may in fact reveal a similar attitude to that of the medical authors. This remains to be discovered.

ABBREVIATIONS AND REFERENCES

Skt. = Sanskrit

AHS — Vāgbhaṭa: *Aṣṭāṅgahṛdayasaṃhitā.* Srikantha Murthy, K.R. (tr.) (1999–2000). *Vāgbhaṭa's Aṣṭāṅga Hṛdayam. Text, English Translation, Notes, Appendix and Indices.* 3 vols. (4th edition). Varanasi: Krishnadas Academy.

ASaṃ — Vāgbhaṭa: *Aṣṭāṅgasaṃgraha.* Srikantha Murthy, K.R. (tr.) (1995–1997). *Aṣṭāṅga Saṃgraha of Vāgbhaṭa. Text, English Translation, Notes, Indices etc.* 3 vols. Varanasi: Chaukhamba Orientalia.

BhaPr — Bhāvamiśra: *Bhāvaprakāśa.* Srikantha Murthy, K.R. (tr.) (1998–2000). *Bhāvaprakāśa of Bhāvamiśra, Text, English Translation, Notes, Appendeces [sic] and Index.* 2 vols. Varanasi: Krishnadas Academy.

BR — Govinda Dasji: *Bhaiṣajyaratnāvalī.* Mishra, B. (ed.). (2006). *Bhaiṣajyaratnāvalī of Shri Govinda Dasji.* Enlarged edition. Commented upon by A. Shāstrī; English translation by K. Lochan; translation technically reviewed by A.K. Choudhary. 3 vols. Varanasi: Chaukhambha Sanskrit Bhawan.

CaS — *Carakasaṃhitā.* Sharma, P. (ed., tr.) (2003). *Carakasaṃhitā, Agniveśa's treatise refined and annotated by Caraka and redacted by Dṛḍhabala, Text with English Translation,* 4 vols. (8th edition). Varanasi: Chaukhamba Orientalia.

CD — Cakrapāṇidatta: *Cakradatta.* Sharma, P.V. (tr.) (1994). *Cakradatta, Text with English Translation, A Treatise on Principles and Practices of Ayurvedic Medicine.* Varanasi, Delhi: Chaukhamba Orientalia.

Ci — *Cikitsāsthāna*

CSS — Vaṅgasena: *Cikitsāsārasaṃgraha.* Saxena, N. (ed.). (2004). *Vaṅgasena Saṃhitā or Cikitsāsāra Saṃgraha of Vaṅgasena. Text with English Translation, Notes, Historical Introduction, Comments, Index and Appendices.* 2 vols. Varanasi: Chowkhamba Sanskrit Series Office.

Dutt 1922 — Dutt, U.C. (1922). *The Materia Medica of the Hindus,* revised edition with additions and alterations by Kaviraj Binod Lall Sen, Kaviraj Ashutosh Sen and Kaviraj Pulin Krishna Sen (Kavibhushan). Calcutta: Adi-Ayurveda Machine Press.

Eliade 1969 — Eliade, M. (1969 [1958]). *Yoga. Immortality and Freedom.* Princeton: Princeton University Press.

Hoernle 1893 — Hoernle, A.F.R. (1893). *The Bower Manuscript. Facsimile Leaves, Nagari Transcript, Romanised Transliteration and English Translation with Notes*. Calcutta: Archaeological Survey of India.

Ka — *Kalpasthāna*

Leslie 2009 — Leslie, C. (2009). 'The Professionalization of Ayurvedic and Unani Medicine.' In Freidson, E. and J. Lorber (eds.), *Medical Professionals and the Organization of Knowledge*, pp. 39–54. New Brunswick, NJ: Transaction Publishers.

Maas 2008 — Maas, Ph.A. (2008). 'The Concepts of the Human Body and Disease in Classical Yoga and Āyurveda.' *Wiener Zeitschrift für die Kunde Südasiens*, 51: 125–62.

Maas 2010 — Maas, Ph.A. (2010). 'On What Became of the Carakasaṃhitā after Dṛḍhabala's Revision.' *eJournal of Indian Medicine*, 3(1): 1–22.

Meulenbeld 1999–2002 — Meulenbeld, J.G. (1999–2002). *A History of Indian Medical Literature*. 5 vols. Groningen: Egbert Forsten.

Ras — *Rasārṇava*, in: Oliver Hellwig: DCS – The Digital Corpus of Sanskrit. Heidelberg, 2010–2012, http://kjc-fs-cluster.kjc.uni-heidelberg.de/dcs (accessed 1 October 2015).

RH — *Rasahṛdayatantra*, in: Oliver Hellwig: DCS – The Digital Corpus of Sanskrit. Heidelberg, 2010–2012, http://kjc-fs-cluster.kjc.uni-heidelberg.de/dcs (accessed 1 October 2015).

Roşu 1978 — Roşu, A. (1978). *Les Conceptions Psychologiques Dans les Textes Médicaux Indiens*. Paris: Collège de France, Institut de Civilisation Indienne.

RS — *Rasaratnasamuccaya*, in: Oliver Hellwig: DCS – The Digital Corpus of Sanskrit. Heidelberg, 2010–2012, http://kjc-fs-cluster.kjc.uni-heidelberg.de/dcs (accessed 1 October 2015).

Scharfe 1999 — Scharfe, H. (1999). 'The Doctrine of the Three Humors in Traditional Indian Medicine and the Alleged Antiquity of Tamil Siddha Medicine.' *Journal of the American Oriental Society*, 119(4): 609–29.

ŚS — Śārṅgadhara: *Śārṅgadharasaṃhitā*. Srikantha Murthy, K.R. (tr.) (1984). *Śārṅgadhar-Saṃhitā (A Treatise on Āyurveda) by Śārṅgadhara*. Jaikrishnadas Ayurveda Series no 58. Varanasi: Delhi: Chaukhamba Orientalia.

Sū — *Sūtrasthāna*

SuS — *Suśrutasaṃhitā*. Sharma, P.V. (ed., tr.) (1999–2001). *Suśruta-saṃhitā with English Translation of Text and Ḍalhaṇa's Commentary along with Critical Notes*. 3 vols. Varanasi: Chaukhambha Visvabharati.

Utt — *Uttarasthāna*

Vi — *Vimānasthāna*

White 1996 — White, D.G. (1996). *The Alchemical Body. Siddha Traditions in Medieval India*. Chicago: The University of Chicago Press.

Wujastyk 2000 — Wujastyk, Dominik. (2000). 'The Combinatorics of Tastes and Humours in Classical Indian Medicine and Mathematics.' *Journal of Indian Philosophy*, 28: 479–95.

Wujastyk 2003	Wujastyk, Dominik. (2003). *The Roots of Ayurveda*. Second revised edition. London: Penguin Books.
Wujastyk 2013	Wujastyk, Dagmar. (2013). 'Perfect Medicine.' *Asian Medicine. Tradition and Modernity*, 8: 15–40.
Wujastyk 2015	Wujastyk, Dagmar. (2015). 'On Perfecting the Body. Rasāyana in Indian Medical Literature.' *AION – sezione filologico-letteraria*, 36.
YR	*Yogaratnākara*. Kumari, A. and P. Tewari (ed., tr.) (2010). *A Complete Treatise on Āyurveda. Yogaratnākara*. 2 vols. Varanasi: Chaukhambha Visvabharati.

Chapter 6

When *Ngülchu* Is Not Mercury: Tibetan Taxonomies of 'Metals'

BARBARA GERKE[1]

> It is written [in an unknown Chinese encyclopedia] that 'animals are divided into: (a) belonging to the Emperor, (b) embalmed, (c) tame, (d) sucking pigs, (e) sirens, (f) fabulous, (g) stray dogs, (h) included in the present classification, (i) frenzied, (j) innumerable, (k) drawn with a very fine camelhair brush, (l) et cetera, (m) having just broken the water pitcher, (n) that from a long way off look like flies.' In the wonderment of this taxonomy, the thing we apprehend in one great leap, the thing that, by means of the fable, is demonstrated as the exotic charm of another system of thought, is the limitation of our own, the stark impossibility of thinking that.
>
> (Jorge Luis Borges, *The Analytical Language of John Wilkins*)[2]

The use of metals such as gold, silver, bronze, lead, et cetera, in Asian medicines is not unusual. Among them, mercury has received special attention

1 Barbara Gerke currently holds a FWF-funded Lise-Meitner senior fellowship at the Department of South Asian, Tibetan and Buddhist Studies at the University of Vienna (Austria).
The research and writing of this article was supported by the German Research Foundation (Deutsche Forschungsgesellschaft), grant nos. 53307213 and 55127201 during a postdoc (Eigene Stelle) at the Institute for Asian and African Studies at Humboldt University of Berlin (Germany) and completed in Vienna during the FWF grant no. M1870.
The author thanks all the Tibetan physicians with whom she discussed mercury classifications, sources and substitutions, especially Dr Tenzin Thaye of the Men-Tsee-Khang in Dharamsala, India, as well as Calum Blaikie, Denise Glover, Florian Ploberger and Jan van der Valk for their suggestions and critical feedback.

2 Cited in Foucault (2002: xv).

in many medical traditions, which have emphasized the rejuvenating and curative powers of this heavy metal in various ways, but only if processed correctly.[3] In Tibetan medicine, also known as Sowa Rigpa (*gso ba rig pa*), processed mercury has been known as the 'king of essences' (*bcud kyi rgyal po*), resembling the semen of Lord Śiva and the vital forces of this powerful Hindu god (White 1996), thus demonstrating Indian origins.

While unprocessed mercury was used early on in India (second century BCE and probably earlier), recipes with processed mercury entered Sanskrit medical literature only around the thirteenth century (Wujastyk 2013). At the same time, more complex mercury compounds reached Tibet via the Swat valley and over the centuries became an integral part of Tibetan pharmacopoeia (Czaja 2013; Simioli 2013). Today, the use of mercury in Tibetan medical compounds is largely limited to the making of certain precious pills (*rin chen ril bu*). In the form of cinnabar (*mthsal* or *chog la*), which is processed in various ways, it is also added as a potent ingredient and used to coat certain herbal multi-compound pills. In Tibetan medical literature, liquid mercury is called *ngülchu* (*dngul chu*),[4] which literally means 'silver water,' an appellation also found in other languages. In Greek, mercury is called ὑδράργυρος (*hydrargyros*) and its Latin variant is *hydragyrum*. The Chinese term is *shui yin* (水银). All of them translate literally as 'water silver,' alluding to mercury's color and mobility at room temperature.

In the English literature on Tibetan medicine *ngülchu* is generally translated as mercury. But is the Tibetan *ngülchu* that is sourced from various substances really equivalent to the elemental metallic mercury that we label Hg in chemical terms? And is this a useful question to ask? In this chapter, I analyze Tibetan medical classifications and sources of this substance, questioning the single identification of *ngülchu*. What exactly is this substance that the early eighteenth century pharmacopoeia *Shelgong* (*Shel gong*) and its commentary *Sheldreng* (*Shel greng*), hereafter called by its brief popular title *Crystal Rosary*, classified as a 'meltable precious substance' and a 'cold inanimate poison'? What do such classifications tell us about Tibetan understandings of *ngülchu*? While looking at the classification of mercury as a poison and at relevant sections in seminal Tibetan pharmacopoeia that describe the sourcing of *ngülchu*, it will become clear

3 See Needham et al. (1976) on the use of cinnabar in Chinese elixirs.

4 Tibetan terms are followed by their Wylie transliteration at first use. Frequently used terms are phoneticized. Primary Tibetan sources are mentioned by their English short title followed by their Wylie transliteration at first use. In the bibliography Tibetan primary sources are listed alphabetically according to their short English title.

that *ngülchu* might not always be what we consider liquid metallic mercury. I argue that instead of simply finding equivalence for *materia medica* substances, we have to consider culture-specific nomenclatures and their underlying epistemologies.[5] This discussion contributes to the current debate on the naming and classification of medical substances in Asian medicine and adds an example from the Tibetan *materia medica* that should caution us when applying distinct European equivalents and classifications of metals, plants and animals to Asian contexts. When analyzing Tibetan classificatory systems and the place assigned to *ngülchu* within them, I prefer to ask how *ngülchu* has been described by authors in certain medical texts and contexts of varied Tibetan medical histories and what these tell us about the socio-historical influences on ways to classify and also source a substance.

I question that *ngülchu* is equivalent to metallic mercury in all instances in Tibetan pharmacopoeias. As we shall see, *ngülchu* is a substance that can be sourced from rocks, plants, animals, ragged clothes and garments of human corpses. Such sources are still cited in modern pharmacology texts in differing orders of priority, although most are not used today. However, contemporary Tibetan physicians, who use liquid mercury at the beginning of the complex alchemical process to make *tsotel* (*btso thal*, lit. 'cooked ash') for use in precious pills, consider *ngülchu* sourced from a specific plant as a potential substitute for liquid mercury bought in an Indian chemist's shop. Such instabilities in identification should be part of the debates on equivalence of technical terms, which in the Tibetan medical contexts have been discussed by several authors;[6] they point to specific challenges, such as widespread practices of synonymy and substitution, regional variations of *materia medica*, and the clashes of Asian epistemologies with modern biological and chemical classifications.

5 I do not view the term epistemology here in an exclusively scientific sense but refer to epistemologies within a broader meaning of how things are known. This is similar to other scholars who have worked on Tibetan medicine, for example Adams et al. (2011: 8).

6 Such discussions refer, for example to the 'three humors' or *nyépa* (*nyes pa*), 'wind' or *lung* (*rlung*) disorders, the anatomy of the channels, and the 'scientification' of medical terms in general. See Gerke (2010: 128–33) for a summary of these debates.

RECONSIDERING EQUIVALENCE

One excellent example on the instability of translations of *materia medica* terms in Asian medicine is Carla Nappi's analysis of the caterpillar fungus called 'summer grass, winter worm' (Tib. *dbyar rtswa dgun 'bu*; Chin. *dongchong xiacao*). Nappi skillfully analyzes the changes in historical identification of this substance, which is listed in both Tibetan and Chinese *materia medica* texts. Apparently, at various times it was classified as a half-plant, half-insect, plant or fungus. Offering a brilliant example in the colonial historiography of medicine, Nappi explores how the caterpillar fungus exemplifies changing classifications of medicinal raw material since late imperial China. As can be expected, the transformations in nomenclature, classification and practice were also determined by certain political and economic factors, as well as foreign relations; since the 1990s, the fruit of the fungus has become a popular tonic with a pedigree and a global business worth millions (Nappi 2010: 23). Drawing a parallel to Latour's work on the identification of Koch's bacillus, Nappi argues that the medicinal substances she questions 'are not trans-historically identifiable things as much as they are bits of discourse that reveal networks of production. Each time one of these names is synonymized with another a new set of relationships and a new series of networks is brought into existence' (ibid.: 29–30).

Reading Nappi's article made me wonder whether we find similar networks of relationships at play in the context of *ngülchu* and mercury. To summarize the key issue here in Nappi's own words:

> *Dongchong xiacao* and *Cordyceps* are not obviously the same object, or even the same kind of thing, despite the current tendency to treat them as alternate names for a single being. The entities denoted by these names arguably occupy very different epistemic spaces in the context in which they emerged. It is instead the practice of synonymy[7] since the eighteenth century that has equated these entities and mashed them into the singularity that they seem to represent today. (2010: 24)

Elisabeth Hsu has shown in the context of Chinese medicine that plant classifications are culture specific and change over time. Her study of the therapeutic preparations of *Artemisia annua* L. or *qinghao* in medieval China

7 Practices of synonymy and polyequivalence are very common in Asian medicines. Tibetan pharmacopoeia often begins with a list of synonyms. The *Crystal Rosary* (105/20–106/12) lists more than fifty synonyms for *ngülchu*.

led her to argue that 'the so-called "herbal" medical practice depends not only on plant classifications that are culture specific, but also on practical interventions that treat the plant as a thing' (Hsu 2014: 1). With her question 'What is in a name?' she explores how the act of naming is influenced by the ways in which people think about what comprises a medicinal drug or an ingredient. In the case of *qinghao*, before the tenth century CE the term did not necessary refer to the plant as such, while since the tenth century the term referred to both a plant in the field and a medical preparation. This means that when analyzing the history of naming Asian medical substances, one also has to consider the polysemy of terms and how a name can mean different things at different times. Hsu has taken this analysis further, redefining what language does with Chinese *materia medica* and how certain substances have a 'Gestalt,' which also emerges from the sound of their names (Hsu 2015).

There are two ideas from the works of Hsu and Nappi that guide the following analysis. In the context of *ngülchu*, when looking at textual records on the various sources, we need to keep in mind that *ngülchu* might have referred to different raw materials altogether. A substance labeled *ngülchu* might be sourced from plant or animal matter and might still be used the same way as liquid metallic mercury distilled from cinnabar, evoking different networks of relationships based on different historical discourses, as in Nappi's example. Thus, an analysis of *ngülchu* has to be open to detect practices of synonymy and indigenous taxonomies that might have influenced its classifications, identifications and sourcing practices over time. When was *ngülchu* metallic mercury and when was it something else? Were the different classifications and sourcing methods linked to trading *ngülchu* in the form of cinnabar rock, which only occurs in geographically distinct places in Tibet and was often imported from China? How was *ngülchu* classified and described in *materia medica* texts? To answer some of these questions, I first outline the basic classification schemes occurring in Tibetan pharmacopoeia. To keep the analysis within the limits of this chapter, I only refer to the *Four Treatises* (*Rgyud bzhi*) – the seminal medical compendium dating back to the twelfth century and partially memorized by medical students to this day – and key Tibetan pharmacopoeia from the early eighteenth and the nineteenth centuries still in use today, as well as some contemporary works.

TIBETAN *MATERIA MEDICA* CATEGORIES

It is important to recognize that classifications of metals and minerals have not yet been studied by scholars of Tibetan medicine. Plants form the largest source of Tibetan medical substances, and to date scholars have focused on plant categories (Boesi and Cardi 2006; Boesi 2005/2006, 2007; Glover 2005, 2010). Glover compares *materia medica* classifications in the *Four Treatises* with the contemporary Tibetan pharmacopoeia known as *Crystal Garland* (*'Khrungs dpe dri med shel gyi me long*; Gawa Dorje 1995). Her results show that *materia medica* texts written and published by Tibetan authors in the People's Republic of China (PRC) post 1959 are heavily influenced by modern biological categories, and classical parameters are no longer the organizing principle of plant substances in these texts (Glover 2005: 194). Classical parameters are largely based on organoleptic experience, such as 'efficacy/potency' (*nus pa*), 'power' (*stobs*) and 'quality' (*yon tan*). Efficacy/potency can depend on 'taste' (*ro nus pa*) or might refer to the natural efficacy of the substance itself (*ngo bo nus pa*). All medicinal substances in the principal *materia medica* section (chapter 20) of the second of the *Four Treatises* are organized according to these principles.[8] Moreover, they are classified according to eight *materia medica* categories.[9]

The introduction of a new hierarchical structure according to 'kinds' or *rik* (*rigs*) in *Crystal Garland* and related literature follow the three modern biological classifications of plants, animals and minerals; *rik* is thus sometimes equated with the problematic category 'kingdom' (Glover 2010: 262). Such categorizations (which are also questioned in current biology) override classical Tibetan taxonomies, of which we have not even understood the principles and historical developments. For example, in the classical *Four Treatises* metals and minerals are variedly found within three of the eight *materia medica* categories, i.e. under 'precious medicine,' 'earth medicine' and 'stone medicine.' *Crystal Garland* groups these three categories

8 See Four Treatises (65/10–75/17) for the Tibetan version; Ploberger (2012: 267–91) for a German translation; Men-Tsee-Khang (2011_a: 194–218) and Clark (1995: 129–83) for English translations.

9 These eight categories are: (1) precious medicine (*rin po che'i sman*); (2) earth medicine (*sa sman*); (3) stone medicine (*rdo'i sman*); (4) woody plant medicine or tree medicine (*shing sman*); (5) essence medicines (*rtsi sman*), which are often scented substances that are extracted from roots and trees as well as animals; (6) medicine from the plains (*thang sman*); (7) herbal medicine (*sngo sman*); (8) animal medicine (*srog chags sman*). They are discussed in more detail in Boesi (2005/2006) and Glover (2010).

together under the neologism 'mineral' (*gter dngos kyi rigs*, lit. 'kinds of real deposits,' also translated as 'ores'), which is also found in other modern texts (e.g. Dawa Ridak 2003), but not in classical pharmacopoeia. This categorization was probably introduced by contemporary Tibetan physicians trained and working under Chinese influence and in an effort to classify Tibetan medicine according to the modern plant, mineral and animal 'kingdoms' (Boesi and Cardi 2006). Pordié calls such pharmacopoeia based on bio-pharmacology a kind of 'deculturated ethnopharmacology' and calls for approaches to look at pharmacopoeia as social expressions (2002: 195). Nappi labels such practices of overlaying Latin binomial nomenclature on classical Chinese plant and animal names 'effectively a kind of textual and nominal imperialism' (2010: 22). One of the results of such 'imperialism' is that traditional categories of medicinal substances, which often overlap, do not necessarily make sense when clustered into modern biological taxonomies. To give an example: Boesi and Cardi (2006) showed that the three *rik* do not account for several of the substances mentioned in the classical Tibetan categories. For example, musk (*gla rtsi*) and bitumen (*brag zhun*), which in classical texts are listed under 'essence medicines,' end up in the 'plant' category under the new scheme.

There are other problematic results of such practices of conformity. For example, Tibetan doctors who are not trained in biology but nevertheless try to conform to presenting Tibetan medicine to the Western world in 'scientific' fashion, have introduced botanical equivalents and English identifications of plants in their Tibetan medical dictionaries (e.g. Drungtso and Drungtso 2005). Such botanical identifications are often copied from the internet, Āyurvedic and other botany books of variable quality, are largely listed without sources, and thus do not only end up often being wrong, but also unquotable. Notably, the botanical names are rarely used by practicing Tibetan physicians in India, who for the most part work exclusively with the Tibetan (sometimes also Chinese) and Hindi names of their medicinal substances.[10] Translations of Tibetan substance names into Chinese also pose problems, and Tibetan physicians have cautioned that erroneous translations might lead to a change in identification and a consequent loss of efficacy in their multi-compound formulas (Gawa Dorje 2009).

This general introduction has elucidated the challenges involved when overriding traditional classifications. These discussions will now be taken further through analyzing examples of metals, specifically *ngülchu*. All

10 I thank the ethnobiologist Jan van der Valk for related discussions and comparison of our ethnographic material with Tibetan doctors in Dharamsala, India.

contemporary Tibetan physicians I spoke with clearly consider *ngülchu* a metal, called *chak* (*lcags*). However, when looking at metals in Tibetan medical literature we notice that *ngülchu* was not always classified as *chak*. The term *chak* is a label for the 'eight metals' (*lcags bgryad*) explained below and is also the specific name of one of them – iron. The term *chak rik* (*lcags rigs*) thus refers to various types of iron as well as to metals in general.[11] As we shall see, the *Four Treatises* and the classical Tibetan pharmacopoeia of the early eighteenth and the nineteenth centuries consulted for this chapter do not list metals within a designated category called 'metals.'

ANIMATE AND INANIMATE TAXONOMIES: CLASSIFICATIONS OF *NGÜLCHU*

When Foucault wrote *The Order of Things*, he tried to trace the cultural experiences that underlie the modalities of creating order, which become the positive basis of knowledge in a given historical time and space. I chose Borges' quote as an opening for this chapter to show the 'uneasiness' that comes with uncommon classifications. It intrigued Foucault that Borges 'simply dispenses with the least obvious, but most compelling, of necessities; he does away with the *site*, the mute ground upon which it is possible for entities to be juxtaposed' (Foucault 2002: xviii, original emphasis). Foucault then asks the crucial question about what comprises the 'ground on which we are able to establish the validity of [a] classification with complete certainty' (ibid.: xxi). This and the following sections present the different taxonomies of *ngülchu* in the *Four Treatises* and selected key classical Tibetan pharmacopoeias of the early eighteenth and the nineteenth centuries, and analyze the varied principles underlying such classifications and what these tell us about what Tibetan authors might have thought about such substances.

The three main classifying parameters I found with relation to *ngülchu* in different contexts are dual opposites: 'animate' (*rgyu ba*)[12] and

11 Metallic substances and metallurgy as a practice were widely known in Tibet (Oddy and Zwalf 1991). On Tibetan alchemical practices to transform metals into gold see Simioli (2013).

12 The literal meaning is 'moving.' So the distinction is made between objects that move and those that do not. Since all objects are comprised of the five elements, in Tibetan Mahāyāna Buddhism inanimate things also might be considered sentient beings, such as stones, trees and sacred sites, all of which can carry life forces (Gerke 2012).

'in-animate' (*mi rgyu ba*), 'wild' (*rgod pa*) and 'mild' (*g.yung ba*), as well as 'meltable' (*bzhu ba*) and 'non-meltable' (*mi bzhu ba*). Briefly, the first set of these parameters refers to classifications of poisonous substances, the second and third set address *ngülchu* as a medicinal substance. The 'wild' and 'mild' categories describe the 'nature' (*rang bzhin*) of a substance and are frequently used to group certain varieties of the same substance. 'Meltable' and 'non-meltable' categories refer to metals and precious substances that can be melted, fused or amalgamated. These categories reveal that in some contexts *ngülchu* is considered a type of poison, which can be used to intentionally harm others, but it is also a raw ingredient for medicines. That *ngülchu* can be a poison as well as medicinal ingredient is not a contradiction but follows Tibetan episteme that the strongest poison can be the best elixir when properly processed.

In the following, I explore the above classifications in more detail and examine how they relate to *ngülchu.*

NGÜLCHU *AS A TYPE OF 'INANIMATE' POISON*

Tibetan medical texts link the origin of classifying parameters such as 'animate' and 'inanimate' to Indian myths. In one of them the gods and demigods search for the nectar of immortality during the creation of the universe. A common theme in this myth and its variations is the appearance of threatening creatures – manifestations of poisons – that are tamed by the gods through mantras, by the power of which their demonic bodies break up into 'inanimate poisons,' such as various types of aconite, and 'animate poisons,' such as poisonous snakes and scorpions (Four Treatises 544/6–10).

The distinction of 'animate' and 'inanimate' is also found in the context of 'natural poisons' (*dngos dug*), and the third part of the *Four Treatises* dedicates an entire chapter to this topic (Four Treatises 544/5–548/5). Inanimate poisons are of two kinds, 'hot' poisons and 'cold' poisons.[13] 'Hot' poisons typically refer to poisonous plants, whereas 'cold' poisons are usually precious metals, gems, rocks and types of soil (Sonam Dolma 2013: 112).[14] *Ngülchu* is thus classified as a 'cold inanimate poison' among the natural poisons. The *Four Treatises* mention *ngülchu* also in the context

13 On the division between hot and cold poisons in a Chinese context see Gould-Martin (1978).

14 Sonam Dolma (2013) also discusses poison classifications and the three poisons of mercury.

of intentionally 'compounded poisons' (*sbyar dug*), which are made of 'precious' substances, among which *ngülchu* is listed (Four Treatises 513/3). There is no specific category of 'poisons' (*dug*) in classical Tibetan pharmacopeia even though some poisonous substances are clearly used as medicines; but these are then classified among other categories, as explained below.

NGÜLCHU *AS A* MATERIA MEDICA *SUBSTANCE*

In the *Four Treatises*, *ngülchu* is not only considered a type of poison that can intentionally be used to harm someone, but it is also an important medicinal substance. The classifications used for the therapeutic applications are different from the poison classifications explained above. As a *materia medica* substance, *ngülchu* is largely mentioned in the form of the natural occurring mercury sulfide rock cinnabar (*cog la ma* or *mtshal*) or the artificially created mercury sulfide vermilion (*rgya mtshal*; lit. 'Chinese' or 'foreign' vermilion).[15] These rock sources of *ngülchu* are classified under 'earth medicines' (*sa sman*). Cinnabar and vermilion are the common forms mercury was traded in, largely from China.[16] Natural cinnabar rock could also be found at various places in Tibet.[17] The demand for mercury was high in Tibet and Nepal, not because of mercurial medicines but because of the fire-gilding techniques required for making Buddhist statues (Lo Bue 1991: 33–34).

The binary nomenclature 'wild' and 'mild' appear frequently in Tibetan *materia medica* and are polysemous. For example, the 'wild' type might refer to a larger size compared to a smaller-sized 'mild' type of a plant. In the context of *ngülchu* they refer to its color and mobility, not its 'size' or quality. In his *Crystal Rosary*, Deumar Geshe Tenzin Phuntsok (born 1672)

15 That Chinese cinnabar was used in Tibet is evident from the frequent use of a Tibetan phoneticized version of the Chinese term for cinnabar, *zhu sha* 朱砂 (lit. 'red sand'; Tib. *cu'u gshag*), which is a synonym for *rgya mtshal* (Jampa Trinlé 2006: 202).

16 According to Turner the trade with China went via Xining (1991: 371, 381). White (1996: 65–66) mentions that that the mercury supply for Tibet came from Yunnan.

17 Cinnabar ores have been reported by various authors to be found in some parts of Kham, e.g. lower Powo in Kongpo in eastern Tibet, near Mount Targo in Central Tibet, in southeastern Tibet, and near Mount Kailash in western Tibet (summarized by Lo Bue 1991: 44).

describes the 'wild' type as more whitish and movable, whereas the 'mild' type is bluish in color and slow in its movements (Crystal Rosary 107/5). Typically, through complex processing, which Tibetan physicians call 'taming' (*'dul ba*),[18] mercury becomes less and less 'wild,' until it is completely 'mild' and immobile.[19] This classification of 'wild' and 'mild' is still found in modern medical dictionaries (Jampa Trinlé 2006: 186) and other contemporary works (Dawa Ridak 2003: 414/11 and Gawa Dorje 1995: 42/23–25).

NGÜLCHU *AS A MELTABLE PRECIOUS MEDICINE*

As already noted, although it might have been perceived as a metal, *ngülchu* was not classified as a 'metal' as such, since there is no such category in pre-1959 Tibetan pharmacopoeia. Common metals, though often collectively labeled *chak*, were either listed among 'meltable precious medicines' or 'stone medicines.' The first category emphasizes their reaction during processing and their perceived value, while the second category refers to rocks and stones as their main source.

The 'eight metals' used to make *tsotel*, the main potent compound for many Tibetan precious pills, are copper, gold, silver, iron, bronze, brass, tin and lead. The group of eight metals as a set of ingredients usually appear side by side with the 'eight elements' (*khams brgyad*), commonly translated as 'eight minerals,' which refer to a set of minerals or mineral-containing substances that help to bind *ngülchu*. Both sets are used in specific mercury processing techniques that came to Tibet from the Swat valley, and were also documented in the Buddhist Canon (Simioli 2013). Notably, *ngülchu* is never counted or listed among the eight metals, maybe because of its ability to form alloys with most of them. This set of eight metals is also not found consistently in all Tibetan medical works. The *Four Treatises*, pre-dating the compilation of the Buddhist Canon, mentions only four metals (gold, silver, copper and iron) among 'precious medicines.' This does not mean that other metals were not known, but only that these four were classified as 'precious medicines' in the *Four Treatises*. In its *materia medica* chapter, gold and silver ore as well as tin are placed among 'stone medicines,' along with cinnabar and Chinese vermilion (Four Treatises 66/12–17, 67/7). None of the eight *materia medica* categories in the *Four*

18 I explore the 'taming' idea in mercury processing in other publications (Gerke in press).

19 Personal communication, Dr Tenzin Thaye, Dharamsala, May 2015.

Treatises mentions *ngülchu* directly. Only the chapter on treating poisons includes *ngülchu* in a list of six 'precious' (*rin po che*) substances used in compounded poisons, along with gold, copper, bronze, iron and lead (Four Treatises 513/2–3). Even in the description of preparing cinnabar ash in the fourth part of the *Four Treatises*, *ngülchu* appears to be a side product of processing cinnabar, but not a specified ingredient of the medicinal cinnabar ash (Four Treatises 597/5–9; Men-Tsee-Khang 2011_b: 112–13). In the few instances where *ngülchu* is listed as a medicinal ingredient, it is always first processed. For example, it is disintegrated through the juice of the sour plant *tarbu* (*star bu*) and mixed with sulfur before being added to a silver ash compound (Four Treatises 597/2–3; Men-Tsee-Khang 2011_b: 112). These examples from just one seminal medical work tell us that in each context *ngülchu* evokes different networks of relationships – as a poison, a therapeutic substance or a side product of processing cinnabar. Holding such different positions, it seems not surprising to find a variety of *ngülchu* classifications.

Several hundred years following the *Four Treatises*, the *Crystal Rosary*, which was written between 1727 and 1737 (Hofer 2014: 227), subdivided precious substances as 'meltable' and 'non-meltable' and cataloged all eight metals as well as *ngülchu* among the seventeen 'meltable precious medicines' (Crystal Rosary 100/15–115/9). This division is still found in the nineteenth century *Stainless Rosary*, by the Mongolian physician Jampal Dorje,[20] and in Gawa Dorje's contemporary *Crystal Garland*, though with some variations in the lists of metals. The common ground on which medical authors grouped these substances together was their ability to be melted or amalgamated as well as their preciousness, the latter possibly influenced by their perceived monetary and/or therapeutic values.

The different priority given to *ngülchu* in the sequence of metals catalogued in these pharmacopoeias is notable. In the *Crystal Rosary*, *ngülchu* is listed in third place after gold and silver (Crystal Rosary 100–05). Jampal Dorje moved *ngülchu* to first place before gold and silver (Stainless Rosary 40/1–7; Lo Bue 1991: 34). It is not clear what influenced the changed sequence. It could indicate a change in its economic or therapeutic value. Tenzin Phuntsok himself wrote three texts on mercury processing,[21] which testifies to a living tradition of medical use of *ngülchu* in eighteenth

20 This work was reprinted in 1971 under the title *Illustrated Tibeto-Mongolian Materia Medica*. See Hofer 2014 on the pharmacopeias of Jampal Dorje and Tenzin Phuntsok.

21 The titles of these texts are listed in Czaja (2013: 91).

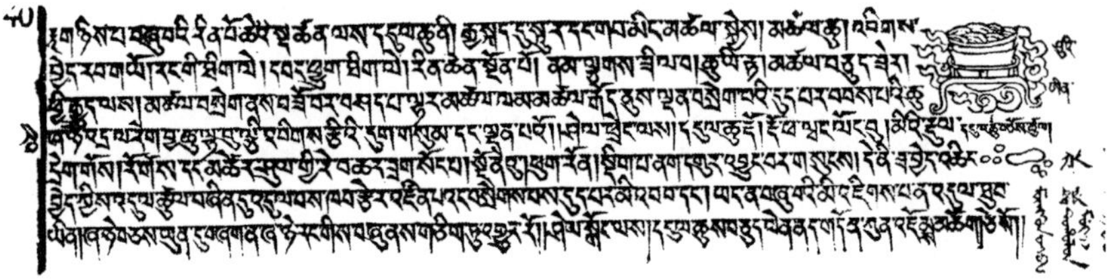

Figure 6.1. The illustrated section on *ngülchu* in Jampal Dorje's *Stainless Rosary*.
This page is from a 1911 print version and is available online at: http://www.wdl.org/en/item/13514/view/1/1/ (accessed 5 October 2015).
Note the sketch of liquid mercury globules underneath the bowl that depicts a processing method. Names are given in classical Mongolian and Chinese script along with a Tibetanized version of the Chinese *shui yin*.

century Eastern Tibet (Czaja 2013: 91). Jampal Dorje describes *ngülchu* as 'an essence supreme in pacifying all diseases and evil spirits' (Stainless Rosary 40/7), which confirms a certain therapeutic 'preciousness.' Post 1959, Gawa Dorje chose to order substances alphabetically within each category, creating a 'modern' ground of indexing, which neutralizes any socio-historical networks that might have played a role in the earlier sequencing of substances.

What do the ways in which Tibetan medical authors classified *ngülchu* over time tell us about their understanding of the substance? The mention of multiple categories and sourcing should make us aware that the identification and categorization of *ngülchu* were not always straightforward; multiple classifications coexisted in wider contexts of *ngülchu* being both a 'meltable precious medicine' and a 'cold inanimate poison.' Categorization probably depended on its perceived qualities and usage, probably historical movements of recipes and substances, as well as the varying popularity of *ngülchu*-containing products, whether in the form of medicine, poisons, or used in religious statues.

The remaining part of the chapter analyzes what the numerous sources of *ngülchu* tell us about Tibetan approaches to resource substances.

SOURCES OF *NGÜLCHU*

As mentioned above, the knowledge of obtaining liquid mercury from cinnabar rocks existed in Tibet at least in written form at the time of the *Four Treatises*. But, there were other sources of *ngülchu*, a comprehensive survey of which is beyond the scope of this chapter. However, the examples presented below suffice to show that *ngülchu* was known to be sourced from substances that question our understanding of sourcing 'metals.' Over time most of these sourcing practices died out, but the ways in which Tibetan authors wrote about them in the past and today raises questions on how to understand and translate *ngülchu*.

What might be termed a 'source' (*'byung ba*) in some cases might also overlap with what is considered a 'substitute' (*tshabs*). Tibetan medical substitution practices often cross 'kinds': e.g. animal and metal substances can easily be substituted by plants. This has provided greater flexibility across time and geographical spaces with varied natural sources, and has allowed physicians to make their medicines locally following a recipe based on the *materia medica* of a different region in Tibet. Since substitution depends on the Tibetan epistemological principles of quality, taste and powers, an animal part can be substituted with a herb if both are similar in taste and power, and thus are considered to have the same potency (see Blaikie 2015; Sabernig 2011).

Around the end of the seventeenth century, a famous physician at the court of the Fifth Dalai Lama, Darmo Menrampa Lobsang Chödrak (Dar mo sman rams pa Blo bzang chos grags; 1638–1710), presents the following sources of *ngülchu* in the *Golden Annotations* (*Gser mchan rnam bkra gan mdzod*; quoted from Dawa Ridak 2003: 413/24–27):[22]

> Then people are saying (*grags so*) on the origins [of *ngul chu*] about sourcing it from springs (*'byung khung*), old felt (?) (*'ching bu*),[23] from rocks (*rdo*), plants (*sngo*), and animals (*srog chags*). In fact, the superior type comes from the mountains [= cinnabar ore], the one sourced from a spring is considered inferior.

Darmo Menrampa himself was not personally familiar with all the sources of *ngülchu* that he lists here, except cinnabar. The phrase 'people are saying' (*grags so*) implies that he has no direct experience of these sources, but that at the same time he does not discredit them.[24] Considering his erudite scholarship and investigative attitude to medicine (Gyatso 2004),

22 TBRC resource ID: W1GS2. I have not yet located this quote in the more than 1000-page work.

23 Also *mching bu*, which actually refers to a trinket or inexpensive jewel (THL 2010). Some Tibetan physicians thought *'ching bu* to be a misspelling of *phying pa*, which means felt. It might thus refer to an old felt blanket, but this is not entirely certain.

24 Personal communication, Dr Tenzin Thaye, Dharamsala, May 2015.

he probably would have voiced a critique if he did not believe these to be valid sources of *ngülchu*.[25]

One could speculate about the reasons behind an apparently more widespread knowledge of *ngülchu* sources beyond medical circles. Maybe it was linked to Buddhist statue making, or to valued medicines. With the reputation of not only being the source of an elixir, but a powerful protection from evil and all kinds of poisoning (Czaja 2013: 84), perhaps people were looking for more common, local and inexpensive sources of the metallic liquid. During the second half of the seventeenth century the therapeutic use of *ngülchu* became more popular in central Tibet. The knowledge of how to make prestigious mercury sulfide-containing precious pills had only just been introduced to representatives of the Gelukpa School and the Fifth Dalai Lama's Government in Lhasa, and the precious ingredients needed to manufacture them had to be gathered from all over, at times from beyond Tibet's borders (ibid.: 84). Their complex and expensive manufacturing processes meant they were largely a medicine for the elite (Gerke 2013). Darmo Menrampa Lobsang Chödrak himself conducted a major mercury processing event in Lhasa in 1678, and was trained in the official lineage that had been passed on by Drigung physicians (Czaja 2013: 83). He was trained in the *tsotel* techniques, which most probably used liquid metallic mercury from cinnabar or vermilion. Further research is required to establish what links might exist between the popularity and spread of precious pills and the occurrence of multiple *ngülchu* sources in the medical texts.

In 1727, in eastern Tibet, Tenzin Phuntsok lists as sources of *ngülchu* the following substances, which partly overlap with those listed by Darmo Menrampa:

> [*Ngülchu* can be sourced from any] type of cinnabar from Tibet or China, whichever is suitable, and silver ore (*rdo dngul*), which is like a rock and when burnt [*ngülchu*] dissipates into smoke; the rock *pawalongbu* (*pha wang long bu*),[26] human sweat (*mi'i rngul*), filthy garments (*dreg can gyi gos*), and clothes of a corpse drenched by rain (*ro gos char zag song ba*), yak fabric drenched in rain on a nomad's camping place (*mtsher shul gyi re ba char zag song ba*), the plant *ne'u* (*sne'u*), a pigeon (*bya phug ron*),[27] and a black scorpion (*'bu sdig pa nag po*). (Crystal Rosary 106/18–107/2)

25 Personal communication, Dr Tenzin Thaye, Dharamsala, May 2015.

26 *Pawalongbu* (also *pha bang long bu*) is one of the 'eight elements,' most probably pyrite, which is an iron sulphide, FeS_2 (Pasang Yontan 1998: 139). It is also identified as galena, which is lead sulfide, PbS (THL 2010).

27 Gurupel in eighteenth century eastern Tibet mentions in his *Essential Drop of Nectar* (*Bdud rtsi'i thig le* 7/4) 'pigeon feathers' (*bya phug ron gyi sgro*).

In another section of the *Crystal Rosary*, Tenzin Phuntsok mentions ten sources from which to obtain *ngülchu*:

> Silver stone (*dngul rdo*), cinnabar, *pawalongbu*, realgar (*ldong ros*),[28] *ne'u*, dirty sweat (*rngul dreg*), worn-out cloth (*re hrul*), garments of corpses (*ro gos*), pigeons (*phug ron*), scorpions (*sdig pa*). (ibid. 576/7–9)

Tenzin Phuntsok presents the above-mentioned substances as valid sources of *ngülchu*, all of which he roughly summarizes into the three groups of rocks, animals and plant sources (skipping the others). He writes: 'whichever source it is, if it is white and mobile it is wild, if it is blue and slowly moving, it is mild' (Crystal Rosary 107/4–5). The contemporary author Dawa Ridak lists several of Tenzin Phuntsok's sources, but divides all these sources simply into 'natural' and 'artificial' types of *ngülchu*, which come either from ores, animals or plants:

> The nature (*ngo bo*) [of *ngülchu*] is two-fold: arising naturally (*rang 'byung*) and being artificially processed (*bcos ma*). There are three [ways to] collect [it]: sourced from ores (*gter dngos*), from animals (*srog chags*), and from plants (*sngo ldum*). ... 'the one appearing from the ocean is inferior, in fact, the [type] from the mountains is the best.' (Dawa Ridak 2003: 414/4–6)

He comments that nowadays *ngülchu* is mainly sourced from rocks, but that the other sources exist as well. He also makes the distinction between the 'wild' and the 'mild' type (ibid.: 414/7–11).

Other contemporary authors introduce different classifications. PRC-based Sonam Bakdrö, for example, classifies *ngülchu* sources as mainly derived from ores. Additional sources are (1) 'old felt or clothes' (*'ching bu'am gyon chas*), specifically clothes filthy with human sweat, rain-soaked garments of a corpse and yak fabric drenched in rain on a nomad's camping place; (2) from 'living things' (*skye dngos*), with the example of the plant *ne'u* and garlic; and (3) from 'animals' (*srog chags*), mentioning pigeons and black scorpions. Sonam Bakdrö notes that these sourcing practices have died out but he does not critique them as invalid (2006: 68/14–21). Both Jampal Dorje's *Stainless Rosary* and Gawa Dorje's *Crystal Garland* maintain that the main source of *ngülchu* is cinnabar. Gawa Dorje states that all other sources 'are explicitly quoted nowadays, [but] the traditions have vanished' (1995: 42/20–21).

28 Gurupel suggests that there are many types of rocks from which *ngülchu* can be sourced, like all types of realgar and others (*ldong ros rnams kyi rdo bzhu rgyu sogs mang zhing*; Essential Drop of Nectar 7/3).

Most of the Tibetan physicians I interviewed did not know much about how to source *ngülchu* from animals and plants, etc., but most believed in its possibility and had heard about these sources. Scorpions apparently needed to be boiled for a long time to secure some drops of *ngülchu*, which they might have consumed from rocks and soil. 'Dirty sweat' was considered a possible source, since spending months without washing, which often happened in the harsh climate of the Tibetan plateau, might lead to all kinds of semi-liquids arising from skin folds, including silvery ones, considered to be *ngülchu*.

The most well-known alternative source of *ngülchu* is the plant *ne'u*, which in itself is not considered poisonous in small quantities, but is considered to have a warming nature, oily and heavy qualities, and sweet and hot tastes. The classical description in the *Crystal Rosary*, which is also cited in modern plant monographs (Kletter and Kriechbaum 2001: 240–45), explains that the entire plant is used in medicines to induce sweating and to bring down fevers (Crystal Rosary 375/10–11). Tibetans also consume it as a green vegetable inside dumplings, although overconsumption might lead to diarrhea.[29] The 'blue Tibetan' type secretes 'nectar dew' on the underside of the leaves (Crystal Rosary 376/2; Kletter and Kriechbaum 2001: 240). We also find a distinction between 'Tibetan wild' and 'Tibetan mild' types of *ne'u*, of which the latter 'releases a colourless liquid when rubbed' (ibid.).[30]

According to Kletter and Kriechbaum (2001: 240), *ne'u* is one of the 150 *Chenopodium* species worldwide, with large morphological variations, of which seven species occur in Tibet. Within Tenzin Phuntsok's nomenclature, *ne'u* is classified specifically under 'the category of those plants where the leaves, stalks, flowers, and fruits need to be gathered together simultaneously' (*lo sdong me 'bras lhan gcig tu btu bar 'os pa'i rim pa bstan pa*; Crystal Rosary XVIII). Tenzin Phuntsok does not describe it as a source of *ngülchu* in this section, and we can only guess that seeing the 'nectar dew' or holding a leaf 'releasing a liquid when rubbed' provided certain characteristics that might have caused physicians to consider it as a source of *ngülchu*. In his modern Tibetan medical dictionary, Pasang Yontan Arya writes that *ne'u* 'is harmful to the eyes because it contains mercury' (Pasang Yontan 1998: 124). It is known that plants can accumulate varying amounts of heavy

29 Personal communication, Dr Tenzin Thaye, Dharamsala, May 2015.

30 This translates more literally as 'If the Tibetan [type of] *ne'u* is rubbed, [it has] the nature of water' (*bod sne bsrubs na chu'i rang bzhin no*; Crystal Rosary 376/6).

metals from the soil and might contain them in certain concentrations; *Chenopodium* species are among these.[31]

When I discussed *ne'u* with Tibetan physicians during a Tibetan medical workshop in Kathmandu in December 2011, I was told that when *ne'u* is processed, it produces a silvery liquid, which they call *ngülchu*. One senior physician described the method as follows:

> From small plants you cannot get *ngülchu*; the plant should be big and the leaves very shiny. You have to put *ne'u* inside a stone-lined hole in the ground, add some water and keep it exposed to sunlight for two months. Then, once you remove the material, small quantities of *ngülchu* will be at the bottom of the hole. This is *ngülchu*. It is poisonous and is the same as *ngülchu*, but the quantity you can get this way is very little. I have heard of doctors doing this, but I have never done it myself.[32]

Other contemporary Tibetan physicians I interviewed in India and Nepal between 2011 and 2013 confirmed that *ne'u* could be a valid plant source for *ngülchu*, but producing only a little. Doctors prefer the liquid mercury readily available in chemist's shops. A senior pharmacist at the main Tibetan medical institute, the Men-Tsee-Khang, in Dharamsala, India, considered *ne'u* a substitute for *ngülchu* in case liquid mercury became difficult to obtain if the global UN ban on mercury affected traditional medical practices in India. The plant *dambukara* (*'dam bu ka ra*) is known to be an actual substitute for *ngülchu* because of its similar properties, but it is considered weaker in potency.[33]

CONCLUSION: WHEN *NGÜLCHU* IS NOT MERCURY

Tibetan medical practitioners used and still use liquid metallic mercury – which they call *ngülchu* – in processed forms, largely as mercury sulfide, in some of their medicines. Its classifications in various medical texts from the twelfth to eighteenth century reveal parallel sets of categories and

31 See Bhargava et al. 2008: 111, who suggest that 'Chenopodium spp. have the ability to accumulate large quantities of heavy metals in the leaf tissues even when they are present in low concentrations in the soil,' based on the assessment of six heavy metals (excluding mercury). Thanks to Jan van der Valk for pointing me to this reference.

32 Personal communication, Dr Tsultrim Gyatso, Kathmandu, December 2011.

33 Personal communication, Dr Tenzin Thaye, Dharamsala, May 2015. So far, I have been unable to find any textual reference for this statement.

networks of relationships with certain ranks and positions in different contexts. For example, in the context of poisons, we find the hierarchy of 'animate' and 'inanimate,' the latter of which breaks down into 'cold' and 'hot' types, among which *ngülchu* finds its place as a 'cold inanimate poison.' In the context of *materia medica* substances, the category of 'precious' substances divides into 'meltable' and 'non-meltable' without further hierarchical subdivisions. In the *Crystal Rosary*, *ngülchu* is simply a 'meltable precious medicine' among seventeen other metals, although its positioning among these substances changed over time, which might or might not present a difference in the priority that Tibetan authors attributed to it. The Tibetan term 'meltable' (*bzhu ba*) has a broader alchemical meaning that is generally attributed to 'metals'; it also means 'digesting' and refers to substances that can be processed, amalgamated and fused. In classical Tibetan medical literature, the term 'metal' does not appear as a name of a category; 'meltable' and 'precious' seem to have fit the categorizations of metals more suitably in Tibetan thought. These varying sets of classification practices of *ngülchu* raise broader questions on what a 'substance' is actually comprised of in Asian medical contexts. We need to pay attention to the complex relational networks that develop around a substance and influence its classifications and sourcing over time to make sense of the specific order and the grounds on which such order was created. The ways in which a substance in certain cases has been classified and sourced might call into question our own definition of categories and 'kinds.'

I have discussed a variety of *ngülchu* sources based on several Tibetan pharmacopoeias. Contemporary medical authors still list these varied sources of *ngülchu*, even though most of the sourcing practices have clearly disappeared. Contemporary Tibetan physicians I spoke with believe all these sources to provide 'real *ngülchu*,' but regret not having received any orally transmitted practical knowledge (*lag len*) on these sourcing practices. None of the contemporary authors, neither from the PRC nor from India, condemn the numerous sources as invalid, but they acknowledge the superiority of cinnabar in the sourcing of *ngülchu*. While today *ngülchu* seems to refer exclusively to metallic mercury (Hg), in the past this might not have been always the case.

Even though we still do not know when these various sources of *ngülchu* first entered the literature and why, the point I wish to make here is that in Tibetan medical thought, a medical substance such as *ngülchu* is believed to be found in or processed from substances across different 'kinds,' including animals, herbs, human sweat and various types of rotten cloth. Clearly, all Tibetan physicians I spoke with consider *ngülchu* to be a metal, but one that can be sourced across different 'kinds.' This is unusual, even within Tibetan approaches to substances. Notably, even if sourced across 'kinds,' all types of *ngülchu* are ranked according to their 'nature' (*rang bzhin*) of either 'wild' or 'mild,' which in this case is determined by color and mobility. Interestingly, these features are those that visibly change when Tibetan physicians process mercury, so their direct experience of processing a substance impacts ways of classification.

Ngülchu holds a special place in Tibetan pharmacopoeias, perhaps because of its changeability, its shiny qualities and quick movements as the only metal liquid at room temperature, its poisonous nature, and the numerous origin myths attached to it – in brief, the relationship that people have had with it as a substance. Understanding such relationships is key to understanding classifications, argues Ralph Bulmer, who while studying zoological taxonomies among the Karam of the New Guinea Highlands asked the question: 'Why is the cassowary not a bird?' and came up with the answer that the cassowary is classified differently because it 'enjoys a unique relationship' with the Karam (1967: 5). Cultural contexts have been emphasized by many ethnobotanists studying 'general and special purpose' taxonomies (e.g. Ellen and Reason 1979). In Tibetan medical thought, at least in the past, *ngülchu* has been more than a 'metal.' It is a poison, a precious substance that can be transformed into a potent elixir, and sourced from animals, plants, rocks and other substances that were available in the immediate surroundings of a nomadic life on the High Plateau. That garments of corpses, yak fabric and cloth had to be rotten, wet, and/or filthy to become a valid source of *ngülchu* might also be linked to perceived networks between poison, dirt and pollution.

Since none of the Tibetan physicians I spoke with had ever experienced sourcing *ngülchu* from anything other than cinnabar or the bottle of the heavy liquid bought from a chemist's shop, it is difficult to determine whether these are different sources of the same substance (liquid mercury) or substitutions for it. We can at least conclude that it has similarities to the practice of substitution, which is very widespread in Tibetan medicine compounding (*sman sbyor*), even though the lists of synonyms can make it difficult to establish when it is a synonym and when a substitute. The plant *ne'u* was explained to me as both a source of and a substitute for *ngülchu*, which makes me conclude that such categories can easily overlap.

While it would be interesting to know whether *ne'u* contains or can attract unusual quantities of mercury (Hg), perhaps our questions should not come from a biological realism perspective of equivalence but acknowledge the culturally distinct ways in which traditional medical practitioners in Asia identify and think about substances. Without implying that all natural-kind classifications are arbitrary or culture-specific, *ngülchu* as a 'precious' substance appears much broader in its definition and can cross the socially constructed barriers of modern 'kingdoms' simply because the nature and quality that it is identified by seem not to be limited to the world classified by such categories. If we are struck by the 'exotic charm of another system of thought' of Foucault's opening quote when we read that very shiny leaves, filthy clothes, the garments of a corpse or the yak fabric drenched in rain at a nomad's camping place, a pigeon and a black scorpion can all be sources of *ngülchu*, perhaps that charm is based on 'the limitation of our own' system of thought. In certain cases we thus have to rethink our methodological approaches when dealing with Asian medical substances and their classifications and find ways to study substances in classical Tibetan texts that can include but are not limited to identifying equivalents. Borges is of inspiration here with his mind-opening short story on classifications and his conclusion that 'it is clear that there is no classification of the Universe not being arbitrary and full of conjectures.'[34]

ABBREVIATIONS AND REFERENCES

Tib. = Tibetan

Adams et al. 2011	Adams, V., M. Schrempf and S. Craig. (2011). *Medicine between Science and Religion. Explorations on Tibetan Grounds*. Oxford, New York: Berghahn Books.
Bhargava et al. 2008	Bhargava, A. et al. (2008). 'Chenopodium: a Prospective Plant for Phytoextraction.' *Acta Physiologiae Plantarum*, 30(1): 111–20.
Blaikie 2015	Blaikie, C. (2015). 'Wish-fulfilling Jewel Pills: Tibetan Medicines from Exclusivity to Ubiquity.' *Anthropology & Medicine*, 22(1): 7–22.
Boesi 2005/2006	Boesi, A. (2005/2006). 'Plant Categories and Types in Tibetan Materia Medica.' *The Tibet Journal*, Vol. xxx No. 4 (Summer 2005) and Vol. xxxi No. 1 (Spring 2006): 67–92.
Boesi 2007	Boesi, A. (2007). 'The Nature of Tibetan Plant Nomenclature.' *The Tibet Journal*, 32(1): 3–28.
Boesi and Cardi 2006	Boesi, A. and F. Cardi (2006). 'Tibetan Herbal Medicine: Traditional Classification and Utilization of Natural Products in Tibetan Materia Medica.' *HerbalGram (The Journal of the American Botanical Council)*, 71: 38–48.

34 See 'The analytical language of John Wilkins' at: https://ccrma.stanford.edu/courses/155/assignment/ex1/Borges.pdf (accessed 23 September 2015).

Bulmer 1967	Bulmer, R. (1967). 'Why is the Cassowary not a Bird? A Problem of Zoological Taxonomy among the Karam of the New Guinea Highlands.' *Man*, New Series, 2(1): 5–25.
Clark 1995	Clark, B. (1995). *The Quintessence Tantras of Tibetan Medicine.* Ithaca: Snow Lion Publications.
Crystal Rosary	Deumar Geshe Tenzin Phuntsok (De'u dmar Dge bshes Bstan 'dzin phun tshogs, b. 1672) (2009). *Shel gong shel phreng.* Dharamsala: Men-Tsee-Khang.
Czaja 2013	Czaja, O. (2013). 'On the History of Refining Mercury in Tibetan Medicine.' *Asian Medicine*, 8(1): 5–105.
Dawa Ridak 2003	Dawa Ridak (Zla ba ri brag) (2003). *Bod kyi gso ba rig pa las sman rdzas sbyor bzo'i lag len gsang sgo 'byed pa'i lde mig*. Delhi: Rig Drag Publications.
Drungtso and Drungtso 2005	Drungtso, Tsering Thakchoe and Drungtso, Tsering Dolma. (2005). *Tibetan-English Dictionary of Tibetan Medicine and Astrology.* Dharamsala: Drungtso Publications.
Ellen and Reason 1979	Ellen, R.F. and D. Reason. (1979). *Classifications in Their Social Context.* San Francisco: Academic Press.
Essential Drop of Nectar	Dege Drungyig Gurupel (Sde dge Drung yig Gu ru 'phel, fl. 18th century) (1985). *Srid gsum Gtsug rgyan Si tu Chos kyi 'byung gnas kyi zhal lung dngul chu btso chen ril bu'i sbyor sde zab bdun bdud rtsi'i thig le. Instructions for the Preparation of Mercury Pellets According to the Teachings of the Great Si tu Pan chen Chos kyi 'Byung gnas/ by Sde dge Drung yig Gu ru 'phel* (Vol. 139). Leh, Ladakh: T. Sonam and D.L. Tashigang.
Foucault 2002	Foucault, M. (2002). *The Order of Things. An Archaeology of the Human Sciences.* London: Routledge.
Four Treatises	Yuthok Yonten Gonpo (G.yu thog Yon tan mgon po, fl. 12th century) (1982). *Bdud rtsi snying po yan lag brgyad pa gsang ba man ngag gi rgyud.* Lhasa: Bod ljongs mi dmangs dpe skrun khang.
Gawa Dorje 1995	Gawa Dorje (Dga' ba'i rdo rje) (1995). *'Khrungs dpe dri med shel gyi me long.* Pe cin: Mi rigs dpe skrun khang.
Gawa Dorje 2009	Gawa Dorje (Dga' ba'i rdo rje) (2009). 'An Investigation into the Advisability of Translating Names of Tibetan Medicine into other Languages.' *Asian Medicine*, 5(2): 394–406.
Gerke 2010	Gerke, B. (2010). 'Correlating Biomedical and Tibetan Medical Terms in Amchi Medical Practice.' In Adams, V., Schrempf, M. and S. Craig (eds.), *Medicine between Science and Religion. Explorations on Tibetan Grounds*, pp. 127–52. Oxford, New York: Berghahn Books.
Gerke 2012	Gerke, B. (2012). *Long Lives and Untimely Deaths. Life-span Concepts and Longevity Practices among Tibetans in the Darjeeling Hills, India.* Leiden, Boston: Brill.
Gerke 2013	Gerke, B. (2013). 'The Social Life of *Tsotel*: Processing Mercury in Contemporary Tibetan Medicine.' *Asian Medicine*, 8(1): 120–52.

Gerke in press — Gerke, B. (in press). 'Buddhist Healing and Taming in Tibet: Ritualised Pharmacology.' In Jerryson, M. (ed.), *Oxford Handbook of Contemporary Buddhism*. New York: Oxford University Press.

Glover 2005 — Glover, D.M. (2005). *Up from the Roots: Contextualizing Medicine Plant Classifications of Tibetan Doctors in Rgyalthang, PRC.* PhD Dissertation, Department of Anthropology, University of Washington, Seattle.

Glover 2010 — Glover, D.M. (2010). 'Classes in the Classics: Historical Changes in Plant Classification in two Tibetan Medical Texts.' In Craig, S., M. Cuomu, F. Garrett and M. Schrempf (eds.), *Studies of Medical Pluralism in Tibetan History and Society (Proceedings of the 11th Seminar of the International Association for Tibetan Studies, Bonn 2006)*, pp. 255–77. Andiast, Switzerland: International Institute for Tibetan and Buddhist Studies GmbH.

Golden Annotations — Darmo Menrampa Lobsang Chödrak (Dar mo sman rams pa Blo bzang chos grags, 1638–1710) (2006). *Gser mchan rnam bkra gan mdzod*. Pe cin: Mi rigs dpe skrun khang.

Gould-Martin 1978 — Gould-Martin, K. (1978). 'Hot Cold Clean Poison and Dirt: Chinese Folk Medical Categories.' *Social Science & Medicine*, 12(1b): 39–46.

Gyatso 2004 — Gyatso, J. (2004). 'The Authority of Empiricism and the Empiricism of Authority: Medicine and Buddhism in Tibet on the Eve of Modernity.' *Comparative Studies of South Asia, Africa and the Middle East*, 24(2): 83–96.

Hofer 2014 — Hofer, T. (2014). 'Illustrated Materia Medica Prints, Manuscripts, and Modern Books.' In Hofer, T. (ed.), *Bodies in Balance. The Art of Tibetan Medicine*, pp. 226–45. Seattle and London: The Rubin Museum of Art, New York in association with University of Washington Press.

Hsu 2014 — Hsu, E. (2014). 'How Techniques of Herbal Drug Preparation Affect the Therapeutic Outcome: Reflections on Qinghao 青蒿 (Herba Artemisiae annuae) in the History of the Chinese Materia Medica.' In: Aftab, T., J.F.S. Ferreira, M.M.A. Khan and M. Naeem (eds.), *Artemisia Annua - Pharmacology and Biotechnology*, pp. 1–8. Heidelberg: Springer.

Hsu 2015 — Hsu, E. (2015). 'Towards a Phenomenology of Language: The Sound of Lightness/Qing in Chinese Medicine.' Unpublished paper presented at 'MAGic2015: Anthropology and Global Health: Interrogating Theory, Policy, and Practice,' University of Sussex (9–11 September 2015).

Jampa Trinlé 2006 — Jampa Trinlé (Byams pa 'phrin las) (ed.) and Bod rang skyong ljongs sman rtsis khang (2006). *Bod lugs gso rig tshig mdzod chen mo*. Pe cin: Mi rigs dpe skrun khang.

Kletter and Kriechbaum 2001 — Kletter, C. and M. Kriechbaum. (2001). *Tibetan Medicinal Plants.* Stuttgart: Medpharm Scientific Publishers.

Lo Bue 1991 — Lo Bue, E. (1991). 'Statuary Metals in Tibet and the Himalayas: History, Tradition and Modern Use.' In Oddy, W.A. and W. Zwalf (eds.), *Aspects of Tibetan Metallurgy*, pp. 33–69. London: Research Laboratory and Department of Oriental Antiquities.

Men-Tsee-Khang 2011$_a$ Men-Tsee-Khang (tr.). (2011). *The Root Tantra and the Explanatory Tantra from the Four Treatises of Tibetan Medicine*. Dharamsala: Men-Tsee-Khang.

Men-Tsee-Khang 2011$_b$ Men-Tsee-Khang (2011). *The Subsequent Tantra from the Four Treatises of Tibetan Medicine*. Dharamsala: Men-Tsee-Khang.

Nappi 2010 Nappi, C. (2010). 'Winter Worm, Summer Grass: Cordyceps, Colonial Chinese Medicine, and the Formation of Historical Objects.' In Digby, A., E. Waltraud and P.B. Mukharji (eds.), *Crossing Colonial Historiographies. Histories of Colonial and Indigenous Medicines in Transnational Perspective*, pp. 21–35. Newcastle upon Tyne: Cambridge Scholars Publishing.

Needham et al. 1976 Needham, J., H. Ping-Yu and L. Gwei-Djen. (1976). *Science and Civilization in China. Volume 5. Part 3: Spagyrical Discovery and Invention: Historical Survey from Cinnabar Elixirs to Synthetic Insulin*. Cambridge: Cambridge University Press.

Oddy and Zwalf 1991 Oddy, W.A. and W. Zwalf (eds.) (1991). *Aspects of Tibetan Metallurgy*. London: Research Laboratory and Department of Oriental Antiquities.

Pasang Yontan 1998 Pasang Yontan Arya. (1998). *Dictionary of Materia Medica*. Delhi: Motilal Banarsidass.

Ploberger 2012 Ploberger, F. (2012). *Wurzeltantra und Tantra der Erklärungen aus 'Die vier Tantra der Tibetischen Medizin.'* Schiedlberg, Austria: Bacopa.

Pordié 2002 Pordié, L. (2002). 'Pharmacopoeia as an Expression of Society: A Himalayan Study.' In Fleurentin, J., Pelt, J.-M., and G. Mazars (eds.), *From the Sources of Knowledge to the Medicines of the Future*, pp. 195–204. Proceedings of the 4th European Congress on Ethnopharmocology. Paris: French Society for Ethnopharmacology, Research Institute for Development (IRD).

Sabernig 2011 Sabernig, K.A. (2011). 'The Substitution of Rare Ingredients in Traditional Tibetan Medicine on the Basis of Classical Tibetan Texts, Their Use in Modern Formularies, and a Case Study from Amdo/Qinghai.' *Curare* 34 (1 and 2): 83–96.

Simioli 2013 Simioli, C. (2013). 'Alchemical Gold and the Pursuit of the Mercurial Elixir: An Analysis of Two Alchemical Treatises from the Tibetan Buddhist Canon.' *Asian Medicine*, 8(1): 41–47.

Sonam Bakdrö 2006 Sonam Bakdrö (Bsod nams bag dros). (2006). *Dngul chu'i byung ba spyi dang bye bgrag btso bkru rig pa'i lag len rgyas par bkral ba mkhas grub rtna shi'i dgongs rgyan*. Lha sa: Bod ljongs mi dmangs dpe skrun khang.

Sonam Dolma 2013 Sonam Dolma (2013). Understanding Ideas of Toxicity in Tibetan Medical Processing of Mercury. *Asian Medicine*, 8(1): 106–19.

Stainless Rosary Jampal Dorje ('Jam dpal rdo rje, fl. 19th century) (1971). *Dri med shel phreng nas bshad pa'i sman gyi 'khrungs dpe mdzes mtshar mig rgyan (An illustrated Tibeto-Mongolian materia medica of Ayurveda of 'Jam-dpal-rdo-rje of Mongolia).* New Delhi: International Academy of Indian Culture. A 1911 print version of the manuscript is available online: http://www.wdl.org/en/item/13514/view/1/1/ (accessed 9 October 2015).

TBRC — Tibetan Buddhist Resource Centre, https://www.tbrc.org (accessed 9 October 2015).

THL — Tibetan to English Translation Tool. (2010). The Tibetan and Himalayan Library, *http://www.thlib.org/reference/dictionaries/tibetan-dictionary/translate.php* (accessed 25 September 2015).

Turner 1991 — Turner, S. (1991 [1800]). *Account of an Embassy to the Court of the Teshoo Lama in Tibet. Containing a Narrative of a Journey Through Bootan and Part of Tibet.* Reprint. New Delhi: Asian Educational Services.

White 1996 — White, D.G. (1996). *The Alchemical Body. Siddha Traditions in Medieval India.* Chicago and London: The University of Chicago Press.

Wujastyk 2013 — Wujastyk, D. (2013). Perfect Medicine: Mercury in Sanskrit Medical Literature. *Asian Medicine* 8(1): 15–40.

Section Three

Power and Devotion

Chapter 7

In Search of the *Sādhu*'s Stone: Metals and Gems as Therapeutic Technologies of Transformation in Vernacular Asceticism in North India

ANTOINETTE E. DeNAPOLI[1]

Clay, wood, metals and gems are the heart and soul of India.
We must respect them and use their power wisely.
(Bhuvneshwari Puri Guru Ma, Rajasthan, 2013)

INTRODUCTION: STONE THERAPY MEDICINE IN EVERYDAY RENUNCIANT PRAXIS

India has long been touted as a land rich in cotton, indigo, silks, spices and textiles. It is also known for, as this anthology documents, its invaluable resources of gemstones, mineral ores and metals. It is not surprising that when the British Parliament declared Queen Victoria's (r. 1837–1901) direct rule over India in 1858, because of its natural wealth and the enormous economic profits that the British East India Trading Company was able to reap by exploiting that wealth, India became known as the 'Jewel in the Crown' of the British Empire (or Raj).[2] The famous diamond known

1 Antoinette DeNapoli is an Associate Professor in the Department of Religious Studies at the University of Wyoming (USA).
All translations are the author's and have been completed with the kind assistance of Vanita Ojha. All interviews have been conducted in Hindi and, occasionally, in Mewari.

2 The characterizing of British (or colonial) India as the 'Jewel in the Crown' is attributed to Benjamin Disraeli. The jewel to which this colonial idiom refers is the diamond, which has been mined in India since the fourth century BCE.

as Koh-i-Noor (meaning Mountain of Light), which belonged to several ruling dynasties over the centuries until it was acquired by the British in 1849, featured as the central jewel in the Queen Mother's Imperial Crown of India, which was made in 1937 for her coronation as Queen Consort Elizabeth, the wife of King George VI. The Koh-i-Noor has been characterized as the 'Heart of India,' and is one of the largest diamonds in the world.[3]

The epigraph that frames this chapter similarly casts precious gems and metals, and clay and wood, as emblematic of the 'heart and soul' of India. A female Hindu renunciate (*sādhu*) by the name of Bhuvneshwari Puri (henceforth, Guru Ma) spoke these words on the night of a Fall Equinox (*Śarad Pūrṇimā*) festival, marking the calendrical transition from summer to fall, which was held at her *āśram* in the north Indian state of Rajasthan. On that evening, hundreds of her devotees (*bhakts*) came together for ritual worship, singing and storytelling, and celebrating the arrival of the new season. As we will see, the idea that stones, generally speaking, give expression to India's heart and soul carries more than symbolic weight. It also accents widely-held views of their 'real' power to heal the mind and body.

Sādhus like Guru Ma have renounced the world, meaning that they have left behind marriage and family, work and the ritual-social responsibilities that typically orient the everyday worlds of householders, to spend their lives rapt in the worship of God. Their radical way of life is known as renunciation (*saṃnyās*). While a variety of *sādhus* and *sādhu* traditions exist in contemporary India, *sādhus* often symbolize the spiritual virtuosos of a religiously pluralistic South Asia. Historically, they have played a crucial role in the teaching and transmission of Hinduism(s) throughout the subcontinent and in the diaspora. Moreover, many Hindus, whether uneducated or educated, rich or poor, high-caste or low-caste, rural or urban, local or transnational, see *sādhus* not only as powerful spiritual mediators between gods and humans, but also consult them for assistance in worldly and other-worldly matters and for ritual and/or medical healing. Thus, *sādhus* could be said to represent the heart and soul of vernacular Hinduism(s).

The extensive field research that I have conducted over the last fifteen years with *sādhus* from the Śaiva and Vaiṣṇava traditions[4] in Rajasthan,

3 For information on the Imperial Crown of India, see http://royalexhibitions.co.uk/crown-jewels-2/royal-regalia/.

4 Generally speaking, the Śaiva *sādhu* traditions ideally regard Śiva as their patron deity, while the Vaiṣṇava traditions regard Viṣṇu and/or his incarnations as

and more recently (2013–2014) in the states of Uttar Pradesh and Gujarat, and in the Union Territory of Dadra and Nagara Haveli,[5] suggests that in the context of the phenomenon which I have characterized as 'vernacular asceticism' – *saṃnyās* as it is imagined and practiced in local contexts (DeNapoli 2014$_a$) – the use of gems and metals for healing purposes demonstrates a vital, and yet underrepresented, aspect of everyday renunciant therapeutic praxis in northern and western India. While a few exceptions exist,[6] much of the academic work that discusses the healing practices of *sādhus* foregrounds their ritual therapies of meditation, yoga and prayer. Thus, this chapter's exploration of the therapeutic uses of gems and metals in vernacular asceticism fills a lacuna in the scholarship on renunciation in South Asia.

I examine here the value, significance and meanings that the *sādhus* (and their spiritual constituencies) attribute to the use of gems and metals to treat a range of illnesses (mental and physical), including the tried-and-true methods that both identify and illustrate a unique healing technique that many of them classify as 'stone therapy.' Following their lead, I refer to the *sādhus*' therapeutic practices as 'stone therapy medicine.' Stone therapy medicine as featured in vernacular asceticism brings to light a type of naturopathic (and ritualized) healing practice that is founded on the integrative principles of the traditional Hindu system of medicine known as *doṣa* (lit., the 'science of life'). The analyses I provide are drawn from ethnographic research methods that included participant observation at the *sādhus*' *āśrams* and temples, interviewing the *sādhus* and their *bhakts*, and documenting on cassette tape the *sādhus*' rhetorical, ritual and therapeutic practices. While I spoke with hundreds of *sādhus* at public gatherings, I worked closely with forty-nine of them whose uses and ideas about stone therapy medicine are representative of the larger *sādhu* population that I encountered.

---

their patron deity. In the context of Śaiva *saṃnyās*, I worked with *sādhus* who took initiation into either the Ādi Śaṅkarācārya Daśanāmi tradition or into the Gorakhnāth Kanpaṭha Yogi tradition, which is also commonly known as the Nāth tradition. The Vaiṣṇava *sādhus* with whom I worked came from the Tyāgī, Vairāgī and Sītā Rām traditions. See Gross (2002) for a detailed discussion on the theologies and practices of Śaiva and Vaiṣṇava traditions.

5 This Union Territory borders the states of Gujarat to the north, and Maharashtra to the south. While traveling in this part of western India with Guru Ma, I conducted fieldwork in its capital, Silvassa.

6 For a comprehensive study of the ritual and medical healing practices associated with Aghor *sādhus*, who specialize in treating stigmatized diseases, see Barrett (2008).

Approaching diseases of the mind, body and soul in a holistic way, stone therapy, according to the *sādhus*, qualifies as 'medicine' (*auṣadhi*), 'science' (*vijñān*) and a 'technology' (*takniq*) of *dharm*.[7] *Sādhus* like Guru Ma emphasize the notion that '*dharm* is the highest technology' and that 'science is the deepest *dharm*.' The post-Enlightenment and, predominantly, Western notion that religion designates an autonomous sphere of knowledge and practice distinguished from, and opposed to, those of medicine, science and technology holds no cultural currency for the majority of the *sādhus* whom I know. Challenging this dichotomy, many of them say that *dharm* and medicine constitute not only compatible, but more accurately, equivalent knowledge systems. Their representing of *dharm* as 'technology' suggests that stone therapy medicine similarly serves as a potent 'therapeutic technology' for contemporary India.[8]

Conceiving stone therapy medicine as 'a technology of *dharm*' has encouraged the *sādhus* to experiment with the dominant definitional boundaries of Hinduism in novel ways that take into consideration the transnational cultural productions, issues and challenges that have oriented the twenty-first century Indian milieu.[9] Their reinventing of *dharm* as a 'therapeutic

7 The term '*dharm*' in the practices of the *sādhus* and their communities signifies the idea of Hinduism as imagined in the classical Sanskrit texts and oral traditions. As we will see, this chapter spotlights the ways that the *sādhus* extend the definitional parameters of Hinduism beyond the more standard understandings of rituals, festivals, rules and texts to include in that category notions of, generally speaking, 'science,' 'medicine' and 'technology.'

8 My use of the concept of religion as a 'therapeutic technology' is drawn from Josephson (2013). In his analysis of the series of transformations that Buddhism underwent as Japan transitioned into the modern period and reworked conceptions of the modern over against Buddhism, Josephson argues that, in premodern contexts, 'Buddhism was [viewed as] medicine and empowered prayer was an important therapeutic technology.' The *sādhus*' idea of stone therapy medicine as empowered by *dharm* parallels nineteenth century Buddhist views of prayer and meditation as empowered by the Buddhist *dharma*.

9 As Nelson (2013) discusses, the transnational flows and exchanges of information, money, ideas, infrastructures, technologies and people have shaped the twenty-first century milieu in ways unlike any other historical period has with respect to cultural constructions of self and personhood, and ideas about agency and authority. He also observes that, citing philosopher Charles Taylor (2004), such processes have given rise to a 'crisis of identity,' which religions across the globe work to address and ameliorate within their religious communities, and in doing so, adapt their traditions as a means to create a sense of wholeness within the larger body politic. The transnational issues to which I refer, and against which the *sādhus* whom I know reconfigure their ideas of

technology of transformation' brings into sharp focus the emergence of a phenomenon which I have termed 'experimental Hinduism.' My use of this concept draws on analytic models of 'experimental religion' as discussed in the context of Ward's 2009 study of eighteenth century American Protestant Christianity, and Nelson's analysis of Buddhisms in contemporary Japan.[10] Experimental religion, generally speaking, describes the emphases that religions place on personal experience, scientific method, pragmatism, intentionality and beneficence as sources of authority for negotiating their conceptual parameters.[11] The concept also calls attention to the dynamic processes by which religions adapt their more 'official' positions as they grapple with, and make meaningful sense of, the dramatic sociocultural changes taking place around them in an epoch of unprecedented global shifts. Thus, experimental Hinduism spotlights the *sādhus*' representations of stone therapy medicine as a therapeutic technology of *dharm* in a creative (and urgent) response to shared perceptions that the current and/or nascent 'social imaginaries' of 'the modern' are pushing the sacred canopy of *dharm* to the periphery of a rapidly changing India.[12]

---

dharm, have to do with those of, as this chapter highlights, medicine, but also technoscience, human rights and feminism, human development and ecological sustainability. I discuss the ways that these and other cultural productions operate as newly emerging contexts for renunciant constructions of 'experimental Hinduism' in my current book project, tentatively titled, *Religion at the Crossroads: Experimental Hinduism and Social Imaginaries of the Modern in Twenty-first Century Transnational India*. I have talked about the context of technoscience in particular in DeNapoli 2014_b.

10 Nelson (2013) develops the concept of experimental religion, and 'experimental Buddhism,' specifically, on the basis of adaptations that the Japanese temple priests with whom he worked have forged in a cultural milieu in which Buddhism has to compete fiercely with multiple social ideologies and institutions, which have compromised its relevance in a country that prides itself on its secular national identity. A fascinating point to which Nelson also calls attention concerns the issue that some of the adaptations wrought by the priests (e.g. fashion shows in the name of Buddhist *dharma*) receive a lot of pushback from the centralized temple bureaucracies, which often promote visions of Buddhism that do not account for the sociocultural shifts happening in contemporary Japan.

11 See introductory remarks in Nelson (2013).

12 The notion of 'social imaginaries' has been developed by Charles Taylor (2004). According to him, across many world societies multiple notions of the modern (or what he refers to as 'modernities') grapple for authority in social constructions of identities, self, personhood and culture. Because cultures imagine and construct collective social life in a variety of competing and complementary

Speaking about India's cultural evolution from the time of Independence onward, Varma (2007) has argued that the period following the liberalization of the Indian economy in the early 1990s witnessed a unique watershed in the development of new (and competing) visions about the modern Indian nation state, which hinged in large part on Western-inspired neoliberal capitalist development models.[13] In addition, Buultjens (1988) has suggested that the socio-cultural changes stimulated by such economic models in the areas of industry and business, national politics and foreign policy, government planning, education, and science and technology have inspired the production and proliferation of the prevailing perception that India consists of two distinct and often opposing worlds. On the one hand, Buultjens contends, there exists the 'rural' India, which symbolizes the stronghold of 'old' cultural patterns and ancient traditions (i.e. *dharm*), and which resists change. On the other hand, there lives the 'urban' India, which not only thrives in the metropolises, but also stands for the 'newness of the nation' and, to that extent, the face of change, progress and innovation.[14]

The implications of Varma's and Buultjens' observations that I want to tease out here in relation to our topic are that 'the modern' and *dharm* as envisioned within standard (and secularizing) neoliberal frameworks constitute antithetical social imaginaries. Accordingly, the modern fields of technology, science and medicine advance change, whereas *dharm* stymies it. Furthermore, unlike *dharm*, which is thought to obscure the modern, the secular mediates it.

---

ways, Taylor refers to these multiple (and multifaceted) visions of life, which include notions of happiness and prosperity, as 'social imaginaries.' He concentrates on the cultural production of a distinct social imaginary, which he characterizes as the 'Western social imaginary,' with respect to its emphases on the notions of economy, the public sphere and self-governance. Taylor argues that these cultural productions have helped to give expression to a specific kind of (Western) modernity, which has monopolized a number of social imaginaries about the modern.

13 See also Buultjens (1988, 2015).

14 Only 20 percent of India's population lives in the cities. The majority of Indians, as Buultjens (1988) and Varma (2007) observe, live in villages. India has over 600,000 villages. It is important to note that some Hindu nationalist organizations, particularly as seen in some recent media-based RSS discourse as featured in newspapers like *The Hindu* and others, have invoked this discourse to construct 'rural' India as representative of the ancient 'Bhārat,' (i.e., 'religious India' and, to that extent, the 'authentic' India) and the 'urban' India as illustrating an increasingly secularized nation.

But the *sādhus* push back against such Cartesian dichotomies. For many of them, technology, medicine and science give expression to the pervasive power of *dharm* and reveal the sacred as much as the gods do. More significantly, they stress that because these knowledge systems arise and develop from the same kinds of insights, intentions and methods that they say can be traced to *dharm*, the latter both signifies and propels the modern. The *sādhus* further suggest that to live in a society in which the modern is distinguished from *dharm* is equivalent to living an emotionally 'broken' existence. Hence, the malaise that many of them associate with the current milieu relates to what they believe constitutes a life lived out-of-sync with *dharm*.

Thus, in this chapter, I argue that the practice of stone therapy medicine in vernacular asceticism creates and communicates notions about the 'fullness of life,' or as the *sādhus* say, 'complete living,' with respect to health, happiness, healing and, of course, the modern. I will show that, for the *sādhus*, the concept of complete living not only concerns actualizing the Āyurvedic principles of balance and harmony, but just as significantly, living in-sync with *dharm*. I will also show that this concept pivots on and reinforces renunciant constructions of *dharm* as a comprehensive technology for the modern age. To that end, analyzing the *sādhus*' 'rhetoric of renunciation,' that is, their religious teachings (*dharm-kathā*), devotional songs (*bhajans*) and personal experience narratives, I suggest that their representations of stone therapy medicine as a potent therapeutic technology of *dharm* foreground three motifs: stone therapy medicine as empowered by *dharm*; stones as 'conscious' agents of complete transformation; and stone therapy medicine as a beneficent technology for treating the malaise of the modern milieu. The emphases that the *sādhus* place on these themes illuminate an emerging ideology of wellbeing and the modern as inextricably tied to the salvific world of *dharm*.

Because the *sādhus*' ideas about the modern are refracted through their constructions of stone therapy medicine as a 'treatment for complete living,' I begin this chapter with an ethnographic scene: the *Śarad Pūrṇimā* festival, which occurred at Guru Ma's *āśram* on 18 October 2013, and which dramatizes in high-relief the ways in which the *sādhus* architect their ideas about the fullness of life through use of the authoritative lens of *dharm*. Although this festival serves as the backdrop for my analysis of stone therapy medicine as practiced in vernacular asceticism, I also weave into the discussion the views and practices of the other *sādhus*, in order to provide my readers a more 'complete' or representative *sādhu* perspective on the value/power of stone therapy medicine and the ideas about health, healing and the modern that it 'performs.' Let us now shift our focus to the *Śarad Pūrṇimā* event.

'LIVING IN SYNC WITH NATURE' – STONE THERAPY MEDICINE AS A TREATMENT FOR COMPLETE LIVING

THEME 1: STONE THERAPY MEDICINE AS EMPOWERED BY DHARM

The *āśram* resounds with the melodious chanting of the *Mahiṣāsuramardinī-stotra* (hymn to the Goddess Durgā as the Slayer of the Buffalo Demon) by Guru Ma's *bhakts* on this festival night of *Śarad Pūrṇimā.*[15] Sweetie, a college-aged female *bhakt*, and Pushpendra, a middle-aged widow with two grown sons who lives at Guru Ma's *āśram*, distribute copies of the *stotra* to members of the audience. Most of the *bhakts* gathered at the *āśram* this evening come from the high-caste (warrior) community of Marwari Rajputs. They constitute a well-educated (most of them, women and men, have college degrees and are employed as college professors, lawyers and businesspeople) and mostly middle-class constituency. While the majority of these *bhakts* live in Rajasthan, a handful of others reside in the adjoining states of Madhya Pradesh and Gujarat. Regardless of location, all of them have come to the *āśram* on this *Śarad Pūrṇimā* for the purpose of receiving the 'fullness' of Guru Ma's blessings and *darśan* (sacred sight). The term '*pūrṇimā*' means fullness, and in this context, 'Śarad Pūrṇimā' refers to the full harvest moon of the Autumnal Equinox. At this time, the moon is closest to the earth, and many Hindus, including those gathered at the *āśram*, perform special rituals (*pūjā*) on this night in order to draw down the moon's healing powers and energies and channel them into those aspects of one's life (physical, mental, spiritual) that require healing. Tonight, as I learn, Guru Ma and her *bhakts* will ritually invoke the moon's energies and celebrate life's fullness by practicing complete living.

On the other side of the *āśram*, adjacent to the Śiva (Bholenāth) temple, another group of *bhakts* prepares a delicious sweet dish of *kheer* (H. *khīr*, rice pudding) in a huge pot made of steel. The *kheer*, as Guru Ma explains, has to cook for four hours under the moonlight. By the time the *kheer*

15 In the *Devīmāhātmya*, the sacred text that contains the *Mahiṣamardinīstotra*, the goddess Durgā vanquishes the demon Mahiṣāsura and ushers in an era in which the qualities of truth, compassion and wisdom dominate, and *dharm*, which runs a hundred percent, is the 'true' sacred canopy.

begins to boil, she adds to the aromatic mixture twelve different herbs,[16] and large clear sugar crystals, the kind that are distributed as blessed food (*prasād*) in temples throughout India, and a mixture of dried fruits and nuts. Guru Ma then takes a bowl that contains pieces of silver metal foil and drops them one-by-one into the *kheer*. Silver metal foil is readily available at Indian grocery shops. It is safe for human consumption and is often used to decorate a variety of sweetmeats. Guru Ma's adding silver foil into the *kheer*, though, appears to have more than only an aesthetic benefit. My curiosity piqued, I ask her, 'Why is silver added to the *kheer*?' 'Silver is a neutral metal. It can be combined with any food and will not react badly with them,' she says. She continues, 'Other metals like aluminum and lead react negatively with food. They have lactic and citric acids. With some foods, the reaction can be poisonous. The *kheer* ought to be made in silver utensil. But it's impossible [for me] to make the *kheer* in a silver pot [that has the capacity to feed hundreds of people]. So, I add the silver [foil] to it instead.'

As she drops the last of the foil into the bubbling mixture, Guru Ma says, 'Silver is a cooling [*ṭhaṇḍā*] metal. It's perfect [to use] for the *kheer*. That, too, is a cooling food. And the moon is a cooling energy. The silver brings the moon's energy into the *kheer*. That's why we make *kheer* on *Śarad Pūrṇimā*. It treats the diseases that come with the autumn season.'

Hearing Guru Ma's explanation, it dawns on me that the making of *kheer* in the context of *Śarad Pūrṇimā* signifies an efficacious *pūjā* for Guru Ma and her *bhakts*. What is more, they see the ritual making of *kheer* on *Śarad Pūrṇimā* as akin to preparing medicine. That is, both ritual and medicinal healing practices converge in the context of this *Śarad Pūrṇimā* festival event. Guru Ma's representation of the *kheer* as a special 'treatment' for the autumn-related diseases not only distinguishes it from other types of *kheer* that are prepared in non-ritual contexts (and served as dessert), but also tacitly constructs the silver-infused *kheer* as a therapeutic technology of *dharm*. The medicinal properties of the *kheer* can be traced to its 'cooling' ingredients of rice and milk and to the cooling metal silver, which, as Guru Ma makes clear, attracts the moon's cooling energies into the *kheer*.

16 Guru Ma only gives the names of six of the twelve ingredients which she adds to the *kheer*. These are: *baṃślocan* (the silica produced in the joints of bamboo stems; *Bambusa bambos* [L.] Voss), *baṛī pīppal* (long pepper; *Piper longum* L.), *kamal patra* (leaves of lotus; *Nelumbo nucifera* Gaertn.), *kālī elāci* (black cardamom; *Alpinia oxyphylla* Miq.), *pīpal kī chāl* (bark of Pipal; *Ficus religiosa* L.), and *soṇṭh* (dry ginger powder; Skt. *śuṇṭhi*; *Zingiber officinale* Roscoe.).

According to her, the combination of cooling substances added into to the *kheer* increases the (cool) *vāta* and *kapha doṣas*, which represent the natural material energies associated with the elements of air and earth, respectively, and neutralizes the (warm) *pitta doṣā*, which embodies the fire element. Also, since autumn-related illnesses stem from an increase of *pitta doṣā*, they must be treated with cooling medicines.

Among the precious metals (e.g. copper, gold and mercury) that may be ingested for medicinal purposes, silver, in particular, as Guru Ma explains, has the capacity to treat *pitta*-related diseases, because only this metal can pull the moon's healing energies directly into the earth. The property of coolness shared by both silver and the moon evokes a 'like attracts like' response. But there is another reason for their magnetic attraction. It has to do with the notion that they have in common the deity, Candra Deva (lit., 'moon god'). Knowing which deities rule over which metals, gems, herbs and foods, illustrates a vital component of Āyurvedic stone therapy medicine. Its efficacy depends on the mastering of such specialized knowledge.[17]

Thus, the therapeutic silver-infused *kheer* balances out the three *doṣas* and brings about the healing of diseases that relate to the increase of the *pitta doṣa*, like asthma and the common cold and flu, which, as I am told, often spreads during seasonal transitions. Guru Ma says:

> Normally, during the fall season, the days are hot and the nights are cold. Due to the changes of hot and cold, a lot of diseases are born. To stop them, Āyurved has developed the festival of *Śarad Pūrṇimā*. It is said that on *Śarad Pūrṇimā*, an elixir is falling from the moon. It will come into the *kheer*. But what is that elixir? It is the *vāta-kapha* measurement. And it will balance out the *pitta* [*doṣa*]. That's why the *kheer* has to be made in the moonlight.... The *vāta* and *kapha* increase in the *kheer*, and eating it decreases the *pitta* in the body. The moonlight will add more coolness.... It is very good for asthma patients. If they take this *kheer* as medicine, they will not suffer from asthma again. I experiment with the ingredients every year. We try new experiments [*prayog*] at the *āśram* all the time. Making this *kheer* is an experiment in Āyurved. What does Āyurved mean? To live in sync with nature. To live a balanced life. Āyurved is the practice of balance. It means to live the good life; to live both earth-wise and soul-wise. Āyurved is not just medicine alone. It is also good living. Āyurved means, at the very least, medicine. But mostly, it means complete living. My point is that we have to think about

17 See also Svoboda (1997) for a discussion of the teachings of Āyurveda and its understanding of the different deities that rule over gems and metals.

the kind of life we are living. Whatever it is, improve it. Live naturally. This is the rule of Āyurved.[18]

Let us consider that the *Śarad Pūrṇimā kathā* performance serves as prescription for complete living. Through *kathā*, Guru Ma lays out specific healing treatments. This prescription begins with the emphasis on *doṣa* as 'the practice of balance,' and by implication, stone therapy medicine as a treatment for balancing the three *doṣas*. But Guru Ma also develops the idea of balance in a holistic way to include the notion of *doṣa* as a therapeutic technology for living the 'good life.' Thus, the *kathā* cues that a 'balanced' life and a 'good' life are the same; that *doṣa* makes 'complete living' possible. The association crafted between *doṣa* and complete living further suggests that, in Guru Ma's view, a good life means living 'naturally.'

This is a point worth considering further. In an age in which the idea of living the 'good life' often connotes in the context of middle-class India the mindless overconsumption of goods and services[19] (Guru Ma likes to tease her *bhakts* by saying that they have to buy all the latest Samsung gadgets in order to be happy), Guru Ma drives its meaning in a different direction. She reconfigures the more standard (and capitalist) interpretations of this idiom from within the authoritative lens of *dharm* to create the alternative meanings of 'living in harmony with nature,' and more broadly, 'living in sync with *dharm*.' The *kathā* she gives after the preparation of the *kheer* accentuates this theme of *doṣa* as a treatment for natural living, and more significantly, for living consciously on the planet. In this respect, complete living involves the practice of living consciously (and conscientiously). Moreover, stone therapy medicine offers an effective treatment for living consciously. No wonder Guru Ma frames her performance this evening with the prescient observation that 'Indians,' in which she includes her own *bhakts*, 'are not happy today.' Their minds have become 'dull' due to 'mental disease.' As the *kheer* cooks under the healing moonlight, Guru Ma talks about the mental malaise that she says plagues the current milieu:

Today, we have more options than ever. We have more schools, more universities and more colleges. We have more doctors, clinics and hospitals. We

18 This statement is part of a larger teaching session on Āyurveda, disease and modern Indian life given on the night of *Śarad Pūrṇimā*. It was recorded on 18 October 2013 at Guru Ma's ashram, Pratap Nagar, Udaipur, Rajasthan.

19 See also Varma (2007) for a discussion and analysis of middle-class Indian values, such as greed, selfishness and, perhaps most importantly, the lack of empathy, that perpetuate a neoliberal capitalist mentality of overconsumption.

have more shops, more books, more cars, more temples and more *kathās*. We have *kathās* on every topic.... But did you notice that we also have more 'depression' and more madness [*pāgalpan*] than ever before? These are the diseases of the heart-mind [*man*]. We are not happy. Our comforts have increased, but our mental happiness has decreased. If we have more, our mental problems should be less. But they have increased. This means that something is wrong in our minds. Earlier, there was no TV, no internet, no mobile phone and no Google. We didn't have so many facilities. Today we have them, but still, we suffer from mental disease. Previously, our hearts were happy. But today, we live in so much mental tension.... Our heart-minds are connected to the moon. The god [*devtā*] of the moon rules over the heart. The sun rules over the body, but the moon [rules] over the mind. The moon is said to be good for the heart-mind. The medicine for our hearts is moonlight. Because of moonlight, our hearts remain quiet and calm. When your body gets ill, you take medicine. But when your heart gets sick, what do you take? ... The definition of a healthy person is that his heart is healthy. This *kathā* is a prescription for sick hearts. Is there anything more important than our heart-minds? The moonlight will make your heart happy. It will treat the emotional diseases of loneliness, anxiety, depression, fear, whatever makes our hearts sick. Today, nobody has a happy heart. We're all suffering.... All of our festivals have a deep meaning behind them. But we don't know the meanings. We just perform them mindlessly. You think *Śarad Pūrṇimā* is for [eating] *kheer* only. But we eat that *kheer* for a reason. Because it is medicine for the heart-mind.

Can I tell you one thing about *doṣa*? There is one rule of *dharm*: You will not find a single thing on this earth which is meaningless. You will not find a single thing that is useless to you. I'm telling you one thing, in the whole world, in the different countries of the world, in the different religions of the world, and in our astrology, too, it is believed that the heart of God is who? Who is the heart of God? It is the moon.

So, the importance of *Śarad Pūrṇimā* is to take [in] the moonlight. Just by sitting in the moonlight, we recover totally from our mental diseases. Why? Whatever negative energy we are having, the moonlight destroys it. Our festivals are supposed to make us healthy. There is no better prescription for health than that of Āyurved. It tells us that the moonlight that falls on *Śarad Pūrṇimā* will make our minds healthy. If our mind is healthy, our body will be healthy. When our body is healthy, our knowledge increases. We come closer to God. God lives where there is [good] health and happiness.[20]

20 *Śarad Pūrṇimā* ceremony, 18 October 2013, Pratap Nagar, Udaipur, Rajasthan.

Figure 7.1. Guru Ma teaches about *doṣa* and complete living on *Śarad Pūrṇimā*. Photograph by the author.

Figure 7.2. Members of Guru Ma's *bhakti* community listening to the *Śarad Pūrṇimā kathā*. Photograph by the author.

Guru Ma's *kathā* details the multiple health benefits of *Śarad Pūrṇimā* by emphasizing the idea that moonlight, particularly as it falls on this powerful night, 'destroys' the negative energies that weigh down the heart-mind and stimulate mental disease. At the same time, her discussion on the healing power of moonlight is woven on the global motif of conscious living. Here, according to Guru Ma, living consciously, in the most basic sense, has to do with knowing the 'deeper meanings' behind the Hindu festivals. In her view, which many of the *sādhus* whom I met shared, the holy days 'are there' not only to worship the deities, but also to create healthy and happy minds, bodies and souls. 'God lives where there is health and happiness,' she says.

Through means of *kathā* performance, Guru Ma crafts medicine and *dharm* as complementary therapeutic technologies for generating health and happiness. Juxtaposing *dharm* and medicine allows her to stretch its standard definitional parameters beyond that of only rituals and rules to include notions of *dharm* as a therapeutic technology, one which can effectively treat the (mental) malaise associated with the modern age. In this context, Guru Ma pushes back against the prevailing capitalist claims that having more options available, including 'doctors, hospitals and clinics,' paves the way for happiness with the counterargument that such promises remain empty. Comparing the contemporary Indian milieu to that of 'previous times,' she calls attention to the notion that the current age is plagued by more mental disease, more broken hearts. Her *kathā* provokes the question, 'How can any society call itself "modern" when it suffers emotionally?' The implication is that the material excess which capitalist consumerism has spawned over the last few decades has produced unconscious humans. Becoming conscious requires a completely different approach to life than what neoliberal narratives are able to provide, and for Guru Ma *doṣa* holds the answer. She makes clear that *doṣa*-ic medicine exemplifies the most effective therapeutic technology for 'real' health and happiness, because, unlike allopathic treatments, it is rooted in the principles and practices of *dharm*. For this reason, the healing and health benefits of *doṣa*, in general, and stone therapy medicine, specifically, are 'total,' rather than, as she implies with respect to allopathic medicine, temporary.

Thus, the 'deeper' message that Guru Ma's *kathā* performs relates to the idea that living consciously requires becoming aware of the ultimate therapeutic power of *dharm*, and more precisely, that medicine situated within the framework of *dharm* becomes empowered by it. Stone therapy medicine works because it serves as a vehicle and agent of *dharm*. Thus, living consciously means realizing that *dharm* represents 'the highest technology' for the current age. Herein lies the first crucial step in the practice of conscious living. Guru Ma's *kathā* prescription for complete living starts with the daily cultivating of this awareness. Only when the mind is awakened to the power of *dharm* as the supreme therapeutic technology can a person live 'in sync with nature' and reap the 'deeper' health benefits of festivals like *Śarad Pūrṇimā*.

The overarching motif that Guru Ma weaves into her *kathā* performance concerns the notion that *Śarad Pūrṇimā* represents more than just a festival for celebrating the arrival of autumn. It illustrates, more importantly, an empowered therapeutic technology of *dharm* that targets, in particular, a variety of mental diseases such as anxiety, depression, fear, loneliness and restlessness. If, as Guru Ma makes explicit, *Śarad Pūrṇimā* operates as a holistic therapy for the mind-body-soul complex, it becomes apparent that stone therapy medicine as practiced in connection with the *kheer* preparation constitutes a specific medicinal treatment that works in conjunction with the other (ritual) treatments also featured within the global *Śarad Pūrṇimā* therapy complex. Practices such as reciting of the *Devī Stotra* that marks the start of the festival, chanting prayers (*mantras*) throughout the event, and singing *bhajans* upon its completion, all have the 'deeper' meaning and purpose of creating 'fullness' with respect to health and wellbeing.

THEME 2: STONES AS CONSCIOUS VEHICLES AND AGENTS OF COMPLETE TRANSFORMATION

The use of silver metal foil to draw down the moon's healing energies into the *kheer* and its ingestion to treat mental disease depicts an important stone therapy medicine practiced in the context of *Śarad Pūrṇimā*. But there is another therapy that many of the *sādhus* prescribe to treat mental disturbances, and whose practice is not limited to *Śarad Pūrṇimā*. This treatment involves the wearing of quartz (*sphāṭik*) or pearl (*motī*) encased in silver on the body (and, as I learned, the stone must touch the body). Guru Ma speaks about quartz therapy medicine, too, during the *Śarad Pūrṇimā kathā*. Since her performance pivots on constructing *doṣa* as a potent therapeutic technology for curing mental dis-ease, her discussion on quartz therapy relates directly to the topic at hand. She teaches her *bhakts* about the efficacy of quartz by telling a story about her own personal experience with being 'short-tempered' (*chiṛchidā*) and how wearing quartz helped her to overcome that condition. She says:

> You people believe that if you wear a particular ring, it will bring you good luck and make you rich. [The members of the audience shake their heads in affirmation.] If you go to a spa, they heal you with stones. They warm the stones and put them on your body. It's a therapy. It's called 'stone therapy.' Whether we wear a pearl, an emerald or a diamond, these are stones, too. [Wearing them] is a therapy, too. In every stone, there is a *moll*. This is what the astrologers say. They believe that when you wear a ring, it will touch you and do its work. Otherwise, it will not. That means the special stone you're wearing is a metal therapy. We call them metals and stones, but they're all a part of nature. They all came out of the mines, the earth and the ocean. A particular stone can affect your physical condition and your mind. A special mathematical system was developed that if you are born at a specific time, then you must wear this stone or that stone. You should know which stones are suitable for your temperaments. I'm wearing a necklace [*mala*] made of quartz. My guru gave it to me. I started to wear it when I was young. I was around eleven years old when I wore it. I liked it a lot. And when I came to know that my *gurujī* has left his body, that's when I realized why he gave me that quartz *māla*. Actually, I was very short-tempered. When I speak too much, or when I'm tired, my body gets warm and my head starts hurting. That's when I get short-tempered. Quartz brings coolness. It takes away all the extra heat from your body. There are very good qualities in quartz. If you touch it, it's very cold. But when you wear it, after sometime, it gets hot. So that means that the stone is affecting you in some way. [By wearing a *māla* of quartz] I also came to know that it is good for fever. I don't wear any other stones except for quartz. For you people who are very depressed, restless and nervous I tell you to wear quartz, pearls or silver. It's good for you. It will take away your restlessness. It might sound funny to you people. But it's true. There are many medicines that have ashes of pearl or gold inside of them. So, of course, when you wear pearl, it affects you.

According to Guru Ma, the wearing of quartz, pearl or silver – or these stones encased in silver metal – relieves short-temperedness, depression, restlessness and anxiety. It also heals fever and general fatigue in the body. Like silver, the cooling properties of quartz, in particular, absorb the excess *pitta doṣa* and raise the *kapha* and *vāta doṣas* in order to create mind-body balance. Importantly, emphasizing the health and healing aspects of quartz, and to use her words, 'stone therapy' more broadly, Guru Ma shifts the prevailing understanding of stones from that of vehicles for generating 'luck' and 'prosperity' to that of stones as powerful therapeutic technologies of complete (or 'full') transformation. That is, the wearing of stones effects changes in the entire mind-body complex. In other *kathā* contexts, Guru Ma has stressed that 'for Indians, [wearing] stones means aligning your lives with the stars. But the stars don't care if you marry or pass your exams. You don't use stones correctly, because you don't know the real power they have for our lives.' In renunciant therapeutic praxis, the power of stones relates to the notion that they serve both as vehicles of and agents for transforming the whole human. The emphasis Guru Ma puts on stone therapy medicine as a holistic treatment buoys her more global claim that *dharm* operates as an integrative, and hence complete, technology for the actualization of complete (balanced/natural) living. Her comment that, before wearing the quartz *mālā* her guru gave her, she 'was very short-tempered' cues that quartz therapy actually works. It creates 'real' and permanent results.

The idea that metals and stones work as potent agents for transforming human consciousness, in particular, articulates a predominant theme in much of the *sādhus*' rhetoric of renunciation. Take, for instance, the lines of a *bhajan* that the late female *sādhu* Ganga Giri of Udaipur often performed in the company of other *sādhus* and householders at her *āśram* in the context of daily *satsang*, and whose composition she attributed to her late brother, the *sādhu* Prithvi Puri:

This body is the shell where you find many pearls.
This body is the shell where many jewels are hidden.

You have to take a dip in the ocean of your heart and
Bring out those jewels and pearls.

Or, consider another *bhajan* that Ganga Giri used to sing in *satsaṅg*:

Oh *sānts*, my tongue is earning *bhajans*.

The bright diamond in the body is my satisfaction.
If it shines in you, it will make your body and mind bright.

We have met the true guru in the form of Ram.
We have me the true guru in the form of *bhajans*.

In all seven seas, the water is deep.
In all seven seas, the water is deep.

The believers fill their cups and drink.
And the non-believers stand by the shore thirsty.

Don't run away from your body.
Someday it will be filled with precious diamonds.
Oh *sānts*, my tongue is earning *bhajans*.

In her commentaries on these *bhajans*, Ganga Giri described that the jewels featured in the *bhajans* symbolize the 'priceless' salvific knowledge, which she characterized as *brahma-jñān*. What is more, in contrast to 'weak' (i.e. non-religious) knowledge, *brahma-jñān* is 'heavy' and 'expensive,' and hence, comparable to the 'unbreakable' (*aṭūṭ*) pearl (*motī*) that grows in the depths of the ocean. In her words: 'When the pearl ripens, the creature in the shell dies. That pearl is more expensive than gold. It is more expensive than any jewel. It is so strong that nothing can break it. Knowledge is something like that.' In this framework, the ripening of the 'creature' into a pearl symbolically corresponds to the awakening of human consciousness. Furthermore, singing *bhajans* signifies an efficacious practice for developing *brahma-jñān* and transforming consciousness. Thus, the *bhajan* rhetorically establishes a linkage between water, pearls and the heart-mind, suggesting the 'cooling' power of stones featured in *doṣa*-ic teachings as agents for awakening consciousness and transforming the mind-body complex. And yet, Ganga Giri's locating their power strictly in the symbolic eclipses their practical efficacy.

Thus, it is significant that in many of the *kathās* that she gives Guru Ma accentuates the tangible, physical efficacy of stones and stone therapy medicine. She traces their power to their possessing an acute intelligence (she says '*moll*'), by which they carry out their healing 'work.' The implication here is that stones have 'brains' and can 'think.' Guru Ma draws out this specific meaning in the *Śarad Pūrṇimā kathā* by representing stones as 'conscious' beings. She says:

> This whole universe is the body of God. We are all the one body of God. We are all connected. There is a single thread that connects everything on this planet. It connects the sun and the moon, the rivers and mountains, and humans all together. That is the life-power [*prāṇ-śakti*]. The most important thing in this world is life [*prāṇ*]. When we make offerings to the gods, we worship the *prāṇ-śakti* only. Of all the gods [worshiped], the *prāṇ-śakti* is chief. This is true since the Vedic times. It is said that the *prāṇ-śakti* is the form of Brahman, who comes to us through the *prāṇ-śakti*. God [*parabrahman*] is life. People love life only. Life is the biggest evidence for the existence of God. If we say that God exists, it is because life exists. The *prāṇ-śakti* is the foundation of all existence. These plants, trees, shrubs, mountains, what's in them?
>
> Life is there. God is there. That which we call life is God. If God is life, it means that this whole expanding cosmos [*brahmāṇḍ*] is alive [*jīvit*]. Everything in this cosmos is alive. It can think and feel deeply. Try to hug a tree. Try to hug water. You will feel that they are alive. You will feel that

> they can think. It means that every creature on this earth, it can be a tree, a plant, the sun and moon, a shrub, a stone [*pattar*], is conscious [*cetan*]. When you start to talk to nature, it will talk back to you. Nature is your true friend. You think only human beings are your true friends because only they can think. But nature thinks, too. Nature is never wrong. All life has consciousness [*caitanya*]; it is alive. It thinks and breathes like we think and breathe. There is nothing on this earth without *prāṇ-śakti*. Nature has *prāṇ-śakti*, so it means it's alive. It is not dull [*jaḍa*]. Human minds have become dull, but not nature's. This is very important. If you see nature as alive you will respect it. See all life as the beauty, creativity, and love of God. These are the 'DNA' of God. We all have this DNA. We are one body of cosmic Brahman.

This segment of the *kathā* performance dramatizes in sharp relief the holistic understanding that 'nature is alive' with what Guru Ma describes as the 'life power' of God. Therefore, everything in nature qualifies as 'conscious.' Distinguishing 'deadened' human minds from the 'awakened' mind of nature, Guru Ma foregrounds the inclusive notion that 'every creature on this earth' can 'think and feel deeply.' Importantly, this understanding, which many of the *sādhus* share, suggests that all types of stones – from the precious gemstone and metal to the ordinary pebble and rock – constitute sentient beings. These 'creatures,' as Guru Ma points out, have the capacity to befriend humans as much as other (non-human) life forms. In addition, recognizing the conscious quality of stones, in particular, engenders respect for nature and the planet. In this way, the *kathā* prescribes the second step for actualizing the fullness of life: Conscious living requires the realization that first, 'the whole expanding cosmos' pulsates with the positive life-energy of God, and second, that the entire natural world is empowered by that divine energy. Being conscious 'creatures' within, and connected to, the 'body of the cosmic god,' stones, too, are empowered to heal disease. Their capacity to heal signifies, to use Guru Ma's apt words, an inherent trait of their 'DNA,' and constitutes a function of God's healing power. Against this backdrop, just as stone therapy medicine is empowered by *dharm*, stones themselves are empowered by the pervasive consciousness of the cosmic Creator (*brahman*). It is significant that as the precious resources of nature, stones, which symbolize the 'heart of India,' serve as empowered agents for healing the hearts, minds and souls of the people of the planet.

Thus, the practical efficacy that Guru Ma attributes to stone therapy medicine as a technology of *dharm* pivots on the perception that stones represent conscious creatures of nature. They have the power to think and, just as significantly, to act. Every stone not only contains within its DNA a unique formula (or 'wisdom') for treating illness, but also actualizes this specialized knowledge as a technique of healing. The notion that stones think and act bolsters Guru Ma's claim that stone therapy medicine illustrates an empowered therapeutic technology for total transformation. But notice that the healing generated by stone therapy medicine occurs only when the stones 'touch' the body. Hence, natural healing depends both on the physical connection created between stones and the body, and on the view of all life as interconnected.

THEME 3: STONE THERAPY MEDICINE AS A BENEFICENT TECHNOLOGY OF DHARM

The widely-touted efficacy of stone therapy in renunciant praxis has given it a reputation among the *sādhus* as 'practical medicine.' But there are two other key reasons they cite for its practicality. Besides the power of stone therapy medicine to generate long-lasting, tangible results, the *sādhus* say that it is based on the principles of 'science' and benefits the planet. Guru Ma threads these overlapping notions seamlessly into the *Śarad Pūrṇimā kathā*. She says:

> Earlier people used to cook in clay dishes, because it had a lot of nutrients in it. If it didn't have nutrients, how would the crops grow? You should cook in clay or copper dishes, but not those made out of aluminum. It is a very heavy metal. It reacts badly in your body. The body is harmed by it. And, if you eat food cooked in aluminum every day, the body gets worse. The people who use aluminum to cook have asthma. They're always sick. Then they have to spend your money on medical treatments that don't work. But you think that aluminum is easy to get, so it's good. It's not good. Clay or copper is good. Clay dishes have nutrients. When the food cooked in those dishes is boiled, it absorbs the nutrients. Everything in this *āśram* is cooked in clay or copper dishes. I never cook with aluminum. I keep telling you that copper and clay are better. But no one listens to me. Potters don't have employment anymore, because people are always buying aluminum vessels. People don't use the *loṭās* [special drinking vessels] anymore. The people who made *loṭās* are called Thateras. But you don't find Thateras these days. *Loṭās* were very scientific. They were very good for health. *Loṭās* are large and round. Their shape is better than if you drink from [the shape of] glasses. Whatever we drink from, the qualities of that vessel get absorbed by our bodies. There was a reason people didn't drink from square *loṭās*. They had to be round. The round design is scientific. The bottom of a *loṭā* has more surface area, which is good for your body, because it can absorb more of its nutrients [from the

vessel]. If you drink water from a copper *loṭā*, your body absorbs the copper. It's good for the body. Copper is a holy [*pavitra*] metal. It works as a medicine. When you drink from a copper *loṭā*, it purifies the water of toxins; and when you drink that water, the copper [in the water] purifies the body of toxic metals. Our bodies contain a lot of metals. Copper rids the body of toxic metals. People with vitiligo [an autoimmune skin disorder that causes depigmentation] have a copper deficiency. I tell them to wear copper rings and bangles, and to drink from copper *loṭās*. To balance the metals in the body, I suggest that they wear copper. It works as a medicine for the body.

But what do you do? You people drink from aluminum [glasses] and cook in aluminum. Aluminum pressure cookers are the worst. There is a rule in *doṣa* about cooking: the food should get the sunlight and air so that the toxins in the food die. Sunlight doesn't let any bacteria get into the food. It finishes them. Today, 90 percent of pressure cookers are made from aluminum. This is not a good metal at all. Even the medicines made by allopathy don't use aluminum. It reacts badly with the medicine. It produces side-effects. When allopathy doesn't use aluminum, why do you use it? The food absorbs the citric acid and lactic acid caused by [use of] aluminum. Food is such a big medicine. And the process by which it is cooked is a big medicine.... The soul of India lives in the utensils of clay, wood and [precious] metals. All of this is written in our texts. You should follow these principles if you want to have good health. You should understand that if you can use clay and copper to make food for the gods, why don't you use it for yourself? They are holy. "Holy" means that they have good qualities. But all this knowledge is finishing, because you people don't understand their benefits for health.

This part of the *kathā* discusses the benefits of clay and copper for everyday life. Both represent 'holy' natural resources, because, as Guru Ma says, they possess 'good qualities,' and thus, are good for the planet. The beneficial properties of clay and copper appear to constitute an underlying condition of their holiness. Not surprisingly, Guru Ma mentions that 'the soul of India' exists in (the use of) clay and copper. In this respect, she indexes their inherent capacities (and role) as potent therapeutic technologies of *dharm* (she refers to them as 'medicine' several times in her *kathā*) and, more precisely, as beneficent therapeutic technologies for the planet.

In the case of copper, it works as an ingenious antimicrobial/antibacterial powerhouse in that, with the help of the sun (and heat from cooking), it purifies food, water and the body of toxins and bacteria. It also removes heavy metals like aluminum from the blood system. Guru Ma says that people who have too much aluminum in their bodies suffer from asthma. 'They're always sick.' This is a valuable point to ponder. Given that the silver-infused *kheer* prepared on *Śarad Pūrṇimā* is used to treat asthma, her statement about the aluminum-fighting properties of copper may also be applied to silver. Both metals act as purificatory agents. Lastly, because of its power to purify and prevent infection, Guru Ma's practice of stone therapy medicine uses copper to treat (and cure) chronic skin conditions like vitiligo, which is commonly seen in India.

Besides detailing the potent benefits of clay and copper, Guru Ma's *kathā* foregrounds that everyday acts as simple as drinking from a copper *loṭā*, or cooking food in a clay dish (and, in other contexts, eating food served on banana leaves), depicts the practical medicine of *dharm*, and as I suggest, stone therapy medicine as a therapeutic technology of transformation. *Dharm*, as Guru Ma suggests, neither has to be complicated in order to be meaningful (or purposeful), nor limited to the practice of rules and rituals. Guru Ma's emphasis on the simplicity of ordinary life practices involving the use of stones as medicine heightens the practical value of *dharm* as a holistic technology for achieving the fullness of life *in any milieu*. And yet, the actualizing of that goal depends on becoming conscious of the deeper health and healing properties embodied within all of the stones that live within the divine body of nature.[21] For this reason, because this *kathā* aims to prescribe effective treatments for the mental malaise associated with the modern period in particular, Guru Ma's accent on the simplicity of *dharm*'s practical medicine amplifies her point that complete living requires an integrative, rather than compartmentalized, approach to life: It demands the cultivation of a worldview that values nature as a conscious healer of disease and an ethos that strives to live in a balanced *and* beneficent way with the natural world.

But the main point that Guru Ma makes to buoy the practical value of *dharm* as a holistic technology for the modern age concerns aligning stone therapy medicine with 'science.' This representation makes it possible for her to bring a seemingly distinct, if not opposing, field of study within the authoritative canopy of *dharm* and to construct *dharm*, technology and science as equivalent knowledge systems. It also helps to draw in the attention of her primarily middle-class audience, many of whom work in medicine and other science-related fields, and, despite their advanced knowledge, continue to live unconsciously and out-of-sync with nature. Guru Ma cues this idea in her critique that they 'drink from aluminum glasses and cook in aluminum pots.'

21 As an aside, after transcribing the tape on which I documented this *kathā*, my research associate Manvendra Singh immediately phoned his mother and told her to throw out all her aluminum pots and pressure cookers. He said, 'From now on, everything we eat will be cooked in copper pots.'

Significantly, by situating science, technology and medicine within the salvific world of *dharm*, Guru Ma not only stresses the notion of *dharm*'s comprehensive efficacy and value for the modern age, but also constructs it as more practical, and more potent, than techno-science and medicine alone. That is, *dharm* can do whatever those knowledge systems can do, but better. Her *kathā* signals the superior and everlasting potency of stone therapy medicine as an empowered technology of *dharm* by distinguishing it from allopathic medicine, which, as she suggests, operates on the logic of materialistic capitalist consumer values and, hence, represents an impotent form of treatment. Take, for example, Guru Ma's prescriptive comment that people with asthma 'are always sick ... and have to spend their money on medicines that don't work.' The 'medicines' to which she refers concern allopathic treatments. She makes this reference explicit later on in the *kathā* with the statement that even 'when allopathy doesn't use aluminum in [the preparation of] medicines, why do you use it?' The implication is that allopathic medicine is weak and offers no real long-term results, because it is not empowered by *dharm*.

As with Guru Ma, many of the *sādhus* with whom I worked put little, if any, trust in the efficacy of allopathic treatments. Because they see stone therapy medicine as empowered by *dharm* (and God), the majority of the *sādhus* emphasize that naturopathic stone therapy is all they require for their healing and health. The female *sādhu* Kailash Das, who operates a bustling temple complex (and cow shelter) in Bhilwara, Rajasthan, talks about her failed experience with allopathic medicine with respect to treatment for a dog-bite in the following personal narrative:

> I feed the cows every day, three, four, times a day. One morning as I was taking them to a nearby field, I noticed a stray dog. I thought, 'it must be hungry. I'll feed it some *capātīs*.' I feed these dogs who come to the *āśram* all the time. I brought it a couple of *capātīs*, but it got scared and bit my [right] leg. I screamed and screamed. When it finally let go, there was blood everywhere. Its teeth tore into my skin and I could see the veins and the tissue. People said, 'Go to the hospital. We'll take you to the clinic.' I went to four, five different doctors, who gave me this medicine and that medicine. But none of them worked. Finally, my guru-brother [a *sādhu* whom Kailash Das knows, and who was initiated by her guru] said, 'You just try this stone medicine instead.' He told me to have a metal worker [*sunār*] make silver rings for my toes and wear them every day until the leg heals. [Author asks: 'Did it work?'] See for yourself! I've been wearing these [rings] for four months now and there is no swelling, no pain. My leg doesn't hurt anymore.... This is all God's [Bhagvān's] grace. *Dharm* is the best medicine [*auṣadhi*].

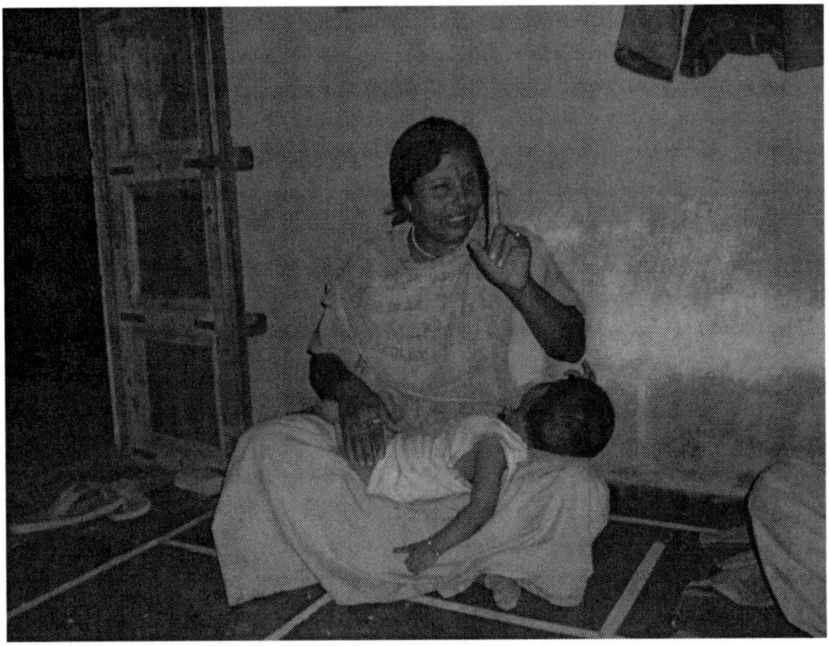

Figure 7.3. While holding a *bhakt*'s child, Kailash Das talks about her success with stone therapy. Photograph by the author.

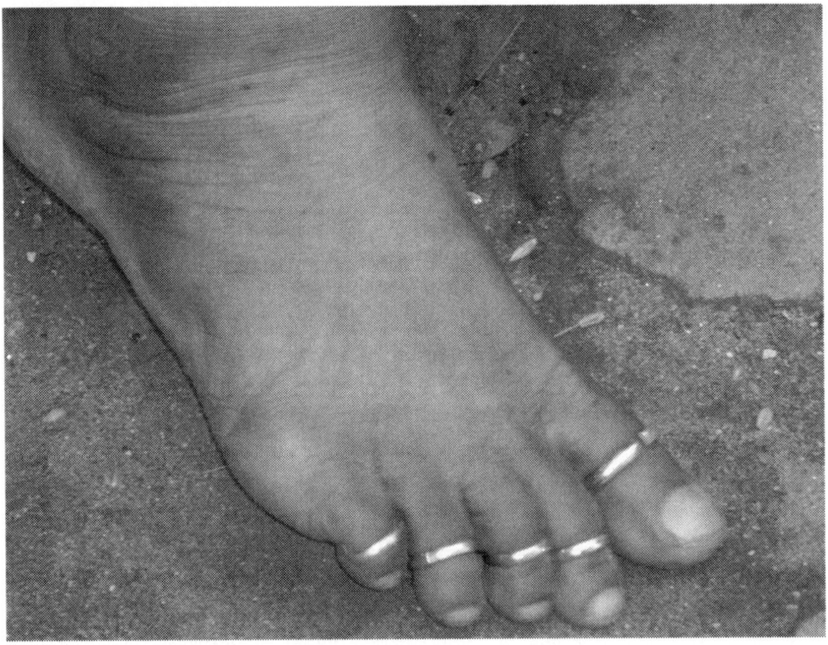

Figure 7.4. Kailash Das displays the silver toe-rings she had made to heal her dog-bite. Photograph by the author.

While Kailash Das does not discuss the physiological (or scientific) reasons behind the healing of her flesh wounds, or those for the silver rings used to treat her injury, she does not have to, as these aspects of her healing are not what she thinks matters in explaining why her treatment 'works.' For her, God has healed the injury, because, as she implies, God works through and, thus, empowers natural medicinal techniques like stone therapy medicine, but not allopathic treatments, which 'don't work.' In a similar vein, the female *sādhu* Tulsi Giri underscores the power of God and, by implication, *dharm*, in connection with the stone therapy medicine she uses to treat (and control) a common navel condition, which is known as *ḍhonḍhī golā ṭalnā* (lit. 'misplaced belly button'). Noticing what I later learn are copper-based metal rings attached to each of her big toes prompts me to ask Tulsi Giri what they are and why she wears them. She says, 'They keep my navel in place. When I had the surgery to remove the stones from my gallbladder [two years ago], this problem happened. If I don't wear [the rings], I have a lot of stomach problems. I have to return to the hospital. I have to take a lot of medicines that I don't like [Tulsi Giri jokes that she lets the rats at her *āśram* eat those medicines instead]. But if I wear the rings, I have no problems at all. Hey Bhagvān! God takes care of all these problems.'

The people on whom *sādhus* like Kailash Das and Tulsi Giri rely to prepare their stone therapy medicine come from a specific caste community known as Sunārs (lit., 'goldsmiths' or 'silversmiths'). Sunārs specialize in the crafting and manufacturing of gold, silver and other precious metals in South Asia. Many of the *sādhus* who use stone therapy medicine as part of their everyday therapeutic praxis work with Sunārs, whose jewelry shops can be seen in many villages throughout India. Since most of the *sādhus* do not keep the materials needed to prepare stone therapy medicine at their *āśrams* (many cite reasons of theft), they send their *bhakts* to local Sunārs with detailed prescriptions of the stones to purchase and their exact methods of application. Against this backdrop, the *sādhus*' stone therapy medicine reveals an entrenched and complex nexus of the people and practices involved in the running and maintaining of local knowledge systems of naturopathic healing as practiced in lived Hinduisms.

And yet, as Guru Ma poignantly observes in the *Śarad Pūrṇimā kathā*, 'all this knowledge is finishing.' The unconscious ways in which the *bhakts* in her community and, as she indicates, Indians more generally, live their everyday worlds by using allopathic medicine to treat illness and purchasing aluminum utensils for food preparation and consumption come with a heavy price tag: the loss of highly skilled craft traditions and the

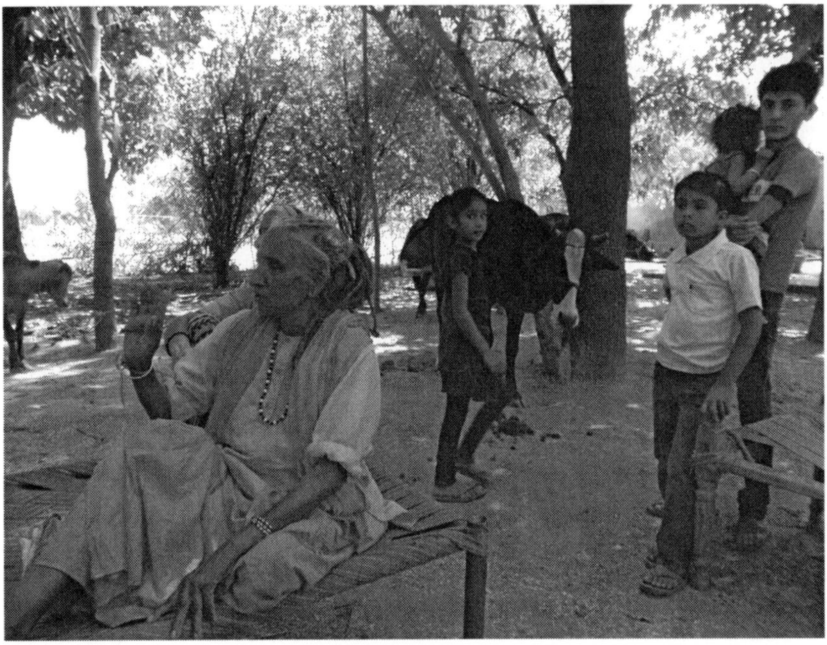

Figure 7.5. Tulsi Giri explains the health and healing benefits of stone therapy for her navel condition. Photograph by the author.

Figure 7.6. Copper ring on Tulsi Giri's right big toe. Photograph by the author.

specialized knowledge systems they embody. In this context, Guru Ma talks about the Thateras, a caste community that specializes in the production of copper *loṭās*. 'But,' according to her, 'you don't find Thateras these days,' because people have stopped using copper *loṭās*. And why have they stopped using copper *loṭās*? For the simple reason that, as Guru Ma stresses, people think it is cheaper and, therefore, 'better' to buy aluminum utensils than it is to purchase carefully-crafted *loṭās* of precious metal. And yet, the predominant capitalist consumer mentality of 'having more but paying less' to which Guru Ma obliquely refers, and which she critiques, is poisoning humans' ability to live in sync with nature and realize the fullness of life. Such an unconscious approach to life creates and perpetuates disastrous results for the planet in its careless destruction of a priceless ecosystem, in which Guru Ma also includes specialized craft and wisdom traditions.

Thus, constructing stone therapy medicine as a powerful treatment for complete living, Guru Ma raises its status and value as a beneficent technology for the current milieu. The *sādhus* suggest that, unlike other industries whose practices tend to produce maleficent 'fruits,' stone therapy medicine is good for the planet. Its benefits directly address the health and healing of the mind-body-soul complex, but also stretch far beyond these goals. The capacity of stone therapy medicine to promote the priceless virtues of respect, empathy and compassion for nature and its diverse natural and cultural ecologies (recall Guru Ma's words: 'Try to hug water'), identifies it as a superior technology for practicing balance and conscious living. Becoming conscious, as the *sādhus* suggest, brings humans closer to the divine. It facilitates a vision of the interconnectedness of all existence, and by implication, fosters a sense of being accountable for the decisions and actions one takes on the planet. Hence, the holistic approach of stone therapy medicine in renunciant praxis distinguishes it as a potent technology for treating the malaise of the modern age. By the same token, its characterization by the *sādhus* as empowered by the divine constructs *dharm*, more generally, as a comprehensive therapeutic technology for human transformation.

CONCLUSIONS: HEALTH, HAPPINESS AND THE MODERN IN EXPERIMENTAL HINDUISM

In this chapter, I have argued that stone therapy medicine in everyday renunciant praxis demonstrates a context in which the *sādhus* rework the dominant definitional parameters of *dharm* to emphasize that it encompasses more than rituals, rules, traditions and festivals. Just as significantly, *dharm* constitutes a comprehensive therapeutic technology for, as *sādhus* like Guru Ma teach, 'complete living,' in an age in which people suffer from all kinds of dis-ease. The notion of complete living as constructed in the *sādhus*' rhetorical, ritual and therapeutic practices brings into sharp focus a tripartite set of themes: balance and harmony in everyday life; natural living; and conscious living. Specifically, stone therapy medicine balances the three material elements (*doṣas*) of *kapha*, *vāta* and *pitta* in the body, and, in this way, causes harmonization of mind, body and soul. Moreover, as the practice of stone therapy medicine involves the selective use of natural ingredients, such as herbs and stones, for treating illness, it enables a person to harmonize his or her life-power with that of the natural world. Furthermore, living in sync with nature inspires the recognition of the interconnectedness of all creatures of the cosmic body of God and, in effect, promotes respect for all forms of life on the planet.

To that extent, the *sādhus* represent stone therapy medicine as a panacea, a superior therapeutic technology, for the malaise that they say describes the modern age. Unlike allopathic treatments, stone therapy medicine delivers on its promises to heal, prevent and cure that malaise, and much more. To the *sādhus*, stone therapy medicine not only produces real results, but it also generates the complete transformation of the mind-body-soul complex in the process.

Importantly, as this chapter has shown, stone therapy medicine as practiced in vernacular asceticism also reveals an embedded ideology about what it means to be a happy and healthy person in the world. In the context of Guru Ma's *Śarad Pūrṇimā dharm-kathā*, being happy and healthy hardly connotes standard neoliberal capitalist notions of the unconscious consumption and accumulation of goods and services; of 'having more but paying less.' That mentality feeds off and perpetuates the deadly poisons of greed, selfishness and lack of empathy, and, consequently, foments destructive (and unconscious) thinking and behaviors on the planet. It ruins not only precious natural ecosystems (one female *sādhu* Chetanananda Swami of Udaipur, with tears in her eyes, gives the example of the mining industry that is destroying the entire local ecosystem), but also the local cultural traditions that embody specialized healing knowledge.

It is significant that Guru Ma directs her critique of, and occasionally her vitriol against, the consumer capitalist approach to life at her predominantly middle-class constituency of *bhakts*. This middle-income group, as a class, generally lauds the value of free enterprise in

India.[22] In the view of the *sādhus*, however, the idea of being healthy and happy has to do with (the practice of) living a life of simplicity (but not of poverty) – that is, of being content with having, as Guru Ma has said in the context of a *dharm-kathā* she gave in Silvassa, 'one dress instead of ten.' Along with this, it requires the surrendering of the ego-self to the wisdom (and beauty) of nature, and to the divine life-energy that pulsates throughout nature, trusting that 'nature never makes a mistake.' Hence, stone therapy medicine represents a beneficent technology for actualizing the values of simplicity and surrender, because one yields to the infinite wisdom embodied in stones as 'conscious' beings.

Finally, the ideologies of health and happiness as refracted through the practice of stone therapy medicine in vernacular asceticism articulate an important (and urgent) message concerning ideas about the modern and its relationship to *dharm*. From the *sādhus*' shared perspectives, a life lived in sync with *dharm* creates everlasting happiness and health (recall Kailash Das' statement that '*dharm* is the best medicine'), but just as significantly, gives expression to the modern in twenty-first century India. The modern human, as the *sādhus* suggest, experiences the whole cosmos as alive with the life-power and energy of the cosmic Creator and looks no further than to nature's precious stones for healing and well-being. She or he understands that living without compassion, respect and empathy for nature as the sentient body of God sets off a domino effect of consequences in a world that is deeply interconnected.

But there is more to what the modern means to the *sādhus*. As the *Śarad Pūrṇimā kathā* has made clear, the modern human no longer straddles the multiple worlds of *dharm*, techno-science, and medicine, but instead positions herself confidently within the sacred canopy of *dharm* alone. As a comprehensive technology, *dharm* both encompasses and empowers these knowledge systems and, thus, represents the authentic face of the modern. Guru Ma's *dharm-kathā* foregrounds that situating oneself within the authoritative world of *dharm* makes possible the fullness of life, because, as a holistic technology, *dharm* promotes the integration of all aspects of a person's life, which, in turn, creates wellbeing and the healing of dis-ease.

22 Many of Guru Ma's *bhakts* work as industrial capitalists and think free enterprise is a good thing for India. However, as scholars like Varma (2007) and Buultjens (1988) have discussed, the free enterprise mentality also has had damaging consequences for the ways that people think about happiness and health, and (access to) planetary resources. Guru Ma's *kathās* primarily function to draw out those connections between an unconscious 'free enterprise' mentality and the unconscious destruction of the earth.

With *dharm* as her or his foundation (and shield), the modern human is not only happy and healthy, but also prepared to deal with the challenges that the contemporary milieu throws her or his way.

In sum, the holistic ideologies of health and happiness that are woven into the *sādhus*' ideas about stone therapy medicine, and the notions of the modern that its practice in vernacular asceticism creates and communicates, demonstrate in high-relief the dynamic processes by which the phenomenon of 'experimental Hinduism' emerges and takes shape in twenty-first century India. As this chapter has shown, the experimental Hinduism to which I have called attention involves the accenting of the values of simplicity and surrender in the *sādhus*' constructions of *dharm* as a therapeutic technology of transformation for modern India. Thus, it becomes readily apparent that the classic ideals of *saṃnyās* play a vital role in renunciant reconfigurations of the dominant definitional boundaries of *dharm*, as well as in renunciant representations of stone therapy medicine as an empowered (and conscious) technology (and agent) of *dharm*.

On this full moon night of *Śarad Pūrṇimā*, Guru Ma's *dharm-kathā*, too, seems to have empowered her *bhakt*s to think and act more consciously as they go about constructing their worlds. As the *kheer* makes its rounds among the audience (it is served in small recyclable paper bowls), I notice more smiles, more laughter and more exuberant exchanges than at the start of the event four hours earlier. The *bhakt*s waste no time in eating the piping hot *kheer* which has been cooking under the moon (Guru Ma jokes that it is what they have been waiting for all along), and as they do, the *āśram* appears to radiate with happier hearts and healthier minds.

ABBREVIATIONS AND REFERENCES

H. = Hindi

Barrett 2008	Barrett, R. (2008). *Aghor Medicine. Pollution, Death, and Healing in Northern India*. Berkeley: University of California Press.
Buultjens 1988	Buultjens, R. (1988). *Windows on India*. New York: Express Books.
Buultjens 2015	Barrett, R. (2015). 'Understanding Modern India.' *Asia Society*, <http://asiasociety.org/asia101/understanding-modern-india> (accessed 25 August 2015).
DeNapoli 2014$_a$	DeNapoli, A. (2014). *Real Sadhus Sing to God. Gender, Asceticism, and Vernacular Religion in Rajasthan*. New York: Oxford University Press.
DeNapoli 2014$_b$	DeNapoli, A. (2014). 'Religion is Technology: Experimental Hinduism through the Lens of Vernacular Asceticism in North

	India.' Paper presented at the Annual American Academy of Religion, San Diego, CA.
Gross 2002	Gross, R. (2002). *The Sadhus of India*. Jaipur: Rawat Press.
Josephson 2013	Josephson, J. (2013). 'An Empowered World: Buddhist Medicine and the Potency of Prayer in Japan.' In Stoler, J. (ed.), *Deus in Machina. Religion, Technology, and the Things in Between*, pp. 117–41. New York: Fordham University Press.
Nelson 2013	Nelson, J. (2013). *Experimental Buddhism. Innovation and Activism in Contemporary Japan*. Honolulu: University of Hawaii Press.
Svoboda 1997	Svoboda, R. (1997). *The Greatness of Saturn. A Therapeutic Myth*. Twin Lakes, WI: Lotus Press.
Taylor 2004	Taylor, C. (2004). *Modern Social Imaginaries*. Durham and London: Duke University Press.
Varma 2007	Varma, P. (2007). *The Great Indian Middle Class*. New York: Penguin Books.
Ward 2009	Ward, P. (2009). *Experimental Theology in America. Madame Guyon, Fenelon, and Their Readers*. Waco, Texas: Baylor University Press.

Chapter 8

'Deg Tegh Fateh!' Metal as Material and Metaphor in Sikh Tradition

ELEANOR NESBITT[1]

Deg tegh fateh, a historic Sikh rallying cry and slogan, translates literally from Panjabi as 'Victory to the cauldron and the sword!'[2] and evokes the iron cooking pot (*deg*) and the steel sword (*tegh*), the weaponry of the *sant sipāhī* (warrior saint). *Deg tegh fateh* encapsulates Sikhs' emphases on hospitality for all and military prowess in protecting victims and challenging oppressors. Weaponry and iron and steel production were important to Gurū Hargobind, the sixth Gurū, and Gurū Gobind Siṅgh, the tenth Gurū. Indeed Gurū Gobind Siṅgh used the epithet '*sarab loh*' (all iron, i.e. pure iron) for Akāl (the Timeless One, God).

Sikh history includes the formulation of the five Ks (external signifiers of Khālsā membership) and the adoption of the *niśān sāhib*. Among the

1 Eleanor Nesbitt is Professor Emerita in the Centre for Education Studies, University of Warwick (UK).

2 The phrase 'the Khalsa's blessing' (Pan. *khālsā dā āshīrvād*, Nabha 1974: 649), occurs in DG (310 and 570, cited in Singh 2014_c: 463) and in the Ardas (cited by Cole and Sambhi 1990: 62). McLeod (1995: 69) translates this as 'Victory by the grace of the Guru to the wielder of the sword,' explaining that '[i]n the eighteenth century the *degh* or *deg* (cooking-vessel) symbolized the *laṅgar* [the *gurdwārā* kitchen where food is prepared and served to all visitors irrespective of their background] which in turn symbolized the grace of God in a casteless society.' Articulating a more usual understanding, Cole and Sambhi (1990: 62) state that these Persian words for 'kettle and sword' signify 'the dual responsibility of the Panth to provide food and protection for the needy and oppressed.' For a detailed exploration of '*deg tegh fateh*' see Singh 1995.

five Ks are the *kirpān* (sword) and the *kaṛā* (a bangle-like circle of iron or, more usually, steel). The *niśān sāhib* is the flagpole and pennant, bearing the Khālsā's insignia, the *khaṇḍā* emblem, which indicates that a building is a *gurdwārā*, a public place of worship. Comprising the *khaṇḍā* itself (a sword that has two concave cutting edges) cupped by two interlocking *kirpāns* (curved swords), and including the circle of a quoit or a *kaṛā*, the insignia conveys Khālsā identity through visual reference to iron/steel. So, iron and steel are implicated in the external evidence that an individual is a Khālsā Sikh and that a building is a place of Sikh congregational worship.[3]

In this chapter the ongoing relationship between Sikhs and iron and steel, their importance to the Khālsā as material and their exaltation in metaphor will be discussed. This will be prefaced by some contextualization: the historical setting of Indian metallurgy and references to iron in Sikh scripture, the *Gurū Granth Sāhib* or *Ādi Granth*. Textual reference to the Indian tradition of the *pāras*, the Indian version of the European philosopher's stone (*lapis philosophorum*) that can transmute base metal into gold, provides a pertinent counterpoint to subsequent Sikh affirmation of iron. Recognition of the contributions of Gurū Hargobind (1595–1644) and Gurū Gobind Siṅgh (1666–1708) introduces the praxis of twenty-first century Khālsā groups in terms of both the physical presence of iron and the concept of *sarab loh*, expressed both physically and metaphysically. A theoretical perspective is offered by discussions of 'material religion' (Houtman and Meyer 2012), of 'materializing religion' (Arweck and Keenan 2006) and of materiality in Sikh tradition more specifically (Murphy 2012). The need to distinguish the material of things (and its supposed properties) from their form (and apparent uses) will be suggested.

SIKHS AND KHĀLSĀ SIKHS

Although the designations 'Sikh' and 'Khālsā' are sometimes used interchangeably, in this chapter the word 'Sikh' encompasses all who self-identify as Sikh whereas the designation 'Khālsā' usually applies more specifically to *amritdhārī* Sikhs, those who have been initiated in a rite variously referred to as *amrit chhaknā*, *khande dī pāhul*, *amrit sanskār* and

3 One other 'K', *kaṅghā* (comb), is made of wood; another, the *kachha* (shorts, also known as *kachhahirā*), is made of cotton. These two materials feature much less prominently in Sikh literature than iron and steel, suggesting that iron and steel themselves, and not only their form and function, are significant.

taking *amrit*,[4] and who thereafter observe a daily discipline. However, many non-*amritdhārī* Sikhs also identify with the term Khālsā, and many Nihaṅg Sikhs (see below) do not emphasize the need for initiation with *amrit*, while regarding themselves as the true Khālsā.

As discussed in Jasjit Singh's analysis of the UK's religiously observant young Sikh adults (2014$_a$), the early twenty-first century Khālsā includes the Nihaṅgs, the Akhaṇḍ Kirtanī Jathā, the Damdamī Ṭaksāl, whose very title has a metallic ring (see below), and Sikh Dharma of the Western Hemisphere. The centrality of iron and/or steel to all these groups will be outlined after an examination of iron/steel in the evolving Sikh tradition.

SIKH IRON IN CONTEXT

While the Sikhs' emphasis on iron is unusual among communities that are defined by their religious allegiance, metallurgy has played a prominent role in Indian civilization for several millennia. Iron-working may have started as early as 1800 BCE at sites in northern India (Tewari 2003). The Greek historian Herodotus (484–425 BCE) mentioned that the Indian and Persian armies used arrows tipped with iron. The Indian King Porus's gift of 40 pounds of steel to the invading Alexander the Great suggests that in the fourth century BCE steel was the most valuable gift that a defeated king could give, and the one that the new ruler would most appreciate (Mushet 1840: 366). Evidence of India's ancient expertise in metallurgy is provided by the over seven-meter-high iron pillar near Qutb Minar, South Delhi, which has resisted corrosion or rust since its production in probably the fourth or fifth century CE (Goswamy 2002).

Steel is an alloy of iron with carbon (and in some cases other metals) and is harder than iron.[5] Relatively unknown in the West is the fact that Wootz steel originated in India in the third century BCE (Srinivasan and Ranganathan 1997). Often known as 'Damascus steel', it was highly prized in Europe for centuries (Smith 1960) but India's British rulers – wanting India to supply only raw materials for manufacture in Britain – subsequently prohibited its production.

4 *Amrit* ('immortal') is the term for healing water and, specifically, for sweetened water ('nectar') that is ritually stirred and administered to initiation candidates.

5 For the many types of iron and steel see http://www.chemguide.co.uk/inorganic/extraction/iron.html (accessed 24 August 2015).

A brief overview of Hindu and Islamic traditions – the two religious cultures in north India contemporary with the evolving Sikh community – suggests that Sikhism is unusual in its focus on iron and steel, although iron is celebrated in the Qur'ān (Anon. 2007).[6] However, a glance at cross-cultural assumptions about the properties of iron reveals multiple parallels and precedents, with iron being widely believed to counter witchcraft and black magic.[7]

IRON IN THE GURŪ GRANTH SĀHIB

Phrases such as 'the Dark Age, the Iron Age of Kali Yuga' occur in English versions of the *Gurū Granth Sāhib*,[8] e.g. ĀG.346 (Ravidās), 406 (Arjan Dev) and (twice) 446 (Rām Dās), 470 (Nānak), 880 (Amar Dās), 1024 (Nānak), 1390 (Amar Dās). Even though textual references to the traditional Indic sequence of ages nowhere mention iron, the Sikh translators' readiness – in line with translators of *kali yuga* in Hindu literature – to term *kal jug*, humanity's least enlightened age, the 'age of iron' suggests a low regard for this metal.

Consistently with this low regard, many of the approximately fifty mentions of iron (*loh*) suggest an unfavorable comparison with more precious metals, notably gold.[9] Gold's value is juxtaposed with the worthlessness of

6 Tension emerges between popular practice and formal teachings of Islam. According to one tradition, the Prophet Muhammad reprimanded a man for wearing an iron ring as protection from disease and said 'Take it off your hand, for verily, it will only increase your weakness.' See http://www.iqrasense.com/muslim-character/healing-using-false-taweez-lucky-charms-and-more.html (accessed 19 August 2014).

7 For example, according to an online explanation of the horseshoe's use in Britain to avert evil/bring good luck, 'because iron could withstand fire, it was also believed that it could ward off evil spirits.' See 'Horseshoe superstition' in http://psychiclibrary.com/beyondBooks/superstition-room/ (accessed 24 August 2015). However, the cup shape of the horseshoe, its association with an animal that was revered in pre-Christian Britain, and its traditional association with St Dunstan (blacksmith turned archbishop) are also factors. On the offering of horseshoes to Śani (Saturn) in India, see Chapter 3 of this collection, p. 60.

8 I have used in this chapter the Khalsa Consensus Translation provided by Dr Sant Singh Khalsa. (Diacritics have been added.) Two websites provide the full text. See www.Sikhs.org and www.srigranth.org.

9 In English translations, 'age of gold' represents, by contrast, *sat jug* (Skt. *satya yuga*), the age of spiritual enlightenment.

dust, iron, lead, copper and glass. Iron features most often in images of the *pāras* (philosopher's stone) (Kabīr, ĀG 481; Amar Dās, ĀG 638; Ravidās, ĀG 973;[10] Rām Dās, ĀG 1245; Nānak ĀG 1311) which transforms iron into gold.[11] In the Gurūs' hymns, the *pāras* represents, variously, Rām ('the Lord') (Rām Dās, ĀG 688 and Nāmdev, ĀG 1351); *rām nām*, i.e. the divine name (ĀG 994, Amar Dās); the Gurū (Rām Dās, ĀG 1114); 'the sanctuary of the True Gurū' (Arjan, ĀG 960); and the *saṅgat*, i.e. the congregation (Rām Dās ĀG 303:7; Nānak ĀG 505). Rām Dās's words read:

> Just as when the iron slag is transmuted into gold by touching the philosopher's stone – when the sinner joins the *saṅgat*, he becomes pure, through the Gurū's teachings. (ĀG 1297)

For Rām Dās (ĀG 1337), as for Kabīr (ĀG 1368), the *pāras* represents the *saṅgat* and for Nānak (ĀG 930) it is the dust of the *saṅgat*'s feet. For the bard, Nall, 'Iron is transformed into jewels when the True Gurū bestows his glance/ grace' (ĀG 1399).

Iron is also repeatedly mentioned as a weighty material which will sink, unless supported by a wooden raft or boat. To quote Gurū Amar Dās:

> The crow does not become white, and an iron boat does not float across. (ĀG 1089)

The iron and wood analogy refers to the human soul and its means of crossing the sea of existence. In this image, wood trumps iron, as an analogy for the salvific. The raft is variously the Lord (Rām Dās, ĀG 309), 'You' (Rām Dās, ĀG 1296), the Gurū (Rām Dās, 1265), the Gurū's *śabad* (words) (Rām Dās, ĀG 1309), and the *saṅgat* (Kabīr, ĀG 335; Arjan Dev, ĀG 896; Rām Dās, ĀG 1309). Thus:

10 A popular story in Ravidās's eponymous community, the present-day Ravidassias, tells how he preferred poverty to wealth which he could have enjoyed had he availed himself of the *pāras* which a visiting saint tried to persuade him to use. For one retelling see http://www.gururavidassguruji.com/stories-of-guru-ravidass (accessed 23 May 2014). This is echoed in the life story of Bhāī Randhīr Siṅgh, founder of the Akhaṇḍ Kīrtanī Jathā, who reported his mental and spiritual turmoil after Bhāī Hirā Siṅgh entrusted him with a proven formula for making gold out of copper; he prayed to God, burned the formula and resolutely dismissed the ideas of wealth that it had induced.

11 The *pāras* (or *pāras pathar* which provided the title of Satyajit Ray's film *Pather Panchali*) corresponds to the *chintāmani* (wish-fulfilling jewel) in Hindu and Buddhist writings and the philosophers' stone in European alchemical writings.

> Just as heavy iron is carried across on wood, so sinners are carried across in the *sadhsaṅgat*, the company of the holy, and the Gurū, the True Gurū, the holy Gurū. (Rām Dās, ĀG 1297)

> One who realizes the Hukam/God's command – his face is radiant and bright, he floats across, like iron upon wood. (Amar Dās, ĀG 1089)

Scripture also evokes the forge: The recasting of iron is an analogy for a wrong-doer's successive rebirths (Nānak, ĀG 752, in Grewal 1986: 204) and for the forging of the mind by the furnace of the body's 'five fires' (Nānak, ĀG 990).

References to attempts at eating iron – especially red-hot iron (Rām Dās, ĀG 317) or by a toothless person (Nānak, ĀG 940) or by someone with 'teeth of wax' (Nānak, ĀG 943) – convey what is humanly impossible, a sentiment also expressed in this verse of Gurū Nānak's:

> Wisdom cannot be found through mere words. To explain it is as hard as iron. (ĀG 465)

Later, the highly respected Bhāī Gurdās (1551–1636) wrote about miracles: 'I may eat iron and make the earth move to my orders' (cited in Prill 2011: 134). Again, through the Gurū's teachings, the impossible is miraculously effortless: 'Through the word of the Gurū's teachings they bite with teeth of wax and chew iron, drinking in the sublime essence of the Lord.' (Rām Dās, ĀG 1324)

It is concentrated meditation on *nām* which makes the impossible possible (Grewal 1986: 205). The capacity to look without bias on gold and iron characterizes the Gurmukh, 'the one who does not slander or praise others' (Tegh Bahadur, ĀG 685), those who are 'the very image of the Lord God' (Kabīr, ĀG 1123), and 'one who is beyond praise and slander' (Tegh Bahadur ĀG 1427).[12]

Viewed positively, iron is, in the scriptures, an element with remarkable potential. It can float, despite its mass, it can be reshaped in the smithy (Nānak, ĀG 752), and can be transmuted into gold through the Gurū's power. Meanwhile, gold is of no use after death, and so it symbolizes *māyā* (illusion) as a distraction from the spiritual path (Arjun Dev ĀG 42, 547; Nānak ĀG 63, 416; Rām Dās ĀG 167; Tegh Bahadur ĀG 718; Kabīr ĀG 343). One passage particularly foreshadows Sikhs' subsequent exaltation of iron/steel and weaponry:

12 Cf. BhG VI. 8 and XIV. 25.

If Truth be the sword and Truth its steel,[13]
Then whatever it cutteth (for oneself) is of infinite Glory,
Yea, if it be sharpened on the whetstone of the Word,
And kept soft in the Sheath of Virtue,
Then if the Sheikh surrenders his head to the sword,
His blood of greed floweth out,
And lo, his life is fulfilled and he's yoked to God,
And mergeth he in God's Vision at the Lord's Gate.
(Nānak, ĀG 956, as rendered by Grewal 1986: 205)

Taken as a whole, scriptural references to iron suggest that iron's high place in subsequent Khālsā Sikh tradition does not stem from the *Gurū Granth Sāhib*. Rather, one needs to reflect on the military leadership of the sixth Gurū, Gurū Hargobind and the tenth Gurū, Gurū Gobind Siṅgh.

GURŪ HARGOBIND AND IRON

On the orders of the Mughal emperor, Jahāngīr, Gurū Hargobind's father Gurū Arjan Dev had been tortured and killed by being boiled in a hot cauldron and then placed on a hot iron plate. On succeeding his father in 1606, Gurū Hargobind was resolved to resist Mughal oppression. He assumed two swords, declaring that one was for *mīrī* (temporal power) and the other for *pīrī* (spiritual authority). He commanded gifts of weapons and war-horses, and his blacksmiths produced weapons for him. When Gurū Hargobind saw the bravery of his youngest son, Tyāg Mal Khatrī (b. 1621), in battle against the Mughals, he renamed him Tegh Bahādur (Pan. 'Brave Sword'), the name by which he is famous as the Sikhs' ninth Gurū. Near Amritsar, Gurū Hargobind had a fort constructed, Lohgaṛh (literally 'Iron Fort') (Fenech 2008: 95).

The steel town of Mandi Gobindgarh in Fatehgarh Sahib district, Panjab, also reputedly originated with Gurū Hargobind, and a local *gurdwārā* commemorates him. Reputedly, during his forty-day stay beside a pool, subsequently renamed Gobindgarh in his honor, the Gurū's men fought against Mughal soldiers.[14] When their weapons were damaged, his men protested that they could not carry on fighting as there was no steel workshop

13 The word *sār* (which occurs only three times in ĀG) is translated as 'steel.'

14 Puzzlingly, this happened in 1646 (i.e. two years after the Gurū's death) according to the above article and elsewhere, e.g. http://mysteelcity.blogspot.co.uk/2011_01_01_archive.html (accessed 4 August 2015).

nearby to repair them. According to local tradition, Gurū Hargobind then declared the place a future center for steel production. In 1902 the local ruler, Mahārāja Hirā Siṅgh of Nabha, started industrial units here; Mahārāja Partap Siṅgh too encouraged the town's industrial development; in 1928 Gobindgarh was declared a free trade zone for steel, and in 1940 land was made available cheaply for blacksmiths.

During the time of Gurū Hargobind we see the elevation of weaponry to being a manifestation of God (Singh 2014_b: 188–89). Satnam Singh quotes the Brahm Kavach, reputedly recited by the Gurū's soldiers:

> [You are] the Kharag (a double edged sword), [you are the] Khandā (another form of long double edged sword), [you are the] Asi (double handled sword).... [You are] the *Dharadhār* (another form of sort sword) and the *Kirpān* (single edged knife) [....] (Singh 2014_b: 189, citing Nihang n.d. 356)

GURŪ GOBIND SIṄGH AND IRON

The rich and complex legacy of Gurū Gobind Siṅgh, Gurū Hargobind's grandson and the tenth human Gurū, similarly includes a site associated with iron. His Lohgaṛh (now replaced by a 1980s *gurdwārā*) was one of five fortresses established by the Gurū to protect Anandpur Sahib (in present-day Himachal Pradesh). Under Gurū Gobind Siṅgh, specialist metal-workers – *mārvāṛīs* and *siklīgaṛs*[15] – produced weapons there, including swords and other arms classified as *amukat* (hand-held), *mukat* (weapons such as quoits which could be hurled) and *muktāmukat* (e.g. javelins and clubs, which could be both wielded and thrown).

It is, however, to Gurū Gobind Siṅgh's legacy of liturgy and literature that the spiritual and symbolic significance of iron and steel for later Sikhs can be most clearly traced. The initiation of Sikhs into the Khālsā re-enacts

15 The word *siklīgaṛ* (A. *ṣaiqal*) means 'polisher.' The Gurū gave these weapon-makers this name because the weapons they made shone so brightly. See 'Sikligar Sikhs' available at www.sikhiwiki.org/index.php/Sikhligar_Sikhs (accessed 6 August 2015). Their descendants, today's Sikligar community, still identify as Sikhs. Today Sikligars probably number about ten million in India, in states including Maharashtra and Andhra Pradesh. Although many Sikligars in Punjab and Haryana identify themselves as Sikhs they do not appear in census returns as Sikh, owing to pressure to record themselves in other categories. They have Scheduled Caste status. They receive some support from diaspora Sikhs concerned at their impoverished condition (oral communications from Kirpal Singh and Pashaura Singh, 16 May 2014).

the 1699 birth of the tradition. Gurū Gobind Siṅgh summoned his followers on the day of the Vaisākhī festival (on the Spring Equinox), challenged anyone ready to die for their Gurū to come forward, and took each of the five volunteers into his tent, apparently to kill them. However, after bringing them out unscathed, he initiated them as his *khālsā* (from A. pure)[16] with water that had been sweetened with *patāse* (sugar sweets) and stirred with his *khaṇḍā*. Initiation into the Khālsā continues to involve sweetened water, stirred in an iron vessel (*bātā*) with a *khaṇḍā*, and initiates are required thenceforth to have five distinguishing markers (the 'five Ks') of which the *kirpān* has a steel blade and the *kaṛā* is made of steel or iron.

In the *Zafarnāmā*, a letter to the emperor Aurangzeb, Gurū Gobind Siṅgh stated: 'When all efforts to restore peace prove useless and no words avail, then the flash of steel is lawful, and it is right to draw the sword.' The sword figures repeatedly elsewhere in the *Dasam Granth* (DG) – also known as *Dasven Pādśāh Kā Granth* – the compilation of poetic works traditionally, but controversially, attributed to Gurū Gobind Siṅgh. God is described as wielding weapons: In *Jāp Sāhib* the Gurū proclaimed:

> Salutations to God who wields the sword,
> Salutations to God who hurls arrows. (DG 52)[17]

For Gurū Hargobind the *tegh* was for the 'defence of the weak' and the 'destruction of the aggressor' (Singh 1995: 544), and – as we have seen – an equation between God and weaponry appears to date from his time. With Gurū Gobind Siṅgh this became a dominant theme:

> Guru Gobind Singh identified the *tegh* or sword with the Lord Creator and thereby gave it a still deeper meaning. He addressed it as Bhagauti (Goddess), Sri Kharag (Lord Sword), Jag Karan (Creator of the World) and Sristi Ubaran (Saviour of the Creation), besides reiterating its role as protector of the good [...] and destroyer of the evil. (Singh 1995: 544)

Quoting from *Śrī Dasam Granth* (1992), Murphy writes:

> The *Shastarnāmmālā* in the Dasam Granth provides the literary apotheosis of this orientation in its praise of weapons and association of them with both worldly and otherworldly power: 'You are the arrow, the spear, the hatchet

16 The term was used by the Mughals for land held directly by the emperor. Similarly, the Gurū's Khālsā would owe allegiance to no intermediaries.

17 The DG pagination refers to the standard printed 1428-page volume.

(*tabar*), and the sword (*talwār*) / Who recites your name crosses the fearful ocean.' (2012: 90, quoting DG 717)

Likewise, Gurū Gobind Siṅgh began his *Bachitar Nāṭak* ('Wonderful Drama') with an invocation to the sword for assistance in completing his writing:

> I bow with heart and mind to the holy sword:
> The sword cuts sharply, destroys the host of the wicked.
> The sword brings peace to the saints,
> Fear to the evil minded, destruction to sin.
> So it is my refuge.[18]

Yet, in the following line, Gurū Gobind Siṅgh calls God not sword but *kharagpānan* ('sword-wielding'). Clearly, for the Guru-poet, the sword and the divine power are one. Gurū Gobind Siṅgh 'consecrated ... a whole spectrum of weaponry':

> The sword, the sabre, the scimitar, the axe, the musket, the shaft.
> The rapier, the dagger, the spear: these indeed are our saints. (Neki and Singh 1995: 322)[19]

Moreover, Gurū Gobind Siṅgh reputedly stated that the sword was one of his own manifestations, alongside the *Gurū Granth* (scripture) and *Gurū Panth* (community) (Singh 2014_b citing Fenech 2008: 205). His signature was a drawing of a sword (Singh 2014_b: 190 citing Singh 2001: 15), and he declared that *darśan* of a sword was *darśan* of the Gurū (Singh 2014_b:197 citing Nihang and Singh 2009: 36).

Jeevan Deol (2001: 32–33) interprets Gurū Gobind Siṅgh as seeing his own mission as 'embedded in a wider cosmological cycle of battles against evil that extends back into mythical time,' consciously referring to Purāṇic tales of divine births/incarnations and to Kr̥ṣṇa's explanation of his *avatāras* (Skt., God's descents to earth to restore righteousness) in *Bhagavad Gītā*. According to Lou Fenech (2008: 45) the reason 'why for the tenth Guru the sword became the ultimate symbol of the Eternal Guru, the *sarab loh*, All Steel, the protector of the oppressed from the scourge of the unrighteous' may, at least in part, lie in his rigorous training through martial

18 For another rendering see Singh and Singh (1999: 109).

19 *As kripān khando kharag tupak tabar aru tīr /*
Saif sarohī saiththi, yahai hamārai pīr.

games and swordsmanship from an early age. As Singh points out (1998: 58), in Akal Ustati Gurū Gobind Siṅgh declared:

> Akal Purakh (Eternal God) you protect me, All-Iron (Sarab loh), you protect me;
> The destroyer of all protects me, Sarab loh always protects me. (DG 11)[20]

Traditionally Gurū Gobind Siṅgh was regarded as the author of the 'All-Iron Volume,' *Sarabloh Granth* (McLeod 2003 and Singh 1998).[21] In part one of *Sarabloh Granth*, Sarabloh is one of the many names given to Devī, the Goddess. In part two, Viṣṇu is 'entreated to become incarnate as Sarabloh (stanza 1167)' and in part five Sarabloh is clearly 'an incarnation of Mahakal or Gopal, the Supreme Deity' (stanza 2386) (Singh 1998: 58).

Thus Gurū Gobind Siṅgh addresses Akāl Purukh (Eternal God) not only as a sword but as Sarab Loh (All Iron) and I would propose that, as an epithet, *sarab loh* is not only a synonym for sword (as Fenech's statement, above, suggests) but also an attribution to God of the qualities of the element of iron itself.[22] These include polished steel's cutting edge and also (as the ĀG verses suggested) hardness, strength and the potential for transformation.

BHAGAUTĪ *AS SWORD*

The sword/*sarab loh* equates to the divine feminine no less than to the divine as (more often) expressed in masculine terms. Neki and Singh outline the 'chequered semantic history' (1995: 319) of a word which is synonymous with the Goddess Durgā, yet also means, in Sikh scripture, a devotee (e.g. ĀG 88) and was then used by Bhāī Gurdās to mean sword. They translate *nāu bhagautī lohu gharāiā* (Var 25: 6) as 'iron (a lowly metal) [which] when properly wrought becomes a (powerful) sword.' In compositions attributed to Gurū Gobind Siṅgh, Śrī Bhagautı means 'the Divine

20 For the original text and another rendering see www.sridasam.org/dasam?Action-Page&p=33 (accessed 26 August 2015).

21 Although the *Sarabloh Granth* continues to be regarded as scripture by Nihaṅgs (see below), most scholars, on substantial internal evidence, now regard the text as largely attributable to others (Singh 1998; McLeod 2003).

22 Similarly, in criticizing material interpretations of Sarab Loh, 'Satnaam Babaji' says 'The Sarab Loh of Guru Gobind Singh ji is where one is impregnated with the Divine Spirit – pure like iron through and through,' which he proceeds to expand as 'being lovingly devoted to Him and doing absolutely no gossiping and slander of anyone or anything.' See www.eternalguru.info/sarab-loh-all-iron-discipline (accessed 26 August 2014).

Sword [which] symbolizes Divine Power' (ibid.: 321). In fact, Bhagautī (Skt. Bhagavatī) is a 'bipolar mystic symbol' (ibid.) which is 'in its symbolic meaning of Divine Power, in contact with the Infinite, and in its concrete form, as a weapon, in contact with the finite' (ibid.: 322). So too, in the congregational prayer (*ardās*), Bhagautī is invoked as a sword, but as a sword that is charged with divine power. As Nikky-Guninder Singh highlights: 'Like the goddess, the sword performs the dual role of preserving the good and demolishing the evil, the negative elements...the sword, like Durgā for the gods, was to be invoked only in self-defense and as a last resort.' (1993: 147)

Vār Durgā Kī graphically describes the sword's/Durgā's attack at the other gods' request on the demon Mahikāshura [Skt. Mahiṣāsura]. As Singh observes, 'besides being the sharp and penetrating power of intelligence, *bhagautī* is also a symbol for crusade against evil' (1993: 146). Such interpretation rather glosses over the contested relationship between Bhagautī and Caṇḍī (the Goddess by another name) and the sword in Sikh literature. As Satnam Singh reports (2014_b), the Singh Sabha scholar, Sahib Singh, strove hard to prove that Gurū Gobind Siṅgh's many references to Caṇḍī and Bhagautī were indeed to the sword and not to the Goddess.

Further to that, despite a frequently 'androcentric' Sikh understanding of the sword 'as a weapon of male heroes' (Singh 2005: 116), we have noted how the name *bhagautī* introduces *de facto* a female dimension. Sikhs generally point to Khālsā symbols and discipline as being unisex. Indeed, for Singh (ibid.: 134), cited by Murphy (2012: 62), part of what the five Ks symbolize is sexual equality, since the *kaṛā*, *kirpān* and other three Ks are to be borne by Khālsā Sikhs regardless of their gender.[23]

THE SWORD IN RAHITNĀMĀS

The term *bhagautī* appears too in *Prem Sumārag* (Pan. 'The Way of True Love'), a *rahitnāmā* or Khālsā disciplinary code possibly dating from the

23 Gender-based distinctions are found in the historic *rahitnāmās* (see below). Thus, a woman candidate for initiation into the Khālsā 'should wear an iron ring on her finger' according to the *Prem Sumārag* (Malhotra 2009: 182, translating Siṅgh 1965: 16–17). Also, the following procedure is only to be followed if a son is born: 'First, [take the child] and touch his forehead on the floor in front of the sword (*śrī sāhib*), the [other] weapons, and the sacred Grānth (*granth pothī*) [...] Touch his *gurhatī* [sweet substance given to a newly born child] with a two-edged sword (*khaṇḍā*) and administer it to him.' (McLeod 2006: 29)

early nineteenth century (McLeod 2003: 151).[24] The prescribed birth ceremonies include a requirement to unsheathe a sword (*bhagautī*), sprinkle red lead on it, and drape a garland over its hilt. Then this is laid on a stool in front of the mother and her baby (McLeod 2006: 31).[25] Indeed, weapons are integral to the rituals of birth, initiation, marriage and death in *Prem Sumārag* (Singh 2014_b: 190) and references to iron recur in the historic *rahitnāmās*. Bhāī Nand Lāl's *rahitnāmā* prohibits showing disregard to iron, the substance of weapons; such disregard results in endless rebirth (Singh 2014_b: 190 citing McLeod 2003: 281).

Mention of iron in the codes of present-day Khālsā groups is instanced below. It is worth mentioning that later *rahitnāmās*, including the *Sikh Rahit Marayādā*, the code that is most widely followed by Sikhs today, specify the preparation of *kaṛāh prasād*, a sweet dish made of wheat flour, sugar, clarified butter and water, which is distributed in the *gurdwārā* as part of congregational worship. It is to be prepared in a *kaṛāhī*, an iron wok-like utensil. 'This linked it to the Khalsa veneration for iron and gave rise to the name *karah prasad*' (Hawley 2014: 322 citing McLeod 2003: 216).

WORSHIP OF WEAPONS

Sikh tradition refers more to *tegh* than to *deg*, to weapons more than to kitchenware. As Murphy explains (2014: 452): 'Weapons are ubiquitous throughout Sikh religious sites, both symbols and actual tools of a larger warrior culture and a commitment to armed defence and protection.' She suggests also that many of these weapons were part of 'the cultural practice of *khil'at*, the formal exchange of gifts between client and patron, ruler and subject, which constituted an important part of community and state formation in pre-colonial and colonial India' (ibid.: 452). Weapons are among other 'relics that substantiate relationships,' for example between a supporter and a guru, that are preserved in private families as well as in some public *gurdwārās*. Such is the case of both Mahārāja Ranjīt Siṅgh's collection of weapons related to the Gurūs (which he worshiped daily) and the Maharaja of Patiala's collection (Murphy 2009 and Singh and Singh 2012). Sometimes weapons' miraculous associations are recalled – for

24 Cf. Malhotra (2009: 182), who suggests a rather earlier date.

25 According to *Prem Sumārag*, baptism of the sword (*pāhul khande kī*) must be administered to a Sikhni who has become pregnant: 'Unsheath a sword and lay it in front of her, together with a bow and five arrows.' (McLeod 2006: 28)

example water gushed out when Gurū Gobind Siṅgh struck the earth with a *barshā* (heavy spear), now in Anandpur Sahib.[26]

Such weapons are not simply housed or displayed but are also enshrined for worship so that *darśan* can be obtained by devotees. Visitors to Gurdwara Nagina Ghat Sahib (near Takht Hazur Sahib in Nanded, Maharashtra) see weapons arranged 'in the form of a lotus, in front of both the Adi Granth and the Dasam Granth' (photographed in Singh 2014_c: 463). At Gurdwara Keshgarh Sahib in Anandpur Sahib, weapons associated with Gurū Gobind Siṅgh are shown to devotees each evening.[27] In the Paonta Sahib Gurdwara too there is an impressive display of weapons.

IRON IN RESURGENT KHĀLSĀ GROUPS

NIHAṄGS

The Nihaṅgs' origins pre-date the reformist Tat Khālsā movement which, from the late nineteenth century, shaped what became the dominant expression of 'Sikhism' (Nihang and Singh 2009; Pamme 2010). The distinctiveness of the Nihaṅgs involves a prominent respect for weapons: the perpetuation of *śastar vidiā* (knowledge of weaponry) and the incorporation of iron or steel martial insignia in their *dumālā* (towering turban) and elsewhere on their person. Their *derās* (centers) in India offer training in traditional martial arts (*gatkā*)[28] (Judge 2014: 378). Moreover, along with the *Ādi Granth* and the *Dasam Granth*, Nihaṅgs honor as scripture the *Sarabloh Granth*.

Traditionally Nihaṅg worship of weapons (*śastar pūjā*) has differed little from other (Hindu) *pūjā*, involving as it does honoring them with flowers and incense. On a more distinctive note, Nihaṅgs' *śastar pūjā* in India may involve anointing them with a decapitated goat's blood collected in an iron utensil. To quote Jathedar Baba Surjeet Singh: 'Weapons are sustained by blood.'[29]

26 Consult http://www.info-sikh.com/GGPage1.html (accessed 11 August 2014).

27 See photographs at http://www.info-sikh.com/GGPage1.html (accessed 11 August 2015).

28 *Gatkā* has soared in popularity among Sikhs worldwide since the 1980s as *śastar vidiā* has been attracting increasing attention (see Singh 2014_c).

29 Quoted from an oral interview in July 2007, cited in 'Jhatka' available at www.nihangsingh.org/website/trad-jhatka.html (accessed 11 April 2015).

On the basis of fieldwork, and reflection on the different cultural contexts of Indian and UK Nihaṅgs, Satnam Singh distinguishes the Indian-raised Nihaṅgs' reverent maintenance of centuries-old *śastar pūjā* from its reinterpretation by UK-raised Nihaṅgs. For the latter, a reverent but less *pūjā*-like acknowledgement of weapons prefaces their practice of *gatkā* and *śastar vidiā*. They regard the practice of *shastar vidiā* as itself worship and see the ritual performance of *pūjā* as unnecessary.

AKHAṆḌ KIRTANĪ JATHĀ

Of more recent origin than the Nihaṅgs, the Akhaṇḍ Kirtanī Jathā (AKJ) draws inspiration from Bhāī Randhir Siṅgh (1878–1961) and takes its name from its practice of continuous all-night hymn-singing (*akhaṇḍ kīrtan*) (Singh 2014_a: 211; Barrow 2001: 100; Nesbitt 2014: 365). The AKJ's disciplinary code (*gurmat bibek*) requires *sarab loh* ('all iron'), in other words iron utensils and iron *thālī*s (trays, platters) rather than the more widespread steel *thālī*, or indeed china crockery. The 'Sarbloh Bibek' section of *Gurmat Bibek* recalls that at birth a child is given *amrit* (sweetened water) '*sarab loh vich*' ('in all-iron,' i.e. from the tip of a *kirpān* that has been dipped into an iron bowl) and 'through Kirpan we are taught to turn eatable food into Amrit food.'[30] The fact that 'as soon as we get out of the Amrit Sinchaar Mandal (place) [i.e. the place where the Khālsā initiation/*amrit saṅskār* took place] we start using same aluminum and brass' is denounced for being as reprehensible as 'wearing the Kachhera to fulfil the task of getting Amrit and then removing the Kachhera and wearing same Dhoti, Tambi and Langoti'[31] – in other words reverting to one's previous garments.

Further to that, iron tools are praised for having various beneficial properties. In a style typical of Sikh 'scientific' defenses of hallowed tradition, Squadron Leader Ram Singh of AKJ, explained: 'We also eat from utensils of iron because iron is a magnetic element and if our women in the villages took their food in that manner there would be no need for iron and B12 supplements during pregnancy' (Barrow 2001: 105).[32] A similar point

30 From Kulbir Singh's 'humble translation' of *Gurmat Bibek*: http://www.gurmatbibek.com/contents.php?id=933 (accessed 6 August 2015).

31 Ibid.

32 While condemning AKJ-style interpretation of Sarab Loh as 'an iron utensil and a strict regime' 'Satnaam Babaji' explains that it was because of the need for iron in food that Gurū Gobind Siṅgh recommended Sarab Loh to his soldiers. See www.eternalguru.info/sarab-loh-all-iron-discipline/ (accessed 26 August 2015).

about the efficacy of the *kaṛā* was made – with hope rather than conviction – by a Coventry female informant who had received *amrit* from the AKJ. She explained that her anemia (which she attributed to her religiously-informed vegetarian diet) was her 'sacrifice' for her faith. Sukhmandir Khalsa's online advertisement of a 'sarbloh spoon' also mentions its value in combating anemia,[33] an example of 'participant believers borrowing the adoptive language of science to articulate personal convictions' (Keenan and Arweck 2006: 2). Elsewhere other analogies are indicated: 'Iron is magnetic in the way in which the Sikh desires to adhere to the Guru. Iron rusts if neglected and must be scrubbed clean immediately after use and regularly maintained. Similarly the mind is easily corrupted if meditation as a spiritual focus is neglected.' (Khalsa n.d.) Khalsa also picks up the *Gurū Granth Sāhib*'s *pāras* theme: 'Spiritual association may transmute the properties of psyche the way the mythical touchstone transmutes the properties of metal.'

DAMDAMĪ ṬAKSĀL

The title of a third Khālsā group, the Damdamī Ṭaksāl, includes the word *taksāl* (Pan. 'mint,' in the sense of a coin factory, although it is translated as 'educational institution'). Although 'there seems to be no difference in the Sikh code of conduct advocated by the Damdamī Ṭaksāl and the SGPC' (Judge 2014: 374), the Ṭaksāl's code of discipline, the *Gurmat Rehat Maryādā Damdamī Ṭaksāl*, is distinctive in its repeated mention of iron (Damdami Taksal 2004):

> The Kara must be of Sarab Loh (pure iron). The Khālsā is not to wear a kara that is made of gold, silver, brass, copper or one that has grooves in it. Only the Sarab Loh is acceptable to Guru Jee. (ibid.: 25)

In this document 'degh' [sic] is used interchangeably with 'karah parshad' and the utensil in which the ingredients are cooked is referred to as 'a Karahi (iron wok)' (ibid.: Annex 1).

> All the utensils used are to be of Sarab Loh. (ibid.)

33 http://sikhism.about.com/od/Sikh-Clothing/fr/Sikhistore-Sarbloh-Spoon-Review.htm (accessed 6 August 2015).

> Before the *kaṛāh parshād* is distributed, 'when the person doing the Ardas says do Bhog[34] to the Degh, at this point the Kirpan is to be placed into the Degh and withdrawn. The Kirpan is symbolic of the Gurū accepting the Degh as weapons are also a form of the Guru.' (ibid.)

With reference to the procedure shortly after a child's birth, the instruction is:

> An amritdharee Singh should get a Sarab Loh bowl and half fill it with water [...] This mixture is stirred with a Sarb Loh [sic] kirpan [...] To break all superstitions a Sarab Loh kara is to be placed on the child's right-hand wrist.[35] (ibid.: 12)

Regarding the *amrit* ceremony:

> All the instruments for the ceremony have to be of Sarab Loh, for example the bowls, bucket, vessel, cauldron, spatula, khanda and a large kirpan [...] (ibid. 15)
> Also 'The Sikhs take Amrit of the Pahul (iron bowl) to become initiated' [...] (ibid. 19)
> Given that iron weapons 'are a form of the Guru' it is unsurprising to read that: 'Weapons are to be shown utmost respect. They are not to be kept in a place towards which your feet will face.' (ibid. 54)

Iron features again in the Ṭaksāl's code, in the punishments described for adulterers after their deaths. Invoking verses in ĀG 546 and 1362, readers learn that: 'Those who have sexual relationships outside of the bonds of marriage go to hell and in the after-life they will suffer the pain of embracing red-hot iron pillars' and 'these individuals are boiled in cauldrons of hot oil.'

34 Bhog is the offering of food to deities (in a Hindu temple) or to the *Gurū Granth Sāhib* in *gurdwārā*s of the Nanaksar tradition. The wording 'do Bhog to the Degh' suggests that food (*kaṛāh prasād*) is being offered to the cauldron, whereas what is meant is that the *kaṛāh prasād* is being offered to God/Gurū as represented by the sword/*kirpān*.

35 The code indicates that the *kaṛā* is to be used instead of the threads and amulets widely used to ward off the evil eye. With reference to iron, a Coventry informant explained that the *kaṛā*, being iron, is believed to counter black magic: 'One of the best ways to prevent evil forces is always wear an iron bangle ... I must make an out of place observation here that an iron bangle also known as the "kara or kada" is one of the five essential items that a true Sikh has to wear. I do not imply that it has to do with warding off evil but certainly some basic unity of thought can be perceived.' (Ray 2013)

Like the AKJ, the Ṭaksāl refers to the beneficial value of the iron *kaṛā*, *kaṛāhī* etc. in counteracting anemia: 'According to scientific research, the Kara adds to the iron levels in the body by rubbing on the skin.'[36]

SIKH DHARMA OF THE WESTERN HEMISPHERE

Uniquely among Khālsā groups, Sikh Dharma emerged as a result of the outreach of Harbhajan Singh Puri (aka Yogi Bhajan) in North America, and consists of mainly *gorā* (Pan. 'white', i.e. of 'European' origin) Sikhs (Takhar 2005). In his recorded English-language lectures, the word 'steel' recurs: 'strong as steel: steady as stone'; 'satkriya naturally helps develop nerves of steel'; 'Hold these hands tight like steel'; 'The time has come when our steel has to be tested'; and 'Forge from this raw steel a tempered sword of clarity about how to be a man.' Yogi Bhajan also envisaged construction of a Temple of Steel.[37]

MATERIALIZING RELIGION

To quote Houtman and Meyer (2012: xv), '[r]eligion cannot persist, let alone thrive, without the material things that serve to make it present – visible and tangible – in the world.' As they point out, probably a Protestant devaluing of the material culture of religions has led to religion being discussed predominantly in terms of beliefs and concepts rather than religious material culture. So, scholarship that focuses on the material aspect of a religion can also draw criticism from an adherent. A Sikh has denounced scholarship which in his view aims to reduce Sikh religion to its material expressions (Mann 2013). Yet such scholarship plays a useful role in addressing an 'opposition between spirituality and materiality' (Houtman and Meyer 2012: 1). Just as objects, no less than books, can be 'read' and disclose much – certainly about science, technology and art – so too they can be read for a fuller understanding of 'religion.'

Nineteenth and twentieth century Europeans who encountered Sikhs tended to understand and explain what came to be termed 'Sikhism' as a

36 This point is also made in the Damdamī Ṭaksāl's disciplinary code. See www.vidhia.com/Rehat%20Maryada/Gurmat_Rehat_Maryada_-_Damdami_Taksa;.pdf (accessed 15 March 2015).

37 See http://fateh.sikhnet.com/sikhnet/Prosperity_Paths.nsf/TOC/5c048599ff60dac085256660001579fa!Open (accessed 8 August 2015).

'Protestant' reaction to Hindu tradition. Certainly from Gurū Nānak's compositions one has a strong sense of a man who, not unlike the drivers of Europe's Reformation, challenged the belief that priesthood, ceremonies and pilgrimages could lead to liberation. Only good works and divine grace could do so. A self-understanding as reformist and 'protestant' tends to attribute preoccupation with material externals to a Catholic (or Hindu) other and to disregard their role in one's own protestant (or Sikh) tradition.

Nevertheless, as Murphy has argued, the Five Ks, markers of Khālsā membership, illustrate how 'Sikh materiality functions to produce the community' (2012: 56). They signify Sikhs who are living out the Sikh code of conduct, the *rahit*, and they are 'symbolically charged' (ibid.: 62). Let us take the case of the *kaṛā*, the most widely observed of the Five Ks which is probably worn by most non-*amritdhārī* Sikhs, as well as by Khālsā Sikhs. In addition to its reputed protective qualities[38] (literally so for the wearer's right arm), the *kaṛā* is a 'handcuff' to God, a reminder to be good, a symbol of God's oneness and infinity. According to Kapur Singh: '[i]n the Sikh symbology [...] the iron bangle represents a view of life that is positive and world-affirming ...' (1989: 113–14). With reference to the *cakra* (Skt.) and *cakka* (P.) of Hindu and Buddhist philosophy, he lists five of its many symbolic meanings: mindfulness of one's 'double role of a spiritual aspirant and a useful citizen'; being compassionately God-centered; being guided by faith in God; remaining 'well-protected against inharmonious influences'; and cooperating with God's will.

Here it is interesting to note that, whereas in four of the five instances above, it is the shape of the *kaṛā*, its likeness – as a circular object – to the *cakra* (Pan. *cakkar*) from which Kapur Singh adduces its symbolic power, in the fourth instance it is to the *kaṛā*'s material substance that he refers, i.e. the fact that it is iron. As he says: 'iron: the world over, is commonly taboo to evil spirits and prevents spells from taking effect' (1989: 114).

Echoing Uberoi's deconstruction of the Five Ks as deliberate inversions of Hindu symbolism (1975), one sees iron – as substance – making a statement, providing a literal hardening of prior tradition. Thus, protection worn around the wrist is no longer a *rākhī* thread but a *kaṛā*. Analogously, with reference to the sacred thread worn across the torso by high-caste Hindus (Pan. *janeu*; Skt. *yajñopavīta*), Muhammad Qasim's *Ibratnāma* reports that, discomfited by the army of Wazir Khan, and grieving for his sons,

38 Nesbitt's eight- to thirteen-year-old informants in Coventry, UK, mentioned that the *karā* brought good luck, protected them in dangerous situations and stopped them feeling scared at night (2000: 198–99).

Gurū Gobind Siṅgh spent his days in mourning. The sacred thread was also given up for the chain of iron (Grewal and Habib 2001: 114).

Neither reliance on iron's protection nor respect for weapons is peculiarly Sikh,[39] nor, as Jewish and Christian scriptures illustrate, is extolling God's protective power through images of weapons and armor, and invoking this imagery to describe spiritual preparedness.[40] However, the significance of iron for the sixth and tenth Gurūs and its centrality in Khālsā initiation and daily discipline, as well as the invocation of God as *sarab loh*, combine both to ensure iron's pre-eminence as an element in Khālsā experience and the Khālsā's uniqueness among religious communities in this respect.

In its capacity for transformation the (lowly) iron, which provides nevertheless a title for God, parallels the (ordinary) water which is transformed into *amrit*, and the human beings who can be transformed into Khālsā. Moreover, as *amrit* can transform by healing and protecting (Nesbitt 1997), so iron can transform – as in the case above of the *kirpān* blessing and transforming food. Much as the word *amrit* expresses *nām*, 'the nectar of the Name,' in the *Gurū Granth Sāhib* too *sarab loh* invokes, and indeed is, God (Nesbitt 1997). Meanwhile, the designation *bhagautī*, the Goddess, dignifies the steel sword. Through daily contact with the *kaṛā* and *kirpān* (and, especially in the AKJ's case, with the ironware involved in cooking and eating) the Khālsā is reminded of the divine. Things – and not words alone – are God's expression for those who have themselves been transformed.

In religious symbolism the material of which things are made may be as significant as their forms.[41] The shapes of the sword (*kirpān*, *khaṇḍā*, *tegh*), the *kaṛā* and the *deg* invite interpretation as powerful symbols and their substance too is freighted with ancient and pervasive belief in supernatural protective properties. Contemporary scientific rationales for *sarab loh*, as counteracting anemia, provide modern expression for iron's intrinsic virtues.

39 For example, the Yoruba people of West Africa venerate Ogun, god of iron and war (Fatunmbi 2000).

40 See for example Ps XXVIII.7 and Eph VI.10–18.

41 Not only shape but also size is significant, as the many discussions of the validity of miniature *kirpāns*, worn as a pendant or attached to a *kaṅghā*, indicate. Although many Sikhs substitute these, the practice is prohibited for *amritdhārī* Sikhs and is a point of difference from Namdharis. See www.namdharitruth-info/fallacies/this-missing-kirpan/ (accessed 27 August 2015).

ABBREVIATIONS AND REFERENCES

A. = Arabic P. = Pali Pan. = Panjabi Skt. = Sanskrit

ĀG — *Ādi Granth* (*Gurū Granth Sāhib*). Translated by Dr Sant Singh Khalsa, http://www.sikhnet.com/files/ereader/Siri%20Guru%20Granth%20-%20English%20Translation%20%28matching%20pages%29.pdf (accessed 25 August 2015).

AKJ — Akhaṇḍ Kirtanī Jathā

Anon. 2007 — Anon. (2007). 'The Miracle of Iron.' Available at www.islamreligion.com/articles/562/ (accessed 29 March 2014).

Arweck and Keenan 2006 — Arweck, E. and W. Keenan (eds.) (2006). *Materializing Religion. Expression, Performance and Ritual*. Aldershot: Ashgate.

Barrow 2001 — Barrow, J. (2001). 'The Akhand Kirtani Jatha: A Local Study of the Beliefs and Practices of its Members.' *Journal of Punjab Studies*, 8(1): 97–115.

BhG — *Bhagavadgītā*. Van Buitenen, J.A.B. (ed., tr.) (1981). *The Bhagavadgītā in the Mahābhārata*. Chicago: The University of Chicago Press.

Cole and Sambhi 1990 — Cole, W.O. and P.S. Sambhi. (1990). *A Popular Dictionary of Sikhism*. London: Curzon.

Damdami Taksal 2004 — Damdami Taksal. (2004). 'Gurmat Rehat Maryada: Sikh Code of Conduct (Summarised Version).' *Vidhia.com, a project of Khalis Foundation*, www.vidhia.com/Rehat%20Maryada/Gurmat_Rehat_Maryada_-_Damdami_Taksal.pdf (accessed 15 March 2015).

Deol 2001 — Deol, J. (2001). 'Eighteenth Century Khalsa Identity: Discourse, Praxis and Narrative.' In Shackle, C., G. Singh and A.P. Singh Mandair (eds.), *Sikh Religion, Culture and Ethnicity*, pp. 25–46. London: Curzon.

DG — *Dasam Granth*

Eph — *Ephesians, The Bible*

Fatunmbi 2000 — Fatunmbi, A.F. (2000). *Ogun. Ifa and the Spirit of Iron*. US: Original Publications.

Fenech 2008 — Fenech, L.E. (2008). *The Darbar of the Sikh Gurus. The Court of God in the World of Men*. New Delhi: Oxford University Press.

Goswamy 2002 — Goswamy, B.N. (2002). 'Enigma of the Iron Pillar.' *The Tribune*, Sunday 14 July, www.tribuneindia.com/2002/20020714/spectrum/art.htm (accessed 18 March 2015).

Grewal 1986 — Grewal, J.S. (1986). *Imagery in the Adi Granth*. Chandigarh: Punjab Prakashan.

Grewal and Habib 2001 — Grewal, J.S. and I. Habib (eds). (2001). *Sikh History from Persian Sources*. New Delhi: Tulika Books.

Gurmat Bibek 2009 — *Gurmat Bibek* (2009). http://www.gurmatbibek.com/ (accessed 24 August 2015).

Hawley 2014 — Hawley, M. (2014). 'Sikh Institutions.' In Singh and Fenech 2014: 318–27.

Houtman and Meyer 2012	Houtman, D. and B. Meyer (2012). *Things. Religion and the Question of Materiality*. New York: Fordham University Press.
Judge 2014	Judge, P.S. (2014). '*Taksals*, *Akharas*, and Nihang *Deras*.' In Singh and Fenech 2014: 372–81.
Keenan and Arweck 2006	Keenan, W.J.F. and E. Arweck. (2006). 'Introduction: Material Varieties of Religious Expression.' In Arweck, E. and W. Keenan (eds.), *Materialising Religion. Expression, Performance and Ritual*, pp. 1–20. Aldershot: Ashgate.
Khalsa n.d.	Khalsa, S. (no date). 'Sarbloh – All Iron.' https://www.google.co.uk/#q=sarb+loh (accessed 18 March 2015).
Malhotra 2009	Malhotra, K.K. (2009). 'Contemporary Evidence on Sikh Rites and Rituals in the Eighteenth Century.' *Journal of Punjab Studies*, 16(2): 179–98.
Mann 2013	Mann, J. (2013). 'Customer Review.' https://pod51036.outlook.com/owa/#path=/mail (accessed 21 August 2015).
McLeod 1995	McLeod, W.H. (1995). *Historical Dictionary of Sikhism*. London: The Scarecrow Press.
McLeod 2003	McLeod, W.H. (2003). *Sikhs of the Khalsa. A History of the Khalsa Rahit*. New Delhi: Oxford University Press.
McLeod 2006	McLeod, W.H. (2006). *Prem Sumārag. The Testimony of a Sanatan Sikh*. New Delhi: Oxford University Press.
Murphy 2009	Murphy, A. (2009). 'The Guru's Weapons.' *Journal of the American Academy of Religion*, 77(2): 1–30.
Murphy 2012	Murphy, A. (2012). *The Materiality of the Past: History and Representation in Sikh Tradition*. New York: Oxford University Press.
Murphy 2014	Murphy, A. (2014). 'Sikh Material Culture.' In Singh and Fenech 2014: 448–58.
Mushet 1840	Mushet, D. (1840). *Papers on Iron and Steel. Practical and Experimental. A Series of Original Communications Made to the Philosophical Magazine, Chiefly on Those Subjects*. London: John Weale.
Nabha 1974	Nabha, K.S. (1974). *Guru Shabad Ratnakar Mahan Kosh*. Third edition. Patiala: Bhasha Vibhag.
Neki and Singh 1995	Neki, J.S. and G.B. Singh. (1995). 'Bhagautī.' In Singh, H. (ed.), *The Encyclopaedia of Sikhism*, pp. 319–22. Patiala: Punjabi University.
Nesbitt 1997	Nesbitt, E. (1997). '"Splashed with Goodness": The Many Meanings of *Amrit* for Young British Sikhs.' *Journal of Contemporary Religion*, 12(1): 17–33.
Nesbitt 2000	Nesbitt, E. (2000). *The Religious Lives of Sikh Children. A Coventry Based Study*. Leeds: Community Religions Project, University of Leeds. Available at http://arts.leeds.ac.uk/crp/files/2014/06/nesbitt2000.pdf (accessed 20 August 2015).
Nesbitt 2014	Nesbitt, E. (2014). 'Sikh *Sants* and their Establishments in India and Abroad.' In Singh and Fenech 2014: 360–71.

Nihang and Singh 2009	Nihang, N.S. and P. Singh. (2009). *In the Master's Presence. The Sikhs of Hazoor Sahib*. London: Kashi House.
Pamme 2010	Pamme, R. (2010). *The Pilgrimage to Takht Hazar Sahib and Its Place in the Sikh Tradition*. PhD Thesis, School of Oriental and African Studies, University of London.
Prill 2011	Prill, S.E. (2011). '"Except the True Name, I Have No Miracle": Modern Sikh Understandings of the Miraculous.' In Singh, P. (ed.), *Sikhism in Global Context*, pp. 130–45. New Delhi: Oxford University Press.
Ps	*Psalms, The Bible*
Ray 2013	Ray, S. (2013). 'How to Break Black Magic?' http://www.metaphysics-knowledge.com/miscellaneous/how-to-break-black-magic.html (accessed 19 August 2014).
Siṅgh 1965	Siṅgh, R. (ed.) (1965). *Prem Sumarag Granth Arthat Khalsai Jivan* (Patshahi Dasvin). Jalandhar: New Book Company.
Singh 1995	Singh, F. (1995). (2nd edition) 'Deg Tegh Fateh.' In: Singh, H. (ed.), *The Encyclopaedia of Sikhism*, Vol. 1, pp. 544–45. Patiala: Punjabi University.
Singh 1989	Singh, K. (1989). *Parasaraprasna*. Amritsar: Guru Nanak Dev University.
Singh 1993	Singh, N.G.K. (1993). *The Feminine Principle in the Sikh Vision of the Transcendent*. Cambridge: Cambridge University Press.
Singh 1998	Singh, G. (1998). '*Sarabloh Granth*.' In Singh, H. (ed.), *The Encyclopaedia of Sikhism*, Vol. 4, pp. 57–58. Patiala: Punjabi University.
Singh 2001	Singh, T. (2001). *The Turban and Sword of the Sikhs*. Second edition. Amritsar: Bhai Chattar Singh Jiwan Singh.
Singh 2005	Singh, N.G.K (2005). *The Birth of the Khalsa. A Feminist Re-Memory of Sikh Identity*. Albany: State University of New York Press.
Singh 2014$_a$	Singh, J. (2014). 'The Guru's Way: Exploring Diversity among British Khalsa Sikhs.' *Religion Compass*, 8(7): 209–19.
Singh 2014$_b$	Singh, S. (2014). 'Worshiping the Sword: The Practice of Sāstar Pūjā in the Sikh Warrior Tradition.' In Jacobsen, K.A., M. Aktor and K. Myrvold (eds.), *Objects of Worship in South Asian Religions*, pp. 182–200. London: Routledge.
Singh 2014$_c$	Singh, K. (2014). 'Sikh Martial Art (*Gatkā*).' In Singh and Fenech 2014: 458–70.
Singh and Fenech 2014	Singh, P. and L.E. Fenech (eds.) (2014). *The Oxford Handbook of Sikh Studies*. Oxford: Oxford University Press.
Singh and Singh 1999	Singh, J. and D. Singh. (1999). *Sri Dasam Granth Sahib Text and Translation*. Patiala: Heritage Publications.
Singh and Singh 2012	Singh, B.S. and R. Singh. (2012) *Sikh Heritage. Ethos and Relics*. New Delhi: Rupa & Co.
Smith 1960	Smith, C.S. (1960). *A History of Metallography. The Development of Ideas on the Structure of Metals Before 1890*. Chicago: University of Chicago Press.

Srinivasan and Ranganathan 1997	Srinivasan, S. and S. Ranganathan (1997). 'Wootz Steel: An Advanced Material of the Ancient World.' In Srinivasan, S. and S. Ranganathan (eds.), *Iron and Steel Heritage of India*, Calcutta: Indian Institute of Metals and Tata Steel.
Takhar 2005	Takhar, O. (2005). *Sikh Identity. An Exploration of Groups among Sikhs*. Aldershot: Ashgate.
Tewari 2003	Tewari, R. (2003). 'The Origins of Iron Working in India, New Evidence from the Central Ganga Plain and the Eastern Vindhyas.' *Antiquity*, 77(297): 536–44.
Uberoi 1975	Uberoi, J.P.S. (1975). 'The Five Symbols of Sikhism.' In Singh, H. (ed.), *Perspectives on Guru Nanak*, pp. 502–13. Patiala: Punjabi University.

Section Four

Body and Embodiment

Chapter 9

A Little Lipstick Goes a Long Way: Chit-Chatting with Women in the *Rāmāyaṇa* and *Mahābhārata*

DEEKSHA SIVAKUMAR

Cosmetics have acquired a bad reputation in recent years. Critics have pointed at the construction of unnatural ideals of beauty, or at the oft-harmful stress on idealized perfection. Discourses on physical ideals of beauty and methods used to achieve these call into question those who use cosmetics as well as those who represent the ideals of beauty. From this perspective, the portrayal of women and their idealization appear to be of utmost importance for a sociological study of cosmetics. Flawless skin, luscious rosy lips, long lustrous hair, and well-outlined eyes form the perfect picture of an attractive woman.

In India, all sorts of skin-whitening creams advertised in countless beauty pageants and by movie stars often with 'botoxed' skin have left the public more than jaded about the industry of cosmetics and makeup. Drawing from the work of several scholars, Ahmed-Gosh notes: 'It is the woman's femininity, purity, submissiveness, mothering, caretaking instincts, compassion, and morality that are evoked by the nation to extol its honor.' (2003: 208) As a matter of fact, nationalism has served a gendered purpose in India, with leaders like Gandhi promoting notions of women being 'guardians of the tradition' and vulnerable to the 'polluting' Western influences of makeup and beauty pageants. More than the cosmetics industry, it is the representation of the moral character of Indian women that really seems to trouble feminists and nationalist leaders alike. As Sutherland (1989) notes, Sītā, unlike Draupadī, was epitomized and cherished for her obedience, not for her beauty. In fact in both the Sanskrit *Rāmāyaṇa* and in its various vernacular versions, very little is said about

Sītā's physical appearance, while all texts agree in privileging her as the epitome of obedience. Seeing Sītā as the ideal wife, daughter or mother fed into her reputation as a submissive and untainted woman. This unsullied reputation is reminiscent of the flawless faces portrayed in contemporary Indian cinema and the advertising industry. Both unquestioningly serve the status quo.

In this chapter I will survey the place of cosmetics within Indian society, moving from an analysis of two conversations recorded in Itihāsa, i.e. Vālmīki's *Rāmāyaṇa* (*Ayodhyākaṇḍa* 109.20ff.) and Vyāsa's *Mahābhārata* (*Āraṇyakaparvan* 222–23). In the historical section preceding this textual analysis, I will survey the use of cosmetics across different regions in the pre-modern world, with special emphasis on the various mineral ingredients that formulate selected unguents and creams. Through this exploration I will show that while cosmetics primarily affect the physical world, their specific components and uses are seldom divulged; if so, only through secrets and in confidence. In the second part of the chapter, I will detail how cosmetics are shared once trust is established. This will be exemplified through a conversation in Vālmīki's *Rāmāyaṇa* between the ascetic *brāhmaṇī* Anasūyā, an elderly wife married to the *ṛṣi* Atri, who asks the young bride Sītā to tell the story (*kathā*) of her miraculous birth from a furrow (*sītā*). To further analyze how cosmetics form a strategy for fostering and maintaining marital relationships, I will then look at an exchange between the wife of five Pāṇḍava brothers, Draupadī, and Satyabhāmā, Kṛṣṇa's favorite wife, at a time when the former's exile in the forest was about to finish. Through this dialog a new rationale emerges for the use of cosmetics. By utilizing makeup, jewelry, creams and flowers both Draupadī and Anasūyā outwardly promote an ethical standard of obedience to one's husband, thus reinforcing the ideal of the *pativratā* (a woman whose vow is to her husband). However, as we will come to see, this relationship is far from subservient and rather shows an active participation by women who use cosmetics to their benefit in social relationships.

THE HISTORICAL USE OF COSMETICS

Cosmetics in modern times are frequently associated with a negative moral characterization of women. Some would argue, however, that the history of cosmetics in the ancient world has more to do with signifying material distinction, cosmopolitan culture, and even medicinal remedies for skin and body ailments. There are many instances where makeup and cosmetics serve real-life purposes without sullying the characters of those who apply them.

Makeup is an essential tool in dance and theatrical performances. It has a transformative and performative power, bringing forth gods, heroes and mythic animals as well as the specific features of human actors. Applying makeup for a performance can take several hours. It may be a tedious occupation requiring numerous extra hands of specialized personnel. Sometimes it even necessitates a personal journey. By ritually transforming the wearer into another being (e.g. a god or a mythical hero), cosmetics create new realities and possibilities for the viewer by enhancing performance and displaying new identities. In the dance and theater performances of Kūṭiyāṭṭaṁ (M.) (Skt. Kūḍiyāṭṭam) and Kathakaḷi (M.) (Skt. Kathākaḷiḥ) from Kerala, this sort of heavy cosmetic application is often cited. In other dance traditions such as Kūcipūḍi (Tel.), makeup along with special contouring techniques, facial expressions, posture and movement, can bend gender norms, allowing a male actor to impersonate female characters and vice versa. While sometimes these types of cosmetics are seen as restyling theatrical conventions, healing the performer, and even entertaining the audience, at other times their purpose is far more sinister: concealing, disguising and masking one's form in order to display another. Our distrust of cosmetics could come from their prolific power in theater, their ability to cloak one's real face and replace it with a false one.

Another problem with cosmetics derives from their composition. In the past, makeup products were hardly considered poisonous because of their unnatural mineral compositions; in fact they have been historically treasured for their unique power to clarify and beautify the human body and face. In order to understand the composition of cosmetics, I will move from accounts recorded in Iranian and ancient Egyptian textual traditions, which provide limited but accessible data regarding what cosmetics would have looked like in the past.

In a study on *haft qalam ārāyish* ('the seven items of cosmetics'), Farmanfarmaian (2000) has examined the cultural and economic importance of cosmetics in Iran and those parts of the world influenced by Irano-Islamic culture. Through ancient textual records to accounts from people in contemporary Iran, including shopkeepers selling cosmetic soaps and rouges, Farmanfarmaian notes the varied perspectives about cosmetics in Iranian culture over the centuries and the techniques for producing them. Of those seven must-have cosmetics, some are obtained from plant-based products, like henna and indigo for dyes. Others, like kohl (eyeliner), whitening powder and rouge, contain minerals. These mineral-based cosmetics were used from at least the twelfth century with very minor changes, to decorate and adorn statues of the gods as well as to beautify human devotees. Pigments obtained from plants and minerals were largely employed for eye shadow colors, cheek stains or rouges and for perfecting lighter skin tones. The luxury and elegance of well adorned and beautified bodies were depicted in paintings and texts now preserved in museum collections. Women in particular were rendered having luscious eyebrows and lashes, large outlined eyes, rosy lips, all achieved by means of advancements in cosmetics. From traces of the contents frequently found in pots in the British Museum, Farmanfarmaian ascribes specific dates and times for the use of various minerals like lapis lazuli, red ochre, black galena and green malachite used in the preparation of eye shadows and eyeliners (2000: 300). The eye in particular was seen as an important and valuable asset of one's face, having the powerful capacity to protect or punish the beholder. In this context, mineral-based eye makeup aided in the treatment of eye disorders and diseases and also rendered the user capable of wielding occult powers or even becoming invisible.

The Egyptians too believed in the curative and protective properties of minerals for eyes, filling their kohl jars with ore of copper, malachite, brown ochre, and magnetic oxide of iron and even lead. While some of the specific cosmetic uses of these minerals are not validated due to lack of evidence on their properties and uses, they are found in considerable amounts in graves – cosmetics were used for the preservation of mummies – and as residues in old cosmetics containers (Lucas 1930: 42–43). Further to that, soot and other vegetable elements were regularly added to kohl, and red ochre accompanied oils to be applied as face pigments and rouge.

The hunger to use cosmetics is recorded as a leitmotif among members of all social classes across the centuries. Usually associated with wealth and luxury, cosmetics have been available to women and men of all socio-economic groups either to display social status, rank and distinction, or as markers of identity. When exotic ingredients were unavailable, other – cheaper – minerals were used. While the upper classes and wealthy elites had access to exotic ointments, the lower classes often made use of powerful substitutes obtained from local resources. In general, the usage and possession of cosmetics was a treasured privilege, as was knowledge of their recipes.

But cosmetics were adaptable too. Popular items on all markets, their original formulas were often altered by producers, depending on availability of sources and prices. Since everyone wished to have them but ingredients were often prohibitive to many, some had to take recourse to cosmetics produced with minerals of lesser potency. On the other hand, with the expansion of the market and a growing clientele, products were enriched with costly ingredients. Such is the case of seventeenth century cosmetics in England, which were basically composed of ceruse, 'the carbonate of lead made by exposing plates of that metal to the vapors of vinegar.' (Gardner 1962: 893) These cosmetic recipes were later altered using Mediterranean herbal ingredients and dyes as well as expensive source materials imported into Venice from merchants all over the world.

The physical self was only part of the cosmetics' purpose. Social relationships could also be enhanced and displayed by effectively using makeup and adorning oneself. Moral character and culture too were distinguishable based on cosmetic usage. In a fascinating study on urban society and Buddhist monasticism in early Indian history, Ali (1998) reports on a cosmopolitan culture indulging on cosmetics and adornment, examining in particular the shift from Vedic society (1500–500 BCE) – a sacrificial culture revolving around *yajña* – to the times of the *Kāmasūtra* (c. second half of the third century CE), when Vātsyāyana adds cosmetics and perfumery in his list of sixty-four fine arts (*catuḥ-ṣaṣṭir aṅga-vidyā*) to be studied by the townsman wishing to seduce maidens:

> The sixty-four fine arts that should be studied along with the *Kamasutra* are: singing; playing musical instruments; dancing; painting; cutting leaves into shapes; making lines on the floor with ricepowder and flowers; arranging flowers; *colouring the teeth, clothes, and limbs* (*daśanavasanāgarāgaḥ*); making jewelled floors; preparing beds; making music on the rims of glasses of water; playing water sports; unusual techniques; making garlands and stringing necklaces; making diadems and headbands; making costumes; making various earrings; *mixing perfumes* (*gandha-yuktiḥ*); putting on jewelry; doing conjuring tricks; practicing sorcery; sleight of hand; preparing various forms of vegetables, soups, and other things to eat; preparing wines, fruit juices, and other things to drink; needlework; weaving; playing the lute and the drum; telling jokes and riddles; completing words; reciting difficult words; reading aloud; staging plays and dialogues; completing verses; making things out of cloth, wood, and cane; woodworking; carpentry; architecture; the ability to test gold and silver; metallurgy; knowledge of the colour and form of jewels; skill at nurturing trees; knowledge of ram-fights, cock-fights, and quail-fights; teaching parrots and mynah birds to talk; skill at rubbing, massaging, and hairdressing; the ability to speak in sign language; understanding languages made to seem foreign; knowledge of local dialects; skill at making flower carts; knowledge of omens; alphabets for use in making magical diagrams; alphabets for memorizing; group recitation; improvising poetry; dictionaries and thesauruses; knowledge of metre; literary

> work; the art of impersonation; the art of using clothes for disguise; special forms of gambling; the game of dice; children's games; etiquette; the science of strategy; and the cultivation of athletic skills. (KS I.3.15, tr. Doniger and Kakar, emphasis added)

Ali recognizes that cosmetics are materials indicative of status as well as wealth. In his analysis he differentiates the monk, who is expected to wear simple clothing, from the townsman who has an extensive knowledge of garments, jewelry and cosmetics (1998: 170). In particular, drawing from *Cullavagga* of the *Vinaya Piṭaka*, it emerges that monks too were concerned with cosmetics. We so learn that the Buddha said:

> Monks, the face should not be anointed, the face should not be rubbed (with paste), the face should not be powdered with chunam, the face should not be smeared with red arsenic, the limbs should not be painted, the face should not be painted, limbs and faces should not be painted. Whoever should do (any of these things), there is an offence of wrong-doing. (CV V.2.5, tr. Horner)

Only if cosmetics were to be used as medical remedies were they allowed:

> Now at that time a certain monk was afflicted by a disease of the eyes. They told this matter to the Lord. He said: 'I allow you, monks, on account of disease, to anoint the face.' (ibid.)

The moral character of a monk – it so appears – was deemed superior to that of a cosmopolitan person who could indulge in fashion and luxury products. Renouncing cosmetics demonstrated disregard for physical beauty and was symbolic of monasticism and ascetic behavior. This strikingly contrasts with gender-specific cosmetics, such as the widespread use in India of vermilion (*kuṁkuṁ*), a concoction of alum, turmeric and lime juice.

Vermilion is a ubiquitous marker of marital status for Indian women. Although *kuṁkuṁ* paste is primarily applied in the parting of hair and on the forehead to indicate the auspicious condition of a married woman, the bright red powder is also used in temple and household *pūjās* to adorn images (*mūrti*) of gods and goddesses as well as the foreheads of devotees upon receiving *darśana*.

Such illustrations bear witness to the range and applicability of cosmetics in India. The ubiquity of cosmetics in early Sanskrit literature further attests its presence and usage among many diverse people and social sectors in pre-modern times. Yet although detailed descriptions are given of miraculous age-defying creams, long-lasting kohl for the eyes, and general makeup tricks which enhance natural beauty, very little is divulged about the making of cosmetics. This field has largely remained a guarded domain, one shared by trusted men and women in the confidence of select spaces, such as pharmacies, harems and theatre greenrooms. It is for this reason that beauty remedies are often termed 'beauty secrets,' implying that knowledge is not necessary to enjoy the miraculous effects they produce on the physical world around us.

COSMETICS ARE SHARED IN CONFIDENCE

We now turn to two exchanges from the Sanskrit *Rāmāyaṇa* and *Mahābhārata*, the main focus of the present analysis. In Vālmīki's *Rāmāyaṇa*, Rāma – soon after Bharata's visit – decides to visit sage Ātri's hermitage, leaving his temporary abode on Mount Citrakūṭa. Here Rāma, Lakṣmaṇa and Sītā meet the *ṛṣi* and his wife Anasūyā, who is renowned for her chastity and ascetic power (*tapas*). Rāma asks Sītā to speak to Anasūyā, whom she approaches in a way appropriate for such a powerful ascetic. Anasūyā is manifestly pleased with Sītā's behavior and, taking her under her wing, she compliments her so:

> How fortunate you have such high regard for righteousness! How fortunate you should abandon your kinfolk, your pride and wealth, proud Sītā, to follow Rāma when he was banished to the forest. A woman who holds her husband dear – whether he is in the city or the forest, whether he is good or evil – gains worlds that bring great blessings. (Rām II.109.20, tr. Pollock)

Upon being lauded by the elderly wife, Sītā returns the favor, paying homage to Anasūyā. Further, she praises Rāma and her mother-in-law Kausalyā for their righteous teachings; and remembers that even ascetic women like Anasūyā and Sāvitrī have always remained obedient to their husbands. Through this simple exchange Sītā wins Anasūyā's confidence, and her response is quite remarkable. Anasūyā's pleasure with the chaste proclamations of Sītā leads her to bestow the gift of a secret age-defying ointment (*aṅgarāga*):

> Here is a choice heavenly garland, raiment and jewelry, and a cream, Vaidehī, a precious salve. This that I give you, Sītā, will beautify your body, it will suit you perfectly, never spoil, and be yours forever. With this heavenly cream

applied to your body, daughter of Jānaka, you will adorn your husband to the same degree that Śrī adorns the eternal Viṣṇu. (Rām ii.110.17–19, tr. Pollock)

While scholars have shown a tendency to focus on the part of the conversation between the two women which consists of Sītā's recounting of the entire story of her marriage to Anasūyā (Hiltebeitel 2015: 66–68), the ritualized gifting from older wife to young bride speaks rather powerfully of the relationship between cosmetics and the physical world.

Cosmetics, it so appears from the above narrative, are protective (from old age) but also transformative, as Anasūyā informs Sītā. The elder wife instructs Sītā and explains that her secret cream, if applied correctly, will make her like a goddess, implying a transformation of her human nature. We do not have any real details about Sītā's appearance, and it is interesting to note how vague descriptions of her have been in comparison to other female characters in this text as well as those in the *Mahābhārata*. The only element we have to evaluate Sītā's appearance depends upon the application of cosmetics, i.e. the comparison with Viṣṇu's consort. Not only does the cream adorn her and make her even more beautiful, it also makes her inseparable from her husband, like Śrī.

Yet, rather than emphasizing the spoken words of obedience between two wives, one must read into the full scenario. This is a private exchange between women; the older wife Anasūyā and the newlywed Sītā are meeting each other for the first time. In creating impressions about one another, this conversation also serves to provide a meaningful exchange about character traits and beliefs. (Later on, when I will discuss Draupadī and her conversation with Satyabhāmā, the reader will recognize similar tropes regarding obedience and subservience to one's husband.) The text, however, should be read carefully maintaining a focus on the type of exchange, that is, of knowledge shared in confidence. Several scholars have identified Sītā's obedience as her core feature. Her relatively quiet demeanor when among elders, her abduction (to which she poses little resistance), and lastly, her selfless immolation when questioned by her husband after enduring several insults, do not do her justice. Sutherland contrasts Sītā's quiet and resilient nature with that of her counterpart Draupadī, and observes that the exchange between Sītā and Anasūyā is precisely the kind of narrative that bears 'terrible results for those who fail in their duty.' (1989: 75) It is Sītā's 'passive behavior' and 'grateful acceptance of advice' that troubles Sutherland because these qualities make Rāma's spouse have a self-denying and 'masochistic' quality (ibid.: 78). While Draupadī is considered vocal, Sītā chooses to bottle up her feelings of anger. Sutherland suggests this as one reason for Sītā being a more chosen ideal for Indian women.

Whichever character one takes as role model, the scene illustrated above between Sītā and Anasūyā has become cited as an encounter where obedience and subservience are discussed, vocalized and lauded by women. Current analyses however do little to explain how cosmetics and adornment are also key factors to attracting, winning and maintaining the attention of one's husband. The same can be said of the second exchange, where Draupadī converses with Satyabhāmā, her cousin-in-law, on the notion that cosmetics are particularly useful in the relationship between husband and wife.

In this colloquy, which is part of *Āraṇyakaparvan* (222–23), Satyabhāmā – the first of Kṛṣṇa's eight wives and his favorite (MBh IV.8.17a) – pays a visit to Draupadī and asks for advice on womanly matters. Most importantly, Satyabhāmā wants to know how Draupadī manages to keep the attention of five husbands who seem to adore and worship her so devotedly:

> Have you followed a vow, done austerities? Is there a special ablution, spells, herbs? A powerful knowledge of roots? Some prayer, or fire oblation, or drug? Tell me the glorious secret of your sexual power, Kṛṣṇā (= Draupadī), so that Kṛṣṇa will always be amenable to me too. (MBh III.222.6–7, tr. Van Buitenen)

Exchanges between female friends are relatable to audiences from different generations. A fairly common concern among women remains how to hold the attention of their husbands and lovers in the face of all the matters that keep them occupied. In this narrative, Draupadī confides in Satyabhāmā about some of the ways in which she keeps her husbands' attention, i.e. by establishing order in her home and familial relationships. Further to that, Draupadī says:

> Well-trained according to the prescriptions, *well-adorned* (*svalaṃkṛtā*) and most eager, beautiful woman, I am bent on what is good for my husband. (MBh III.222.31, tr. Van Buitenen, emphasis added)

Draupadī's advice too can be read along the lines of her obedience and subservience towards the men in her life. However one should take into account her personality as well as the context when making such assumptions. Draupadī is by no means a subservient queen. In fact, we often find her contrasted with Sītā for the main reason that her outspoken personality and manipulative strategies win her alliances with her male counterparts. Sutherland aptly notes that Draupadī's depiction shows her suffering great insults and enduring much hardship while accompanying her husbands faithfully into exile (1989: 68). Draupadī's assertive behavior is evident in her many encounters with male characters, such as when she convinces Bhīma to kill the *sūta* Kīcaka after he has dishonored her in Virāṭa's court (MBh IV.13–23), or when she is described as the main female character to maintain a friendship relation with Kṛṣṇa. Her dynamic and aggressive personality is contrasted, as Sutherland suggests, to the subservient nature of her husbands (1989: 71). It so appears that Draupadī's temperament is incongruent with the notion of obedience, and yet we find her repeatedly telling Satyabhāmā how ardent devotion and subservience to the will of her husbands enable her to keep their attention. If we are to read the dialog in light of this analysis, it is unlikely that we would understand why she divulges adornment strategies and household techniques to maintain order and win her husbands' attention. I thus propose to read Draupadī's dialog as yet another of her dynamic and opinionated strategies to make situations work towards her end.

In this light, Draupadī's advice to Satyabhāmā can be seen quite differently. When she tells Kṛṣṇa's wife that she never complains about her mother-in-law 'no matter how aggravated I am' (MBh III.222.36) and that she worships the elders (*guruśuśrūṣaṇa*; ibid. 37), this should not be read as merely a proclamation of appropriate behavior. Rather it is a plea for right conduct towards good results. Further to that, Draupadī adds that she never contradicts Kuntī on matters of clothes, jewelry or food (*vastrabhūṣaṇabhojana*; ibid. 39). This particular statement should already be a clear indication that we must listen to this section in light of Draupadī's specific personality. Since Satyabhāmā has requested her advice on matters of household management, she suggests a possible strategy for peaceful coexistence with one's in-laws and husband. The extent of detail she provides clearly alludes to her position. Draupadī tells Satyabhāmā about all the chores she has to take care of in the household, i.e. feeding and serving eight thousand Brahmins, eighty-eight thousand bath-graduates (*snātakas*), ten thousand ascetics (*yatis*) and one hundred thousand horses and elephants in the stables as well as keeping track of the one hundred thousand serving maids of the king. Further to that, she also takes care of all the retinue in the palace:

> The treasury, an ocean that even Varuṇa couldn't manage, this treasury of my so very virtuous husbands, I alone knew how full it was. (ibid. 54, tr. Van Buitenen)

Her humorous tone here suggests the magnitude of her awareness on household matters in comparison to her husbands. So she concludes:

> I am the first to wake up, the last to bed down-that, Satyā, is my charm, all the time. Yes, I know how to put a charm on my husbands, and no, I do not practice the ways of bad women (*asatstrī*), nor do I want to. (ibid. 56–57, tr. Van Buitenen)

In the last verse, we learn that Draupadī is against the use of manipulative charms (cf. Patton 2007: 104). She argues that if spells and powders are used, these will render the husband impotent, possessing ill-luck and having diseases or symptoms of old age. Such a strong condemnation by an opinionated Draupadī must be given its due, and must not just appear as a proclamation of chastity and obedience to her male and dominant husbands. Alternatively, it can be read as her advocating that fair means should be used to attract the attention of one's husband, that spells and charms contradict that goal, and further that they are wicked and ineffective. This is not so for makeup. Patton urges us to read this section in alliance with Draupadī's personality, displaying her awareness of the 'power dynamics' inherent in social relationships between men and women (2007: 102).

At the end of this dialogue, Draupadī communicates to her friend – in the form of a short poem – a collection of various strategies to worship one's husband:

> Let me talk of the way, unopprobious way,
> For a woman to hold the heart of her husband:
> You walk that way, friend, the proper way,
> And you'll cut your man from his mistresses.
> There is no such deity, Satyā, here
> In all worlds with all their divinities
> Like a husband: you're rich in every wish
> If you please him right; if he's angry you're dead.
> You get children and all kinds of comforts,
> A place to sleep and to sit and marvelous sights
> And clothes and flowers and certainly perfumes
> And the world of heaven and steady repute.
> No bliss is easily found on earth,
> A good woman finds happiness through hardship:
> So worship Kṛṣṇa with happy heart
> With love and always the acts of affection.
> When from tasty dishes and beautiful garlands,

From domestic adeptness and various perfumes
He reasons that he must be dear to you,
He himself will embrace you with all his love.
When you hear the sound of your man at the door,
Rise up and stand in the middle of the house;
When you see he has entered make haste with a seat
And receive him with water to wash his feet.
And send your serving woman away,
Get up and do all the chores yourself;
Then Kṛṣṇa will surely know your heart:
She loves me completely, Satyā, he'll think.
Whatever your lord may say in your presence,
Even though no secret, keep it a secret:
A co-wife of yours will surely report you
To Vasudeva and he'll turn away.
Invite for a meal, by hook or by crook.
Your husband's favorites, faithfuls, and friends,
And cut forever his foes and opponents,
His ill-wishers, blackguards, and the insolent rude.
If you find your man either drunk or distracted,
Control your temper and hold your tongue;
Though Pradyumna and Sāmba are your sons,
Don't ever attend to them secretly.
Strike up a friendship with highborn ladies
And women of virtue and without vice;
The bellicose, bibulous, gluttonous ones,
The bad, thievish, fickle you must avoid.
This is the glorious secret of sex
That leads to heaven and uproots the foes;
Worship your husband while wearing your costly
Flowers and jewels and make-up and scents.
(MBh III.223.1–12, tr. Van Buitenen)

This entire dialogue reads much like the rest of the colloquy. But, while the impulse would be to focus on obedience and service, one should turn towards the emphasis on strategy in order to win the attention of the husband. In particular, self-adornment – makeup and jewelry – stand out for our purposes. Minerals as precious stones or as bases for cosmetics are pivotal to arranging and maintaining social relationships between man and wife. Draupadī reasons that a man will be faithful as long as the household is kept in order and unguents and perfumes are used. This is indicated when she states:

> When my husband sleeps out on some family business,
> I go without flowers and make-up, and follow a vow.
> (MBh III.222.29, tr. Van Buitenen)

Using or not using cosmetics to maintain and magnify her natural beauty seems to be one of Draupadī's major strategies towards winning the attention of her husbands. Moreover, in the context of the text, she chooses to pass down this information, much as Anasūyā had earlier, to another woman *only* in confidence. This seems more akin to some of the arguments made by scholars on the pre-modern use of cosmetics wherein cosmopolitan people utilized cosmetics to display their social status and rank. A Draupadī without cosmetics, without her jewelry and without her auspicious signs of womanhood appears without husband and hence affects her world through those actions. Her advice to Satyabhāmā hinges on an understanding of cosmetics as effectively shaping the physical world around them.

Whether we focus on Sītā or Draupadī, both display and narrate personal and confidential stories where cosmetics play a significant role. Sītā receives secret information from a well-wisher and older wife. Draupadī offers this information to a friend in need. The act of sharing secrets on cosmetics and their power in confidence is one that requires more thorough examination. The two accounts herein examined occur in moments of privacy strictly among women. While it is clear that Satyabhāmā visits Draupadī just to seek her advice on this matter, Sītā is asked by Rāma to talk to Anasūyā who then rewards her with the age-defying cream. In both situations women establish a trust and a bond with each other in order to receive precious information. Sītā wins over Anasūyā through her pious words and her awareness of other devoted wives and their actions. Draupadī is aware of Satyabhāmā through her friendship with Kr̥ṣṇa, but only once trust is gained – by noting she won't use spells or charms of any sort – does the Paṇḍavas' wife reveal the information requested. These exchanges are not merely about obedience or subservience, as suggested by most, but rather indicative of a space from which women share information privately understood and revealed to them using the means they find effective.

COSMETICS AFFECTING THE COSMOS

Since information about cosmetics appears to be shared in confidence in Itihāsa texts, the ingredients are largely unknown. Creams and ointments are often said to be medicinal, to have life-changing properties, and even to cause miracles. In the narratives of Sītā and Anasūyā (Rām II.109.20ff.) and Draupadī and Satyabhāmā (MBh III.222–23), cosmetics alter the relationship between man and wife, transforming women and making them more appreciated for their chastity and obedience. Rather than being an act of males imposing dominance upon their wives, these exchanges show feminine strategies to effect physical and psychological changes in their world. While in the performing arts we have seen how makeup can be used to manipulate and perform different social and physical realities, in real life makeup and adornment can cause positive changes, especially in the relationship between man and wife.

Men in *Mahābhārata* and *Rāmāyaṇa* display affection towards women who are well dressed and well adorned. These women and their adornments help men to gauge their status and their rankings. Respectable women and mothers are all well adorned. Pious and virtuous women have glowing skin, and chaste women shine through their obedience. Conversely, unkempt women sporting unwashed hair, ungroomed women without flowers and makeup, all signal ill omens, bad fortune, and ill health, and so are meant to be avoided by men. Beauty itself comes from adornment.

Such beliefs are not unique to Itihāsa; they appear to be prevalent in the pre-modern world. The vermilion used by married women in India still remains an auspicious and positive symbol of marriage and a thriving relationship between a couple. Similarly, *wasma*, which was used to dye Iranian women's hair, was a compulsory product during important occasions and among elites. Men too used these dyes for their beards and eyebrows (Farmanfarmaian 2000: 320). Kohl and black liner for the eyes and eyelashes, in addition to protecting the wearer, also prevented the dreaded evil-eye or curses from jealous onlookers. Even prior to colonial influences, women of lighter complexion were praised and sought after, and skin-lightening cosmetics were used. Such markers indicated youth, wealth and, above all, elite status.

In addition to the riddles mentioned in *Gandhasāra*, McHugh notes the transformative capacity of perfumes in pre-modern times. Narrating the inherent relationship between mundane and theological worldviews, he argues that bodily experience could be transformed through odors. Having employed one of the only Sanskrit texts of its kind, McHugh's analysis is critical to our understanding of cosmetics in early Indian culture. He describes the relationship between ethics and aesthetics in regard to perfumery, quite similar to what we saw in the exchanges among women in the epic texts regarding cosmetics. While some flowers' odors could attract one to come closer, some incenses could be powerful enough to be erotic aphrodisiacs (McHugh 2012: 130).

CONCLUSION

We return again to the question we began with: If makeup is so ubiquitous and has been used for centuries, why does it evoke such criticism today? If cosmetics could prove so helpful in the context of marital relationships, as in the case of our discussion on Sītā and Draupadī, how can they be portrayed as superficial, fake and even manipulative today? In Indian texts the use of cosmetics is often associated with the concept of *māyā*, or performed illusion – as in how physical reality could be cloaked or enhanced using magical techniques. In order to understand this contradiction we have to go back to the origin of cosmetics and their use.

Kelly Olson has studied cosmetic substances and poisons in the classical world. She says: 'Scholars have pointed out that there was a difference in antiquity between *kosmêtikon tês iatrikês meros* (the preservation of beauty) and *to kommôtikon* (the "unnatural embellishment" of looks), a distinction clearly found in Galen, for instance.' (Olson 2009: 294, citing Gibson) According to Olson, it is *to kommôtikon* (Gr. τὸ κομμωτικόν) that informs current understandings of cosmetics. Makeup is often cited as an unnatural form of embellishment in contrast to a more natural beauty. Though the influence of colonialism and, more recently, globalization and corporate marketing have fostered a heavily charged narrative on the cosmetic manipulation of the body, such portrayals are equally intertwined with questions about ethnicity and global politics. The substantial loss of pre-modern cosmetic knowledge and traditions is a major gap. Our reading of textual sources is inevitably affected by current beliefs and fashions, so that the original – mineral-based – recipes of ancient Egyptian, Indian, Greek, Roman, Iranian, etc. eyeliner, hair dye, age-defying creams and complexion foundation are dismissed vis-à-vis their more recent, and less expensive, herbal versions. In the name of health and environmentalism, Indian women are informed of the ill-effects of, for instance, mineral vermilion and are invited to purchase herbal – i.e. safer/healthier – alternatives to take care of their natural beauty. More generally, mineral substances are associated with wasteful and potentially dangerous indulgence and luxury.

These critiques, however, are more harmful than the mineral-based substances they condemn, since they favor a disturbing association between moral character and feminine beauty. Olson informs us that both cosmetic and anti-cosmetic traditions hold true in their disgust of 'women in their raw state' (2009: 304). Many of the uses of cosmetics that we have seen throughout this chapter paint rather a different picture of (mostly mineral) makeup. Its transformative power is shared only in confidence and among trusted women who employ specific cosmetics for specific purposes. The multifaceted nature of cosmetic usage displays the inherent ambiguity of its mineral composition and the rare power it possesses to effect change in the relationships one cultivates through physical appearance. It is true that beauty lies in the eyes of the beholder, but while applying cosmetics one often beholds oneself as well, witnessing a glimpse of how one is perceived by others. The notion of adorning oneself shows self-reflexivity in understanding how one is perceived. Cosmetics have had, and continue to have, real-world influences effecting change upon the social, material and hence physical relationships that surround us.

ABBREVIATIONS AND REFERENCES

Gr. = Greek M. = Malayalam P. = Pali Skt. = Sanskrit Tel. = Telugu

Ahmed-Gosh 2003 — Ahmed-Gosh, H. (2003). 'Writing the Nation on the Beauty Queen's Body: Implications for a "Hindu" Nation.' *Meridians*, 4: 205–27.

Ali 1998 — Ali, D. (1998). 'Technologies of the Self: Courtly Artifice and Monastic Discipline in Early India.' *Journal of Economic and Social History of the Orient*, 41(2): 159–84.

CV — *Cullavagga*. (1) GRETIL e-text based on: Oldenberg, H. (ed.) (1880). *Vinaya-Pitaka. Vol. 2: Cullavagga*. London: Pali Text Society, <http://gretil.sub.uni-goettingen.de/gretil/2_pali/1_tipit/1_vin/vin2cuou.htm> (accessed 26 August 2015), input by the Dhammakaya Foundation, Thailand, 1989–1996. (2) Horner, I.B. (tr.) (1963). *The Book of the Discipline. Book V (Cullavagga)*. London: Luzac & Co.

Dhand 2008 — Dhand, A. (2008). *Woman as Fire, Woman as Sage. Sexual Ideology in the Mahābhārata*. Albany: State University of New York Press.

Doniger 1997 — Doniger, W. (1997). 'Sita and Helen, Ahalya and Alcmena: A Comparative Study.' *History of Religions*, 37(1): 21–94.

Farmanfarmaian 2000 — Farmanfarmaian, F.S. (2000). '"*Haft Qalam Ārāyish*": Cosmetics in the Iranian World.' *Iranian Studies*, 33(3/4): 285–326.

Flueckiger 2013 — Flueckiger, J.B. (2013). *When the World Becomes Female. Guises of a South Indian Goddess*. Bloomington, IN: Indiana University Press.

Gardner 1962 — Gardner, T.L. (1962). 'The Cosmetic Industry.' *Journal of the Royal Society of Arts*, 110(5076): 892–903.

GS — Gangādhara: *Gandhasāra*. Vyas, R.T. (ed.) (1989). *Gangādhara's* Gandhasāra *and an Unknown Author's* Gandhavāda *(with Marathi Commentary)*. Vadodara: Oriental Institute.

Hiltebeitel 2015 — Hiltebeitel, A. (2015) 'Dialogue and Apostrophe: A Move by Vālmīki?' In Black, B. and L.L. Patton (eds), *Dialogue in Early South Asian Religions. Hindu, Buddhist, and Jain Traditions*, pp. 37–78. Farnham, Surrey: Ashgate.

Kamath 2012 — Kamath, H.M. (2012). *Aesthetics, Performativity, & Performative Maya. Imaging Gender in the Textual and Performance Traditions of Telugu South India*. PhD Dissertation, Emory University, Atlanta.

KS — Vātsyāyana Mallanāga: *Kāmasūtram*. (1) (1900). GRETIL e-text based on Paṇḍit Durgāprasād (ed.) (1900). *Kāmasūtram. With Yaśodhara's Jayamaṅgalā Ṭikā*. Second edition. Mumbai: Nirṇayasāgarayantrālaya with reference to:
Śrīdevdutta Śāstrī, D. (ed.) (2049 saṁ) *Kāmasūtram*. Varanasi: Chaukhambha Sanskrit Sansthan, <http://gretil.sub.uni-goettingen.de/gretil/1_sanskr/6_sastra/6_kama/kamasutu.htm> (accessed 24 August 2015), input by Mizue Sugita, 1 September 1998. (2) Doniger, W. and S. Kakar (trs.) (2002). *Kamasutra*. Translated with an Introduction. New York: Oxford University Press.

Lucas 1930 — Lucas, A. (1930). 'Perfumes and Incense in Ancient Egypt.' *The Journal of Egyptian Archaeology* 16(1/2): 41–53.

Mankekar 1999 — Mankekar, P. (1999). *Screening Culture, Viewing Politics. An Ethnography of Television, Womanhood, and Nation in Post-Colonial India*. Durham, NC: Duke University Press.

MBh — *Mahābhārata. Āraṇyakaparvan*. (1) GRETIL e-text based on the critical edition of of the Bhandarkar Oriental Research Institute, Pune, <http://gretil.sub.uni-goettingen.de/gretil/1_sanskr/2_epic/mbh/mbh_03_u.htm> (accessed 25 August 2015), input by Muneo Tokunaga et al. Revised by John Smith et al. (1999). (2) Van Buitenen, J.A.B. (tr.) (2007). *Mahābhārata. The Book of the Assembly Hall; The Book of the Forest*. Chicago: University of Chicago Press.

McHugh 2012 — McHugh, J. (2012). *Sandalwood and Carrion. Smell in Indian Religion and Culture*. New York: Oxford University Press.

Olson 2009 — Olson, K. (2009). 'Cosmetics in Roman Antiquity: Substance, Remedy, Poison.' *The Classical World*, 103(2): 291–310.

Patton 2007 — Patton, L.L. (2007). 'How Do You Conduct Yourself? Gender and the Construction of a Dialogical Self in the *Mahābhārata*.' In Brodbeck, S. and B. Black (eds.), *Gender and Narrative in the Mahābhārata*, pp. 97–109. London: Routledge.

Rām — Vālmīki. *Rāmāyaṇa. Book 2. Ayodhyākaṇḍa*. (1) GRETIL e-text by Muneo Tokunaga et al. Revised by John Smith, Cambridge (2001), <http://gretil.sub.uni-goettingen.de/gretil/1_sanskr/2_epic/ramayana/ram_02_u.htm> (accessed 25 August 2015). (2)

	Pollock, S. (trans.) (2005). *Rāmāyaṇa. Book Two: Ayodhyā by Vālmīki.* New York: New York University Press and JJC Foundation.
Ram 1992	Ram, K. (1992). *Mukkuvar Women. Gender, Hegemony and Capitalist Transformation in a South Indian Fishing Community*. New Delhi: Kali for Women.
Shah 2012	Shah, S. (2012). 'On Gender, Wives and "Pativratās.' *Social Scientist*, 40(5/6): 77–90.
Sutherland 1989	Sutherland, S.J. (1989). 'Sītā and Draupadī: Aggressive Behavior and Female Role Models in the Sanskrit Epics.' *Journal of the American Oriental Society*, 109(1): 63–79.

Chapter 10

Ratna: A Buddhist World of Precious Things

MATTIA SALVINI[1]

There is no jewel like the Jewel that is the Buddha.[2]
Yaṁkiñcītigāthāvaṇṇanā

Like the Wish-Fulfilling Gem, he is unshakable by the winds of any concept,
And yet, he fulfills all wishes, for all sentient beings.[3]
Advayavajra

Non-existent, non-arisen, and non-ceased, a reflection is seen thanks to the complete purity of the great beryl. In the same way, Mañjuśrī, thanks to the complete purity of the mind, and due to having practiced proper cultivation, sentient beings see the body of the Tathāgata.[4]
The Ornament of the Light of Awareness

1 Mattia Salvini is Director of the International PhD Programme in Buddhist Studies, Faculty of Social Sciences and Humanities, Mahidol University (Thailand).
The author thanks Harunaga Isaacson and Giuliano Giustarini for precious suggestions and corrections. Any mistake that may remain is the responsibility of the author. The research that resulted in this chapter was made possible by the Numata Foundation, which sponsored the author's stay in Hamburg University as Numata Visiting Professor in Buddhist Studies from October 2014 until January 2015. Deep gratitude is owed to the benefactors of the Numata Foundation. Finally, and importantly, thanks are due the Faculty of Social Sciences and Humanities of Mahidol University, in particular Ajahn Wariya, Ajahn Nopraenue and Ajahn Pagorn, for facilitating and encouraging the reserach trip to Hamburg.

2 *ratanaṃ buddharatanena samaṃ natthi*. YGV.

3 *cintāmaṇir ivākampyaḥ sarvasaṃkalpavayubhiḥ / tathāpi sarvasattvānām aśeṣāśāprapūrakaḥ* | KN, v. 4.

4 *sa cāsan notpanno na niruddhaḥ pariśuddhatvād mahāvaiḍūryasya pratibhāsaḥ saṃdṛśyate* | *evam eva mañjuśrīḥ pariśuddhatvāc cittasya subhāvitatvād bhāvanāyāḥ sattvānāṃ tathāgatātmabhāvadarśanaṃ bhavati* || JS.

Buddhist literature, imaginaire and 'material' culture are largely made of gems, jewels and precious metals; these appear as figures of speech, offerings, ornaments, magical implements, ideal landscapes and suggestive narrative elements. The daily devotions of Buddhist monks, philosophical texts or the precints of a Buddhist sacred site belong to a jeweled world of interrelated complexity, lustre and, most importantly, beneficial efficacy.

A Buddhist is 'one who takes refuge in the Three Jewels.' A successful Buddhist practitioner may accumulate merit to reach Indra's divine realm, inlaid with four precious gems; or s/he might reach 'the diamond-like concentration' as a prelude to final awakening. Mindfulness of Amitābha, the Buddha of Infinite Light, brings about rebirth in his pleasant realm, filled with gold, silver, crystals and other remarkable gems. Those of sharpest faculties may obtain Buddhahood in one lifetime, by relying on the swifter 'Diamond Vehicle.'

Such bright paths contrast with the absence of precious or beautiful materials in the hells, which are filled with less attractive substances like ice, fire, iron and molten copper, and, of course, with an intense profusion of blood and other side-products of bodily pain. Lower realms of rebirth exemplify the opposite of 'something which delights' (*ramayati*, one of the etymologies of the term *ratna*), and demonstrate the relationship between precious beneficial substances and collective karma (P. *kamma*).

In the realm of human activities, from the name of textual collections like the 'Heap of Jewels' to the contemporary Burmese and Thai Buddhist practice of covering statues with layers of gold-leaf, Buddhist culture is everywhere adorned by imagined or actual gems and precious metals, 'outer' and 'inner': even the mind's highest aspiration, and its emptiness, are 'gems.'

I shall discuss in this chapter the place of *rat(a)na*s in the South Asian Buddhist traditions, emphasizing the interrelated nature of different branches of pre-modern Indian knowledge, without claiming to be comprehensive.

WHAT IS A *RATNA*?

Among the various possible Sanskrit and Pāli terms rendered in English as 'jewel,' the Sanskrit *ratna* (P. *ratana*) is arguably the most relevant for the Buddhist tradition. Its usual translation, however, does not do perfect justice to its range of meanings, as this term can refer to a wide variety of substances. *Ratna*s include precious stones like diamonds or emeralds;

precious metals like gold and silver; and, significantly, anything deemed to be the foremost of its own type. It may therefore be useful to keep in mind that *ratna/ratana* is translated as 'jewel' as a conventional approximation, and it may furthermore be desirable to investigate its meaning as discussed in the primary sources.

I propose to start from the tradition of the *Amarakośa*, an extensive lexicon (or perhaps, 'the' lexicon) of Sanskrit, authored by a Buddhist, and omnipresent in the commentarial literature of India. The *Amarakośa*, furthermore, influenced the composition of Pāli lexicons and is known even in the Tibetan tradition, thanks to the translation of Subhūticandra's commentary.[5]

At least three commentators explain the term *ratna* in the *Amarakośa* as being related to the root *ram*, which has the sense of 'delighting in, being fond of,' et cetera. The more restricted meaning of *ratna* is given in the *Amarakośa* as a synonym of *maṇi*, hence gem or jewel, precious substance; a broader meaning is found in another section, where it is said that '*ratna* is also used in respect to what is the best in its class' (*ratnaṁ svajātiśreṣṭhe 'pi*), which itself offers a useful clue to understand why the Buddhist tradition relies on *ratna* to qualify what it considers to be the highest.

In both instances where *ratna* occurs in the *Amarakośa*, Kṣīrasvāmin tells us that 'people take delight in it, thus it is a *ratna*' (*ramante'smin ratnam*).[6] When *ratna* is explained as 'whatever is best in its own class,' Kṣīrasvāmin offers a quote from the *Cāṇakyaśataka*: 'Whatever is excellent within a specific class is called *ratna*.'[7]

Commenting on the first occurrence of *ratna*, Maheśvara offers the same etymology, and adds three interesting verses that offer a link to another branch of knowledge which, as we will see, turns out to be especially useful:

> Gold, diamond, sapphire, ruby and pearl;
> The experts in Ratnaśāstra know these to be the five *ratnas*.
>
> Gold, silver, pearl, *rājāvarta* and coral,
> Are said to be the five *ratnas*: the rest is called a 'thing.'
>
> Pearl, gold, *vaiḍūrya*, ruby, topaz, *gomeda*, sapphire, emerald and coral,
> Are said to be the nine great *ratnas*.[8]

5 See Vidyābhūṣaṇa (1912) and Deokar (2014).

6 Amara(K), p. 155.

7 *jātau jātau yad ukṛṣṭaṁ tad ratnam abhidhīyate*. Amara(K), p. 206.

8 *kanakaṁ kuliśaṁ nīlaṁ padmarāgaṁ ca mauktikam | etāni pañca ratnāni ratnaśāstravido viduḥ || suvarṇaṁ rajataṁ muktā rājāvartaṁ pravālakam |*

The example offered by Maheśvara in reference to *ratna* being used in the sense of 'the best of its class' is the 'woman-gem' (*strīratna*),[9] which, as we shall see, is one of the main accoutrements of a 'Wheel-Turning Monarch' (*cakravartin*), the primary royal simile for a Buddha.

Bhānujīdīkṣita offers some additional grammatical information about the formation of the term *ratna*: it comes from the root *ram* in the sense of delighting, enjoying (*ramu krīḍāyām*). He also quotes from *Haima*, another lexicon, where the term *ratna* is mentioned either as a synonym of *maṇi* or as referring to 'the best in its class' (*svajātiśreṣthe*).[10] While commenting on the second occurrence of *ratna*, Bhānujīdīkṣita also offers the alternative etymology 'it delights' (*ramayati*).[11]

The commentary on the *Abhidhānappadīpikā*, a Pāli lexicon showing remarkable familiarity with Sanskrit sources, glosses the term *ratana* by quoting the *Amarakośa* and hinting at a list of *ratana*s that are of particular relevance for the Buddhist tradition:

> It diffuses delight (*ratiṁ tanoti*), thus it is a '*ratana*' [...]. '*Ratna* also refers to the best in its class', says the *Amarakośa*. Accordingly, the best among elephants is the elephant-*ratana*, and the best among women is the woman-*ratana*.[12]

The next two verses in the *Abhidhānappadīpikā* then offer two alternative lists of *ratana*s, reminiscient of the Ratnaśāstra quotes that we earlier found in the *Amarakośa*.

All these passages present similar ideas about the sense of the term *rat(a)na*, and perhaps what we should note is that the extended sense of 'the best of its type' recurs in all the lexicons, and that the etymological link with the idea of 'bringing delight' is also very well attested. And what, in the Buddhist world of gods and men, could be the very best, and bring the greatest delight?

---

ratnapañcakam ākhyātaṁ śeṣaṁ vastu pracakṣate iti vā | muktāphalaṁ hiraṇyaṁ ca vaiḍūryaṁ padmarāgakam || puṣparāgaṁ ca gomedaṁ nīlaṁ gārutmakaṁ tathā | pravālayuktāny uktāni mahāratnāni vai naveti || Amara(M), p. 234.

9 Amara(M), p. 314.

10 VS, p. 332.

11 VS, p. 418.

12 *Ratiṃ tanotīti ratanaṃ, kammādimhi ṇo, tilopo. 'Ratnaṃ sajātiseṭṭhepī' ty amarakose* [Amara 23.126], *tena gajaseṭṭho gajaratanaṃ, itthiseṭṭho itthiratananti.* (ADṬ 489)

BUDDHIST *RAT(A)NAS*: A PĀLI INTRODUCTION

We may now turn to the commentarial literature on the Pāli Piṭakas, where we find additional etymologies of the term *ratana*, and rather detailed explanations of why the Buddha, the Dhamma and the Saṅgha should indeed be called *ratana*. We will see that these discussions are clearly related to how Sanskrit and Pāli lexicons discuss the term *rat(a)na*, while at the same time they re-contextualize the link to the root *ram* so as to best address Buddhist intellectual and religious concerns.

A number of commentaries on different Pāli Suttas connect the term *ratana* to the root *ram*, not unlike the lexical literature that we saw in the previous section. The following passage represents a recurrent way of explaining how the etymology fits the 'Three Jewels' (i.e. the Buddha, the Dhamma and the Saṅgha):

> It procures, generates, brings delight (*rati*), thus it is called *ratana*, and *ratana*s can be of seven or ten types; those (Three Jewels) are just like this – such is the etymology.
>
> Even without considering their similarity to something else, simply in accordance with the meaning of the words as they have been spoken, the quality of being *ratana* is fit for the Buddha, etc. Since in their case, for someone who becomes intent on their real, good qualities, in accordance with 'Thus he is the Bhagavat,' etc., a joy and happiness, which is not small and is the cause for the attainment of Deathlessness, arises.[13]

This brief passage makes several (more or less explicit) points: A *ratana* is so called because it produces delight; therefore, just like we apply this term to the standard lists of seven and ten jewels (that we know from Ratnaśāstra) we should apply it to the Buddha, the Dhamma and the Saṅgha (whether through a comparison with the mineral word, or by directly considering that the Buddha, Dhamma and Saṅgha generate delight). It is fit

13 *Ratiṃ nayati, janeti, vahatīti vā ratanaṃ, sattavidhaṃ, dasavidhaṃ vā ratanaṃ, tam iva imānīti neruttikā. Sadisakappanam aññatra pana yathāvuttavacanattheneva buddhādīnaṃ ratanabhāvo yujjati. Tesañhi 'iti pi so bhagavā' tiādinā* (*dī. ni.* 1.157, 255) *yathābhūtaguṇe āvajjantassa amatādhigamahetubhūtaṃ anappakaṃ pītipāmojjaṃ uppajjati.* (DN *Sīlakkhaṇḍavaggaṭīkā*, introductory sections) I thank Harunaga Isaacson for correcting my understanding of the syntax of this sentence. The first etymology is here explaining that the Buddha, Dhamma and Saṅgha are *ratana* because they are *like ratana*; the second etymology explains that we can directly say that they are *ratana*, simply because they generate delight, even without comparing them to jewels.

to do so, since when a Buddhist practices the 'mindfulness of the Three Jewels,' recollecting their good qualities through well-known formulas like '*Iti pi so Bhagavā*,' et cetera, this recollection causes a remarkable level of delight. This delight, in turn, is an initial cause for the attainment of final liberation – technically, it is 'the faculty of conviction' (*saddhindriya*), the first of the 'five faculties' that a Buddhist practitioner should perfect in order to progress on the path.

In other words, one obtains delight thanks to the good qualities of the Buddha, Dhamma and Saṅgha, and in this appreciation of good qualities the practitioner becomes perhaps a metaphorical gemologist, an image that, I hope, is starting to become by now more plausibly appropriate. This conviction about the Buddha, Dhamma and Saṅgha having good qualities, and the good disposition of the mind which follows, are the recurrent and standard description of the term *saddhā* (Skt. *śraddhā*) and *pasāda* (Skt. *prasāda*), that constitute a fundamental mental attitude upon which the attainment of the entire Buddhist path is predicated. The Three Jewels are therefore *ratanas* because their genuine good qualities are the basis of a supreme delight (*rati*) based on the unmistaken recognition of their worth. 'Good quality' (*guṇa*) is itself a rather significant term, especially when we keep in mind the context of the 'examination of gems'; the good qualities of a *ratna*, in fact, are not exhausted in their physical traits, but they consist (primarily) in their ability to bring about beneficial results. This sense of the term *guṇa* as 'beneficial power' seems contextually appropriate when *rat(a)na* is a qualification of the Buddha, Dhamma and Saṅgha – and the list of *guṇas* in the standard formulas of praise to the Three Jewels confirm that here 'good quality' has to be something beneficial/virtuous.

The sense of 'producing delight,' however, is not exclusively applied to the Three Ratanas; a passage of the *Abhidhammavatara*, for example, describes discipline (*sīla*), as the 'supreme *ratana*,' and the commentator offers a precise rationale, distinguishing *sīla* from other types of more 'worldly' *ratanas*:

> Even compared to the Wheel-*ratana*, etc., of the Cakkavattis, as this generates superior delight, it is the unexcelled *ratana*. Since, the Wheel-*ratana* etc. only generate a delight based on occurrence (i.e. *saṁsāra*), while this generates a delight that is based on non-occurrence (i.e. *nibbāna*) as well. Thus, it is something that generates delight in a superior way.[14]

14 *Ratananti cakkavattīnaṃ cakkaratanāditopi savisesaṃ ratijananato anuttaraṃ ratanaṃ. Cakkaratanādikañ hi vaṭṭanissitam eva ratiṃ janeti, idaṃ pana vivaṭṭanissitam pi janetīti visesato ratijanakaṃ hoti.* (Abhidhammāvatāra-Purāṇaṭīkā 793–97)

When offering a more extensive explanation, Pāli commentaries often quote a verse that lists five reasons why something might be called *ratana*: '(1) worshiped, (2) of great worth, (3) with no equal, and (4) whose sight is hard to obtain; (5) enjoyed by superior sentient beings; due to this, it called *ratana*.'[15]

The *Yaṃkiñcītigāthāvaṇṇanā* discusses how, if we apply these five reasons rather strictly, the term *ratana* should only apply to the Tathāgata, and argues so by offering references to well-known narratives and to Suttas where the Buddha himself hinted at how these five reasons should be applied:

(1) In the sense of 'being worshiped' the Tathāgata can be said to be the only genuine *ratana* because while the Buddha is alive, all the deities worship only him; and once he has reached *parinibbāna*, his heritage, i.e. the many monasteries where his monks live, and the four holy sites connected to his life, continue to be worshiped.

(2) Similarly, nothing has a worth comparable to that of the Tathāgata. This is proven by the fact that Emperor Aśoka became what he became thanks to a very small offering that he had made to the Tathāgata, which shows the amount of merit that one obtains by making offerings to the Tathāgata. This, in turn, demonstrates the extent of the Tathāgata's worth.

These reasons are not only discussed in relation to the Buddha, Dhamma and Saṅgha, but are also partly applied to the 'Seven Ratanas' that make for a 'Wheel-Turning Monarch' (*cakkavatti/cakravartin*). A 'Wheel-Turning Monarch' is a rare being (although not as rare as a Buddha), well-known in the Pāli and Sanskrit Buddhist traditions (the *Abhidharmakośabhāṣya*, AKB, has an extended description of its sub-types, turning a 'golden wheel,' a 'silver wheel' or an 'iron wheel'). This simile is so important that the first teaching of the Buddha is also called *Dhammacakkappavattana*, 'The Turning of the Wheel of Dhamma': just like a Wheel appears to the Wheel-Turning Monarch, and as he follows that Wheel he easily conquers all the lands that he comes across, so also the Buddha conquers the entirety of the Sahā world through the turning of his Dhammacakka/Dharmacakra. (The similarity between the Cakravartin and the Buddha extends to their exceptional physical marks, which are nonetheless more beautiful in the case of a Buddha.)

(3) The *Yaṃkiñcītigāthāvaṇṇanā* offers a lengthy discussion of the Seven Ratanas of a Cakkavatti when explaining the sense of 'having no equal,' the third reason for calling something *ratana*. It is to be noticed that 'having no

15 *Cittīkataṁ mahagghañ ca, atulaṃ dullabhadassanaṃ; Anomasattaparibhogaṃ, ratanaṃ tena vuccati*. This verse recurs in a number of commentaries. See for example, DN *Mahāvagga Aṭṭhakathā*, *Mahāpadānasuttavaṇṇanā*.

equal' is akin to the sense of 'being the best in one's class' (*svajātiśreṣṭha*) that we found in the *Amarakośa*, and it is in this sense that each of the Seven Ratanas is so called:

A Wheel: It is a *ratana* because it has no equal, appearing without a maker just due to the king's accumulation of *kamma*; made of precious substances in all its parts, it shines so intensely that people become curious: 'I wonder, has a second moon or sun risen?'

An Elephant: He is a *ratana* because he has no equal, in terms of his strength and speed and all other elephantine virtues.

A Horse: He is a *ratana* because he has no equal, for reasons not unlike those given for the elephant.

A Gem: It is a *ratana* because it has no equal, as it shines so brightly that during the night people start thinking 'it is daytime' and start attending to their daytime chores.

A Woman: She is a *ratana* as she has no equal, coming from an exceptional family (perhaps even from Uttarakuru), being free from any flaw like being too tall or too short, bringing warmth in winter and coolness in summer, and having many other good qualities that include a body fragrant like sandalwood and a mouth perfumed like a lotus.

The Lord of the House: He is a *ratana* as he has no equal, being the main officer of the king, and having divine sight from birth; thanks to this divine sight, he finds treasures and ensures the wealth of the kingdom.

The Supreme Leader: He is a *ratana* as he has no equal; he is the eldest son of the king and, thanks to his exceptional ability to understand the mind of the king's retinue, he can rule in the king's stead with astonishing success.

All of the seven *ratanas* are unmatched but, as the *Yaṃkiñcītigāthāvaṇṇanā* (from which these descriptions are taken) tells us, 'there is no *ratana* like the Buddha-*ratana*,' and therefore, if we are strict in applying 'having no equal' as a criterion for something to be a *ratana* 'then only the Tathāgata is a Jewel' (*tathāgatova ratanam*), because:

> [...] It is not possible to compare and examine the Tathāgata in terms of discipline, concentration, higher cognition, or in other respects, against any one else, and then determine that 'this good quality is the same or similar with the case of this person.' Therefore, in the sense of 'having no equal' too, there is no *ratana* that is equal to the Tathāgata.[16]

16 *Yadi hi atulaṭṭhena ratanaṃ, tathāgatova ratanaṃ. Tathāgato hi na sakkā sīlato vā samādhito vā paññādīnaṃ vā aññatarato kenaci tulayitvā tīrayitvā 'ettakaguṇo vā iminā samo vā sappaṭibhāgo vā' ti paricchindituṃ. Evaṃ atulaṭṭhenapi tathāgatasamaṃ ratanaṃ natthi.* (*Yaṃkiñcītigāthāvaṇṇanā*, part of RSV)

(4) When we consider the fourth reason, that a *ratana* is so called because its sighting is hard to obtain, nothing matches the Tathāgata in that respect: Even the *ratanas* of a Cakkavatti arise several times in a single aeon (*kappa*); but since the world can remain empty of a Tathāgata even for an 'immeasurable aeon' (*asaṁkhyeyyepi kappe*), the Tathāgata is to be considered the foremost among those 'whose sighting is hard to obtain.'

(5) The last reason given in the verse is that a *ratana* is to be 'enjoyed by superior sentient beings' (or more literally, not by inferior sentient beings). We see indeed that the Tathāgata is not 'enjoyed' (*aparibhogo*) by inferior beings or by those with a distorted view (like Pūraṇakassapa and the rest), while those who can benefit from the Tathāgata are the Great Sāvakas (disciples), some of whom are even 'born in high families' (*mahākulappasuta*).

When explaining the initial etymological reason to call something *ratana*, i.e. that something is *ratana* in the sense of generating delight (*ratijananaṭṭhena ratanaṁ*), the YGV first of all offers a lengthy list of the types of delight that the Tathāgata generates in innumerable gods and men:

> the delight of the first meditation, the delight of the second, third, fourth, and fifth meditations, the delight of the sphere of space, the delight of the spheres of consciousness, nothing whatsoever, neither notion nor no-notion; the delight of the path of Attaining the Stream; the delight of the result of Attaining the Stream; the delight of the paths and results of Once Returning, Non-Returning and Arhat-hood.[17]

Obviously, no other *ratana*, including the various *ratanas* of the Cakkavatti, can match the type of delight caused by the Buddha – as this includes the supreme happiness of liberation and all the intermediate steps leading to that.

The YGV also offers a detailed subdivision of *ratanas*, meant to be omni-comprehensive, establishing the Buddha as the highest:

Ratanas can be with or without consciousness; for the example, the Wheel and the Gem are without consciousness, while the other *ratanas* of the Cakkavatti starting from the Elephant and ending with the Leader have consciousness. Among these two types, the *ratanas* with consciousness are superior, and this can be established by noticing that inert *ratanas* (like gold, silver, gems, pearls, etc.) are employed to adorn sentient *ratanas* like the Elephant and so forth.

17 *devamanussānaṃ paṭhamajjhānaratiṃ dutiyatatiyacatutthapañcamajjhānaratiṃ, ākāsānañcāyatanaratiṃ, viññāṇañcāyatanaākiñcaññāyatananevasaññānāsaññāyatanaratiṃ, sotāpattimaggaratiṃ, sotāpattiphalaratiṃ, sakadāgāmianāgāmiarahattamaggaphalaratiñca.*

The sentient *ratanas* can also be understood as twofold: animals and humans. Among the two, the human *ratanas* can be considered to be superior, because the animals function as means of conveyance for human beings (hence, they are subordinate).

The human-being-*ratana* is also of two types: woman and man. The man is here considered superior because woman functions as an assistant/servant to man (*paricārikā*).

Men also are of two types: having a home, or homeless. The homeless are superior, because we see that even the highest among householders, i.e. the Cakkavatti, by extensively worshiping a worthy homeless ascetic endowed with discipline and other virtues, eventually obtains *nibbāna*.

The homeless *ratana* is also of two types: the Ariya and the ordinary. The YGV does not state this explicitly, but the progression makes it clear that the Ariya is the superior type. The Ariya is also divided into 'still to be trained' and 'needing no more training,' and here it is implicitly understood that the second type is higher. The *ratana* 'needing no more training' can in turn either be a 'dry *vipassaka*' or 'one who has employed the Vehicle of Samatha,' the second being higher. This second type of Arhat can in turn be divided into the type that has obtained the perfections (*pārami*) of a Sāvaka, and the type that has not; the first type is higher because he has greater good qualities (*guṇamahantatāya*).

> Even compared to the type of *ratana* who has obtained the perfections of a Sāvaka, the *ratana* that is a Paccekabuddha is said to be superior. Why? Because, he has greater good qualities. Since, many hundreds of Sāvakas even resembling Sāriputta and Moggallāna do not approach the one-hundreth-portion of the good qualities of a single Paccekabuddha.
>
> Even compared to the *ratana* that is a Paccekabuddha, the *ratana* that is a Sammāsambuddha is said to be superior. Why? Because, he has greater good qualities. Since, even Paccekabuddhas filling the entirety of Jambudīpa, sitting after having crossed their legs, do not approach any part, portion, or part of a portion of the good qualities of a single Sammāsambuddha.
>
> Indeed, the Bhavagā said: 'As many sentient beings are there, monks [...], the Tathāgata is said to be the foremost amongst them,' etc. In this way, by whichever means possible, there is no *ratana* that is equal to the Tathāgata. Due to this the Bhagavā says: 'There isn't anything equal to the Tathāgata.'[18]

18 *Sāvakapāramippattaratanatopi paccekabuddharatanaṃ aggam akkhāyati. Kasmā? Guṇamahantatāya. Sāriputtamoggallānasadisāpi hi anekasatā sāvakā ekassa paccekabuddhassa guṇānaṃ satabhāgampi na upenti. Paccekabuddharatanatopi sammāsambuddharatanaṃ aggam akkhāyati. Kasmā? Guṇamahantatāya. Sakalampi hi jambudīpaṃ pūretvā pallaṅkena pallaṅkaṃ ghaṭentā nisinnā paccekabuddhā ekassa sammāsambuddhassa guṇānaṃ neva saṅkhaṃ na kalaṃ na kalabhāgaṃ upenti.*

This passage offers one of the most detailed hierarchies of *ratanas* that I am aware of. It highlights significant differences between the term *rat(a)na* and its corresponding English renderings 'jewel' or 'gem.' Apart from the extension to precious substances in general even within the mineral world, the usage of *ratana* for what is sentient seems to be more than just a metaphor: it is part of the direct semantic range of the term. In the classification presented in the YGV, it is clear that insentient *ratanas* occupy the lowest level in the hierarchy, which becomes more complex and rich in its subdivision of sentient beings, culminating in the Tathāgata, the *ratana* whose good qualities (*guṇa*) are unmatched by anything or anyone else.

RATNAŚĀSTRA, THE SCIENCE OF BENEFICIAL GEMS

At this point, I propose a brief complementary tangent into a branch of South Asian learning that Buddhist authors seemed to have remarkable interest in. This is Ratnaśāstra, i.e. the science or technical branch of learning that teaches about *ratnas* in the more restricted sense of precious materials such as gold, silver and a number of gems. Some of the discussions to be found in this branch of knowledge are especially useful in understanding passages which refer to specific gems or to how gemological expertise may work.

I will restrict my discussion of Ratnaśāstra to a few passages found in a relatively large treatise by a Buddhist author, the *Ratnaparīkṣā* by Buddhabhaṭa, and to a smaller 'Examination of Gems' (*Ratnaparīkṣā*), being a chapter of a Kāmaśāstra treatise called 'Everything about the Urbane Man' (*Nāgarasarvasva*) and also authored by a Buddhist, Padmaśrī.

Buddhabhaṭa's *Ratnaparīkṣā* begins (appropriately) with a Homage to the Three Jewels (*ratnatraya*), i.e. the Buddha, Dharma and Saṅgha. This is of course not unusual, but the author's choice to explicitly employ the term *ratna* in the introductory verse is most likely significant:

> Having bowed down to the Three Ratnas,
> that are saluted in the Three Worlds,
> and having looked at the Ratnaśāstra thoroughly,
> in respect to the choicest Ratnas and discarding the worthless,
> Buddhabhaṭa here presents no more than a compendium.[19]

Vuttañ hetaṃ bhagavatā – 'yāvatā, bhikkhave, sattā apadā vā...pe... tathāgato tesaṃ aggam akkhāyatī' tiādi (a. ni. 4.34; 5.32; itivu. 90). Evaṃ kenaci pariyāyena tathāgatasamaṃ ratanaṃ natthi. Tenāha bhagavā – 'na no samaṃ atthi tathāgatenā' ti.

19 *ratnatrayāya tribhuvanavanditāya namaḥ kr̥tvā samavalokya ca ratnaśāstram* | *ratnapravekam adhikr̥tya ca vimucya phalgu saṁkṣepamātram iha buddhabhaṭena dr̥ṣṭam* || (RP v.1)

The author then offers an introductory narrative that explains the origin of all precious *ratnas*:

The was once a powerful King of the demigods, called Bala ('Strength'), who was keeping the Three Worlds under his control, thanks to his great strength. As the gods were unable to defeat him in a fair battle, they resorted to deceit, and asked him as a boon to offer himself as the sacrificial animal for a rite. Due to his lofty character, he accepted; and thanks to his pure karma (*pariśuddhena karmaṇā*), when his body was torn by the gods, the parts became seeds of *ratnas*. The gods, *yakṣas*, *gandharvas* and *nāgas* went quickly through the sky to seize a large quantity of those *ratna*-seeds; and when these seeds fell here and there – in the ocean, in a river, on a mountain or in a forest – the place became a mine, because of the seed's inconceivable weight.

At this point the narration introduces an important distinction that is also the main rationale for the existence of Ratnaśāstra: some *ratnas* have beneficial powers, others don't. Good *ratnas* can vanquish a demon or *rakṣas*, poisons, snakes, illnesses and even sins; bad *ratnas* (born during bad astrological junctures or other bad days) are full of flaws (*doṣa*) and have no good qualities (*guṇa*). Finot (RP) perceptively remarked that the terms *guṇa* and *doṣa* are ambivalent: they can refer to the physical features that make a gem perfect/imperfect, or to its beneficial and maleficent powers. However, these appear as two sides of the same coin, and this match between beauty and beneficence is indeed best exemplified by the supreme *ratna*, the Buddha, whose physical marks of perfection are also a (thoroughly narrated) testimony of the beneficence of his past lives' endeavors.

Ratnas are, therefore, not only beautiful but also (and, perhaps, primarily) powerful. A recurrent line in Sanskrit says that 'mantras, gems and herbs have inconceivable power' (*mantramaṇyauṣadhīnām acintyaprabhavaḥ*); we should also remember that *ratnas* figure rather prominently in traditional medicine (Āyurveda) as antidotes to a number of physical imbalances. It is therefore even in the interest of the king to collect good *ratnas* and avoid the bad ones. This is where even a king should turn to the gemologist – not just for aesthetic reasons, but for his wellbeing and power.

The *Ratnaparīkṣā* explains in two verses the main qualities of a gemologist:

> Expert in the śāstra, and skilled; such is the Examiner of Ratnas.
> Only he is said to be the one to determine the measure of the price.
> Those who know the price of *ratnas* by following specific places and times,

and who do not follow the guidance of the śāstra, are not considered acceptable by the experts.[20]

One of the main qualities of the *ratnaparīkṣaka*, the 'Examiner of Gems' is indeed the ability to recognize good qualities and flaws (*guṇa*s and *doṣa*s), thanks to his acquaintance with the *śāstra*s and with the proper procedures to do so.

The same idea is emphasized in the short *Ratnaparīkṣā* chapter found in NāSa. The focus of this treatise is (primarily sexual) pleasure (*kāma*), and here gems appear to be an important accoutrement of the urbane townsman (*nāgaraka*), who seeks to heighten his own pleasure through varied means of refinement. The author, Padmaśrī, emphasizes that he will explain about good and bad gems, and their effects on their bearers 'for the benefit of sentient beings' (*sattvahitāya*). In verse 3.2, he offers a list of misfortunes that follow from wearing flawed gems:

> If someone, not understanding that a gem is tainted by flaws, wears it,
> Hosts of flaws like captivity, pain, illness, loss of kinsmen and wealth, etc., will fall to the share of that person.[21]

Verse 3.5 explains, on the other hand, the benefits:

> A good type of *ratna* causes the increase of the treasury, wealth, sons and servants,
> And also of good health, luck, results and lifespan;
> A bad type (i.e. a counterfeit), on the other hand, is the seed of quick loss.[22]

Padmaśrī takes Buddhabhaṭa's text as an authoritative precedent, and quotes from it (calling it *ratnaśāstra*). The half verse he quotes corresponds to verse 27cd of the *Ratnaparīkṣā*, in the chapter on diamonds (*vajra*):

> One with good qualities is the origin of riches of good qualities;
> The opposite is the cause of the arising of plight.[23]

20 *śāstravit kuśalaś cāpi ratnānāṁ sa parīkṣakaḥ| sa eva mūlyamātrāyāḥ paricchettā prakīrtitaḥ* ||14||
vettāro ratnamūlyasya deśakālāntarānugāḥ| na śāstravaśagā grāhyā vidvadbhis te 'pi nepsitāḥ ||15||

21 *doṣāpamr̥ṣṭaṁ maṇim aprabodhād bibharti yaḥ kaścana kañcid eva* | *taṁ bandhaduḥkhāmayabandhuvittanāśādayo doṣagaṇā bhajante* ||

22 *sujāti ratnaṁ parivr̥ddhihetur niddhānalakṣmīsutasevakānām* | *ārogyasaubhāgyaphalāyuṣāṁ ca vijātikaṁ tv āśu vināśabījam* ||

23 *tathā ca ratnaśāstre – guṇavad guṇasampadāṁ prasūtir viparītaṁ vyasanodayasya hetuḥ* | *iti* ||

Both texts emphasize that the primary reason to choose *ratnas* carefully, collect them and wear them, is their beneficial power, their ability to bring about an improvement in the favorable circumstances of one's life. This idea should be kept in mind when we observe the omnipresence of *ratnas* in Buddhist culture, as their role is not only, and perhaps not even primarily, aesthetic.

'Good health' is one of the purposes of wearing proper gems and jewels; that these may have medical qualities is confirmed by their usage in traditional South Asian medicine (Āyurveda), which I earlier hinted at. Vāgbhaṭa, one of the great ancient authorities on Āyurveda, and a Buddhist, recommends, as part of a lay person's daily routine: 'He should always wear *ratnas*, *siddha*-mantras, and great medicinal herbs.'[24]

A commentator explains that:

> *Ratnas* are gems (*maṇi*) such as diamonds, rubies, etc.; a '*siddhamantra*' is one such as the 'Aparājita,' etc.; a 'great medicinal herb' is the Sahadevī, etc.[25] (*Sarvāṅgasundarā*)

He also tells us that one should always wear these 'on the arms, neck, etc.' Such advice is part of a set of holistic prescriptions towards a healthy daily routine, and it seems rather clear that, falling within the same category as mantras and medicinal herbs, jewels are here primarily prized for their positive effect on a person's health. It is also significant to notice that the commentary lists *diamonds* as the first among beneficial gems.

Vāgbhaṭa explicitly mentions gems (*maṇi*) as a type of medicine (*auṣadha*), of the sub-category considered 'based on the divine' (*daivavyapāśraya*), in a list which starts (once again) with mantras and medicinal herbs. Verses 12.12–18 describe the beneficial effects of jewels:

> Ruby, sapphire, topaz, and *vidūraka*; pearl, coral, the best of diamonds, *vaiḍūrya*, crystal, etc.;
>
> A precious gem (*maṇiratna*), fluid, cool, astringent, pleasant to touch, pleasing to the eyes, when it is worn removes sin, misfortune and poisons;
>
> It brings wealth, long-life, and energy; it causes joy and enthusiasm, and it is auspicious.[26]

24 *dhārayet satataṁ ratnasiddhamantramahauṣadhīḥ* | (AH II.3)

25 *ratnāni maṇayo vajrapadmarāgādayaḥ* | *siddhamantro 'parājitādi* | *mahauṣadhiḥ sahadevyādikā* | (SSun)

26 *padmarāgamahānīlapuṣparāgavidūrakāḥ* | *muktāvidrumavajrendravaiḍūryasphaṭikādikam* ||12.16||

The idea is identical to what we found in Ratnaśāstra, suggesting a complementarity between traditional medicine and gemology: since gems have beneficial medical properties, the Āyurveda specialist must turn to Ratnaśāstra in order to choose fine and genuine jewels: starting, most probably, from diamonds.

DIAMONDS AND THUNDERBOLTS

Buddhabhaṭa's *Ratnaparīkṣā* offers us some indications of why diamonds should hold a special place within the Buddhist symbolic landscape. The *Ratnaparīkṣā* starts in fact with a discussion of diamonds (and so does NāSa); the reason is given as follows:

> Since the diamond is declared by the experts to have great power,
> The diamond should be examined first; therefore, I will explain about it.[27]

Once again, the main feature of the diamond, the one that makes it foremost amongst gems, is not beauty, but power (*mahāprabhāva*). Buddhabhaṭa, going back to the initial narrative setting, tells us about the origin of diamonds:

> In whichever place of the Earth a piece of bone of that One Wishing to Conquer the Diamond-Bowed One somehow fell,
> There are diamonds of many different types.[28]

The reference to bones must be considered somewhat significant in a Buddhist setting, as anyone familiar with the recurrence of bone-relics in the Buddhist tradition will notice. 'Bones' were not necessarily considered as suitable objects of veneration in non-Buddhist Indian traditions. The special devotion towards the bone-relics of the Buddha (and of other Buddhist masters) earned the Buddhists the sarcastic epithet of 'bone-worshipers' (*asthi-pūjaka*), bestowed upon them by their adversaries to humor them during rather intense philosophical debates. Although the narrative setting makes this reference to bones sensible and acceptable

---

maṇiratnaṃ saraṃ śītaṃ kaṣāyaṃ svādulekhanam | cakṣuṣyaṃ dhāraṇāt tat tu pāpmālakṣmīviṣāpaham ||12.17|| dhanyam āyuṣyam ojasyaṃ harṣotsāhakaraṃ śivam ||12.18 ab|| (ASaṁ)

27 *mahāprabhāvaṁ vidvadbhir yasmād vajram udāhṛtam | vajraṁ pūrvaṁ parīkṣeyaṁ tato 'smābhir nigadyate* || (RP 16)

28 *tasyāsthileśo nipapāta yeṣu bhuvaḥ pradeśeṣu kathaṁcid eva | vajrāṇi vajrāyudhanirjigīṣor bhavanti nānākṛtimanti teṣu* || (RP 17)

against probably any other religious backdrop, a Buddhist audience would most likely have some added mental associations when listening to this story of a meritorious person's bones turning into diamonds.

The *Ratnaparīkṣā* makes it very clear that one of the special features of diamonds, which allows recognition of a genuine one, is its ability not to be scratched by any other material, and its ability to scratch any other material (vv. 48, 49). This feature is particularly important in the Buddhist usages of the term *vajra*, as we shall see. 'A diamond scratches all; a diamond is not scratched by others' (*vajraṁ vilikhati sarvān nānyena vilikhyate vajram*).

The chapter on diamonds ends with a verse that highlights, in a rather triumphant cadence, how a king will benefit from wearing a diamond of good quality:

> Lightning-like, flashing, delightful –
> A king who wears the kind of diamonds explained above,
> With his supreme splendor attacking them boldly,
> Enjoys the entire land of his neighbors.[29]

Two of the terms most employed (even in Buddhist texts) for diamonds are *vajra* and *kuliśa*. These may also refer to a thunderbolt (associated from ancient times with the deity Indra).

The etymology of vajra is usually linked to the root *vaj*, in the sense of 'moving' or 'going,' the idea being that a *vajra* is that which moves about without any impediment. Although this suggests a primary application to a thunderbolt, the unbreakability of a *vajra* – its being 'unimpeded' by any other material – may also be compatible with the commentarial expansions on the word.

When we compare the Sanskrit commentaries on the *Amarakośa* with the Pāli commentary on the *Abhidhānappadīpikā*, we see a rather fascinating continuity between the two languages, suggesting that they may be understood as part of the same tradition of lexical studies:

Amarakośodghāṭana: 'It moves (*vajati*), it just goes, it is not prevented, thus it is a *vajra*.'[30]

Vyākhyāsudhā: 'It moves (*vajati*). "The root *vaj* is in the sense of 'going'"; according to "*ṛjrendra*" we have the suffix *<ran>*.'[31]

29 *saudāminīvisphuritābhirāmaṁ rājā yathoktaṁ kuliśaṁ dadhānaḥ* | *parākramākrāntaparapratāpaḥ samastasīmāntabhuvaṁ bhunakti* ||

30 *vajati yāty eva na pratihanyate vajram* | Amara(K), p. 11.

31 *vajati* | *vaja gatau (bhvā.pā.se.) ṛjrendra* – (u.1.18) *iti ran* || VS, p. 20.

Jātarūpa: 'It only moves (*vrajaty eva*), its activity of going is not prevented by anything whatsoever, thus it is a *vajra*.'[32]

Abhidānappadīpikāṭīkā: '"The root *vaj* is in the sense of going"; it only moves (*vajateva*), its activity of going is not prevented by anything whatsoever, thus it is a *vajira*; the suffix is *ira*.'[33]

In the original, the wording of Jātarūpa's Sanskrit commentary and that of the Pāli *Abhidhānappadīpikāṭīkā* are practically identical.

The other recurrent term for a diamond/thunderbolt, i.e. *kuliśa*, is also similarly explained as something with the power to thin down or destroy:

> It thins (*śyati*) the mountains (*kulinaḥ*), i.e. it makes them thinner by cutting away their wings, thus it is '*kuliśa*'; or, it hurts (*liśati*) what is despicable (*kutsita*), i.e. it cuts down the enemies.[34]

BUDDHIST DIAMONDS

I will now turn to specifically Buddhist usages of the term *vajra*, highlighting what most readers would have already recognized as the main argument of this essay: gems and precious substances are not only symbols of what is precious and beautiful but also, and perhaps primarily, of what is powerful, i.e. supremely effective and beneficial.

I can only hope to muster a representative list, as comprehensiveness would be beyond the scope of a single article. This will be the case even though I have restricted myself to diamonds/thunderbolts; let alone if I wished to include a number of other precious substances, like gold, *vaidūrya*, crystals and so forth, that also figure prominently in Buddhist texts (and on statues, inlaid temple-doors, etc.). I feel, though, that even this rather incomplete selection of Buddhist *vajras* will make it clear that diamonds are supreme because they are unhindered in their beneficial power, and not solely because they shine.

32 *vajaty eva na pratihanyate 'sya gamanaṁ kenāpīti vajram* | Amara(J), p. 42.

33 *vaja gatiyaṃ, vajateva na paṭihaññate yassa gamanaṃ kenacīti vajiraṃ, irapaccayo.*

34 *kulinaḥ parvatāñ śyati pakṣacchedena tanūkaroti kuliśaṁ kutsitaṁ liśati takṣṇoty arīn vā* | Amara(K), p. 10.

THE PATH OF THE ARHAT IS LIKE A VAJIRA

The 'Root Commentary' (*mūlaṭīkā*) on the *Dhammasaṅganī*, one of the seven basic books of Abhidhamma in the Pāli tradition, explains the logic of comparing the Arhat's Path to a 'Vajira':

> For a *vajira*, wherever it falls upon, there is no gem or stone, etc., that is unbreakable, nor is its path of movement blocked by it; in the very same way, wherever it arises, in that (mental) continuum there is no affliction that cannot be broken by the Path of Arhathood, nor, being broken, is it blocked; thus, its quality of being comparable to a *vajira* is to be understood.[35]

This passage is easy to compare with the etymologies of *vaj(i)ra* that we found in the lexicons, and with the principle, mentioned in the Ratnaśāstra, that a diamond can break all other materials and that it is broken by nothing else. Both ideas, as we will see, recur in other contexts; and both have primarily to do with the *efficacy* of diamonds/thunderbolts.

SUPREME CONCENTRATION IS LIKE A VAJRA

The Buddhist path is recurrently described in terms of progressive stages of mental cultivation, often in the rather technical sense of 'concentration' (*samādhi*), where the practitioner is single-mindedly aware of a specific, often lofty, point of reference.

Among the *samādhi*, the highest, to be attained as a prelude to liberation, is the 'Concentration that is Like a Vajra' (*vajropamasamādhi*). This appears in a number of texts of rather different traditions; it figures in the Sarvāstivāda fold and is discussed in the *Abhidharmakośabhāṣya*; and in the Mahāyāna too, it is the mental accomplishment that immediately precedes complete, full Awakening (*anuttarasamyaksambodhi*), as mentioned, for example, in the *Bodhisattvabhūmi* (BBh).

In some of the texts, it seems clear that the common quality between supreme concentration and a *vajra* is the power to pierce through anything else, including the nature of reality; 'piercing through' is a recurring Buddhist image referring to the quality of genuine awareness of

35 *Vajirassa yattha taṃ patati, tattha abhejjaṃ nāma kiñci maṇipāsāṇādi natthi, na ca tena gamanamaggo viruhati, evam eva arahattamaggena yattha so uppajjati, tasmiṃ santāne abhejjo kileso nāma natthi, na ca bhinno puna viruhatīti vajirupamatā veditabbā.* (DSMṬ 101–08)

something. The following passage from the *Suvikrāntavikrāmiparipṛcchā* perhaps captures this well:

> Just like, Suvikrāntavikrāmin, into whatever a *vajra* is cast upon so as to pierce through that thing, it does pierce through it; in the very same way, wherever a monk places and in whichever things he applies his mind to, once he has seized that mind with a higher cognition that has the piercing quality of the Concentration that is Like a Vajra, he pierces through all those things.[36]

BODHICITTA IS UNBREAKABLE, LIKE A VAJRA

The *Abhisamayālaṁkārakārikā* is a crucial text of the Mahāyāna tradition devoted to the exegesis of the textual corpus of the *Prajñāpāramitā* (the 'Complete Mastery of Higher Cognition' or the 'Perfection of Wisdom,' etc.). A substantial portion of Abhisamayālaṁkāra literature is devoted to a discussion of the 'Arising of the Mind' (*cittotpāda*), i.e. to Bodhicitta, understood as the 'wish to obtain Perfect Awakening for the sake of others' (*parārthāya samyaksambodhikāmatā*). In accordance to what accompanies this arising of the mind, it may be subdivided into twenty-two varieties, comparable to a number of different things (including a 'mine of jewels' and, as we will see later, a 'Wish-Fullfilling Gem'). When the Arising of the Mind is accompanied by the Pāramitā of Valor (*vīrya*, defined as enthusiasm for what is wholesome), its term of comparison becomes a *vajra*:

> The eighth, accompanied by the Pāramitā of Valor, is comparable to a *vajra*. As he says: 'A Bodhisattva Great Being who wishes to establish in valor those who are lazy, and who wishes to surpass the valors of all the Śrāvakas and Pratyekabuddhas with the arising of a single mind of rejoicing, should train in the Pāramitā of Higher Cognition.' Thus, it is in the sense of its being unbreakable, thanks to the firmness of one's conviction.[37]

36 *tadyathāpi nāma suvikrāntavikrāmin vajraṃ yasminn eva nikṣipyate nirvedhanārtham, tat tad eva nirvidhyati, evam eva bhikṣur vajropamasamādhinairvedhikyā prajñayā parigṛhītaṃ (cittaṃ?) yatra sthāpayati yeṣu ca pracārayati, tān sarvān nirvidhyati* | Suv, p. 5.

37 *aṣṭamo vīryapāramitāsahagato vajropamo yad āha – kusīdān vīrye pratiṣṭhāpayitukāmena | sarvaśrāvakapratyekabuddhavīryāṇy ekānumodanācittotpādenābhibhāvitukāmena bodhisattvena 'mahāsattvena prajñāpāramitāyāṁ śikṣitavyam' iti | tasya sampratyayadārḍhyenābhedyatvāt* || AAV, p. 18.

Here Ārya Vimuktisena, the great Abhisamaya commentator, links the comparison found in the root text with a passage from the 'Perfection of Wisdom in Twenty-Five Thousand Lines,' showing that the point of commonality is that this type of Bodhicitta is unbreakable – a quality of *vajra*s that is emphasized, as we have seen, even in Ratnaśāstra.

DIAMOND-WORDS

Ulrich Pagel (2007) has offered an in-depth study of the concept of *vajrapada*; the term is itself difficult to translate, because while *pada* does mean a 'word' or an 'expression,' the explanations that we find in the texts also link it to its etymology as meaning a 'locus.' So, a *vajrapada* is a key expression of an important Buddhist teaching, and at the same time it is the locus or basis for something.

*Vajrapada*s are found in a variety of texts; they are prominent in the Mahāyāna Sūtras and also in a Mantranaya context. I will here consider the occurrence of *vajrapada* in two important *śāstras*.

The first is the *Ratnagotravibhāga* ('The Subdivision of the Jewel-Lineage'), the main Sanskrit treatise expounding the doctrine of Tathāgatagarbha, the 'Buddha-Nature' that is present in all sentient beings. The very first verse offers an index of the whole work through seven *vajrapada*s:

> The Buddha, the Dharma, the Assembly, and the Dhātu;
> Awakening, Good Qualities, and, finally, the Activity of the Buddhas;
> This is the body of the entire treatise:
> In brief, the seven *vajrapadas*.[38]

In the commentarial explanation we find, once again, the idea of 'piercing through' as a cognitive metaphor:

> The *pada*, i.e. place, of the referent of realization, which is comparable to a *vajra*, is the '*vajra-pada*.' In that context, due to its being hard to pierce through with an awareness consisting of listening and contemplation, the referent, which has an ineffable nature and is to be experienced by oneself, is to be understood to be like a *vajra*. The letters that express such a referent by elucidating the path that is conducive to its obtainment are said to be the *pada* because they are its basis. Thus, in the sense of 'difficult to pierce

38 *buddhaś ca dharmaś ca gaṇaś ca dhātur bodhir guṇāḥ karma ca bauddham antyam*| *kṛtsnasya śāstrasya śarīram etat samāsato vajrapadāni sapta* | (RGV I.1)

through' and in the sense of a 'basis' the quality of being *vajra-pada*, of both meaning and expression, should be understood.[39]

Vajrapadas figure prominently in another great Mahāyāna *śāstra*, the *Madhyāntavibhāga* ('Differentiation of the Middle from the Extremes'). Sthiramati's subcommentary offers a detailed discussion of the *vajrapadas* appearing in the root text, and he also gives three possible explanations of the term:

> In that context: it is a *vajra* in the sense of being like a *vajra*; because it breaks all mental distortions and is not broken by them. The awareness without distortions explained in the contiguous section of this text is the *vajra* that is understood (*pratīyate*), i.e. comprehended, by these: hence they are *vajra-padas*, meaning 'an explanation of *vajra*.'
>
> Alternatively, placing in the mind the letters and the meanings, etc., is itself expressed by the term '*pada*'; thus, the *padas* of the vajra means its 'points of reference.'
>
> Or, since they are difficult to break, these *padas* are like *vajras* and hence are called *vajrapadas*.[40]

We can easily notice that both treatises explain *vajra* in a way that is parallel and compatible with what we found in the lexicons and in the *Ratnaparīkṣā*. However, while the *Ratnagotravibhāga* focuses on the idea of 'piercing' (*vyadh*), Sthiramati speaks in terms of 'breaking' (*bhid*), just as Vimuktisena does while describing Bodhicitta, and just as the *Hevajra Tantra* does, as we will now see, while explaining its own name.

39 *vajropamasyādhigamārthasya padaṃ sthānam iti vajrapadam | tatra śruticintāmaya-jñānaduṣprativedhād anabhilāpyasvabhāvaḥ pratyātmavedanīyo 'rtho vajravad veditavyaḥ | yāny akṣarāṇi tam artham abhivadanti tatprāptyanukūlamārgābhidyotanatas tāni tatpratiṣṭhābhūtatvāt padam ity ucyante | iti duṣprativedhārthena pratiṣṭhārthena ca vajrapadatvam arthavyañjanayor anugantavyam* | (Commentary on RGV I.1)

40 *tatra vajram iva vajraṃ sarvaviparyāsabheditvāt taiś cābhedyatvāt | anantaroktam aviparyāsajñānaṃ tac ca vajram ebhiḥ pratīyate avagamyata iti vajrapadāni | vajravyākhyānam ity arthaḥ | atha punar vyañjanārthamanaskārādaya eva padaśabdenocyante tato vajrasya padāny ālambanānīty arthaḥ | athavā durbhedatayā vajravad etāny eva padāni vajrapadānīty ucyante* | (MVBhṬ on 5.23ab)

EMPTINESS IS THE UNBREAKABLE VAJRA

Emptiness (*śūnyatā*) is a key concept of Buddhist thought, explained in a number of different ways according to the specificities of the philosophical school. Within a Mahāyāna context, two distinct interpretations of emptiness are to be found in the Yogācāra and Madhyamaka schools, respectively. The first explains emptiness as the fact that mind is devoid of an object and of agent of perception, and projects both as the illusory appearance of sentient beings and the world, from life to life. For the Madhyamaka, that mind too never really arose, and any predication of existence or non-existence is empty, since it is never a self-sufficient concept.

The Mahāyāna tradition includes a swifter path, usually known as 'the Method of Mantra' (*mantranaya*), and texts that expound such a path (generally) presuppose the Yogācāra or Madhyamaka understanding of emptiness. The following verse is found in two (related) commentaries to the greatly important *Hevajra Tantra*: the first (Ratnākaraśānti's 'Garland of Pearls') is from a Yogācāra author; the second (Kṛṣṇācārya's 'Garland of Jewels of Yoga') is from a Madhyamaka perspective.

> Firm, solid, without cavities, being by definition uncuttable and unbreakable,
> Not burnable and not destructible, emptiness is called '*vajra*.'[41]

The adjectives used here to describe a *vajra* make perhaps better sense when we think of a diamond and its examination; for example, having cavities would be perceived as a remarkable flaw for a precious stone. Furthermore, what I have translated as 'not burnable and not destructible' could be translated actively as 'not causing one to burn and not causing destruction,' as causing fires and causing overall destruction are two features of a flawed or fake diamond according to Ratnaśāstra.

Both commentaries attribute the verse to the *Vajraśekhara*,[42] but they quote this to explain different parts of the *Hevajra Tantra*. The *Muktāvalī*

41 *dṛḍhaṁ sāram asauṣiryam acchedyābhedyalakṣaṇam | adāhī avināśī ca śūnyatā vajram ucyate* || The verse appears with some orthographical variation for sauṣirya in different transmissions and, according to the readings reported by the Sarnath edition of the *Yogaratnamālā* (YRM), at least one manuscript has adāhi and avināśi, which would (perhaps) make the last two adjectives apply to *vajra* rather than *śūnyatā*. However, even the latter variation does not change much, as all the adjectives are supposed to be applicable to both.

42 I thank Harunaga Isaacson for pointing out that the verse is in fact found in the *Vajraśekhara/Vajraśikhara* (Tib.), and for offering the relevant reference: see

quotes this verse while explaining the expression 'Higher cognition is called Vajra' (*vajraṁ prajñā ca bhaṇyate*, v. 1.7), while the *Yogaratnamālā* quotes it to explain the sentence 'Being unbreakable, it is called Vajra' (*abhedyaṁ vajram ity uktam*, 1.4). The *Muktāvalī* is therefore explaining the sense of the name of the Tantra itself (*he-vajra*), while the *Yogaratnamālā* is expanding on the epithet 'Vajra-sattva,' a synonym for ultimate Buddhahood (also called 'Vajra-dhara'). Both commentators agree that, although the passages are not explicit on this point, what is actually being referred to by *vajra* is here emptiness.

While (in this context) Ratnākaraśānti does not elaborate on the reasons for the unbreakability of emptiness, Kṛṣṇācārya offers a reference to a (difficult) verse from Nāgārjuna's *Mūlamadhyamakakārikā*:

> The Bhagavat said: 'Unbreakable,' etc.: it cannot be broken, thus it is 'unbreakable'; it cannot be defeated by debaters, whether they may be from on one's own faction or from the opponent's faction. What is that? The emptiness of all the dharmas. Thus the Ācārya, Devapāda, said:
>
> When a refutation is done in accordance with emptiness,
> If someone were to speak a rebuttal,
> For him nothing has been rebutted
> And it all becomes the same as what is to be established.[43]

The verse is actually from the chapter of the *Mūlamadhyamakakārikā* dealing with the refutation of the five aggregates. What I have translated as 'rebuttal' could perhaps also be rendered as 'exception.' Candrakīrti clarifies the sense of the verse as follows: If the Madhyamaka proves that form (the first of the five aggregates) is empty, then the opponent may wish to retort that feeling (the second aggregate) has not been proven to be so, and thus form too might be considered non-empty. But nothing whatsoever escapes the Madhyamaka basic argument that whatever is dependently arisen, is empty; and Candrakīrti seems to imply that the entire *Mūlamadhyamakakārikā* (with the exception of a few chapters, where

---

Derge (Tōhoku 480) f. 149r7–149v1, Peking (Ōtani 113) f. 170r3–4. The verse is also found in Advayavajra's *Pañcatathāgatamudrāvivaraṇa* (PTMV).

43 *bhagavān āha – abhedyam ity ādi | na bhedya iti abhedyam | svaparavādibhir ajayyam | kiṁ tat ? śūnyatā sarvadharmāṇām | tathācāryadevapādaḥ – vigrahe yaḥ parīhāraṁ kṛte śūnyatayā vadet | sarvaṁ tasyāparihṛtaṁ samaṁ sādhyena jāyate ||* (YRM i.4) I read (according to Harunaga Isaacson's suggestion) *jayyam* for *japyam* in the Sarnath edition. The verse corresponds to *Mūlamadhyamakakārikā* 4.9.

he marks the difference) functions according to the following structure: Nāgārjuna refutes some Abhidharmic categories; the opponent proposes a rebuttal/exclusion; Nāgārjuna points to the fact that the next category is also not excluded, by applying the general principle that dependent arising proves emptiness, through the formulation of specific additional arguments.[44]

In Kṛṣṇācārya's exegesis, emptiness is a diamond because no amount of debate will refute the empty nature of anything whatsoever; we find here perhaps another echo of the *Ratnaparīkṣā*, where the chapter on diamonds ended with a verse that depicted a king, wearing a *vajra* and thus invincible in battle.

THE WISH-FULFILLING GEM

While among the jewels within the range of (ordinary) gemologists none surpasses the diamond, the South Asian tradition speaks of another jewel, far more powerful than the *vajra* itself, to be retrieved from some non-human realm: this is the Cintāmaṇi, the Wish (*cintā*)-Fulfilling Gem (*maṇi*). Its main feature is that, whatever one may wish for, it will be obtained thanks to the exceptional power of the Cintāmaṇi. Not unlike the diamond/thunderbolt, the Cintāmaṇi has been employed as a term of comparison for a number of different things, of which I will here offer a small sample.

DISCIPLINE FULFILLS ALL WISHES

The *Abhidhammāvatāra* tells us that *sīla* ('discipline,' i.e. wholesome habits) is, among other things, 'the best of ornaments' (*alaṁkāro anuttaro*), a jewel (*ratana*), a refuge (*saraṇa*), the best of vehicles (*yānam anuttaram*), and a Wish-Fulfilling Jewel. Commenting on the last identification, the *Purāṇaṭīkā* explains:

> Because it brings about the specific accomplishment that one wishes for, it is equal to a Cintāmaṇi, thus it is called 'Cintāmaṇi.'[45]

44 See PP, p. 127.

45 *icchiticchitassa sampattivisesassa nipphādanato cintāmaṇisamanticintāmaṇi.* (AAPṬ on 794)

BODHICITTA, THE ASPIRATION THAT ACCOMPLISHES ALL ASPIRATIONS

After comparing Bodhicitta to a *vajra* (as we have seen above), the *Abhisamayālaṁkāra* offers the Wish-Fulfilling Gem as another term of comparison, referring to Bodhicitta accompanied by the Perfection of Aspiration (*praṇidhāna*):

> The twelfth, accompanied by the Pāramitā of Aspiration, is similar to a Cintāmaṇi. As he says: 'A Bodhisattva who wishes to bring forth the Buddha-Body, who wishes to address with the Dharma-Teaching each of the infinite world-spheres in the ten directions, who wishes not to cut off the family of the Buddhas (…).' The sense is that, in respect to it, results are accomplished in accordance to one's aspiration.[46]

Bodhicitta is like the Cintāmaṇi for a Bodhisattva who has perfect aspirations, because those aspirations will be fulfilled without fail, by relying on the 'Arising of the Mind.'

TWO WISH-FULFILLING GEMS FROM A MEDITATION MANUAL

One of the great Buddhist systems of (Vajrayāna) meditation is the 'Wheel of Time' (Kālacakra), which is the name of a form of the 'primordial Buddha' and of the main text wherein related cosmology, inner physiology and meditative practices are expounded (i.e. the *Kālacakra Tantra*).

The stages of Kālacakra practice are usually divided into a six-fold yoga; the following two Cintāmaṇi-similes are taken from a commentary on a celebrated text on the six yogas. Raviśrījñāna describes two important facets of the Buddhist path as being Wish-Fulfilling Gems:

> It diffuses delight (*ratiṁ tanoti*), thus it is a *ratna*; the ratna that is the good path is a *ratna* like the *ratna* that is the Wish-Fulfilling Gem, because of its being difficult to obtain and because it bestows the wished-for result.[47]

46 *dvādaśaḥ praṇidhānapāramitāsahagataś cintāmaṇisadṛso yad āha – buddhakāyaṁ niṣpādayitukāmena daśasu dikṣu pratyekam anantān lokadhātūn dharmadeśanayā vijñāpayitukāmena buddhakulam anupacchettukāmena bodhisattvena 'mahāsattvene' ti | tasya yathā praṇidhāṇaṁ phalasaṁṛddheḥ* || (AAV, p. 19)

47 *ratiṃ tanotīti ratnaṃ cintāmaṇiratnam iva ratnaṃ durlabhatvāc cintitārthaphalapradatvāc ca sumārgaratnam* || (GuBh, p. 88)

> [...] the arising of non-conceptual awareness, because it accomplishes the good of the world, in all its aspects, is like a Wish-Fulfilling Gem.[48]

The first passage once again reminds us of the lexicons and their definitions of *ratna*; here the path and (perhaps) its most important result (the arising of non-conceptual awareness) are extolled as the supreme *ratna*, the Cintāmaṇi.

THE BUDDHA, SUPREME CINTĀMAṆI

> Mañjuśrī, there exists, in the Great Ocean, a Great Gem-Ratna, called 'The Fulfillment of All Wishes,' tied to the top of a standard. Whichever wish a sentient being has, sentient beings will hear sounds coming out from that Gem and conforming to that wish. Yet, that great, precious jewel does not form mental constructs or concepts. It does not think or ponder. It is inconceivable, free from what can be conceived, and free from mind, mentality, and consciousness.[49]

A number of texts compare the Buddha to the Cintāmaṇi; the two qualities that are common to the Cintāmaṇi and to the Buddha are the ability to fulfill all the wishes of sentient beings, and the absence of mental constructs.

The Cintāmaṇi exemplifies how a Buddha could act (primarily, teach) while at the same time being completely free of concepts (this being a topic of complex and prolonged reflection in the Mahāyāna tradition). This Gem is something that we know to be without any concept (in this case, because it is insentient), and yet its efficacy makes it seem as if it may have a concept of what people wish for.

A Buddha's activity, not unlike the Cintāmaṇi's power, is a reflection of the mind of the sentient beings – which means their wishes in the case of the Wish-Fulfilling Gem, but means their aspirations (*āśaya*) and faith (*adhimukti*) in the case of the Buddha. Both the Cintāmaṇi and the Buddha, moreover, depend for their appearance on the accumulated meritorious karma of sentient beings. The other necessary cause for the appearance

48 [...] *sa ca nirvikalpajñānotpādaḥ sarvākārajagadarthasampādakatvāc cintāmaṇir iva* || (GuBh, p. 118)

49 *tadyathā mañjuśrīr asti mahāsāgare sarvābhiprāyaparipūraṇaṃ nāma mahāmaṇiratnaṃ tad dhvajāgrāvabaddham | yasya sattvasya yādṛśo 'bhiprāyo bhavati, tādṛśaṃ tataḥ śabdaṃ niścarantaṃ sattvāḥ saṃjānanti | tac ca mahāmaṇiratnaṃ na kalpayati na vikalpayati na cintayati na vicintayati, acintyaṃ niścintyaṃ cittamanovijñānāpagatam* | (JS)

of a Buddha is, of course, his own previous accumulation of merit and wisdom during the Bodhisattva path, and the necessary aspirations towards Buddhahood.

The quote at the beginning of this section, from *The Ornament of the Light of Awareness*, highlights an important feature of the Cintāmaṇi that helps to explain its use as a comparison for the Buddha. The Wish-Fulfilling Gem, in fact, has the quality of reflecting the minds of sentient beings; its apparent 'knowledge' of the sentient beings' wishes (that turn into sounds thanks to the Gem's presence) is not the result of conscious effort or conceptual deliberation. I would propose that this reflective quality is part of a mineral metaphor about the purity of the gem itself. That this may be the case is supported by another jewel-simile in the same Sūtra (which I have translated at the beginning of this article): the ground of *vaiḍūrya* (beryl/cat's eye gem/sapphire) is capable of reflecting Indra's palace, just like the mind of sentient beings is capable of reflecting the image of a Buddha, and, in both cases, it is thanks to their purity. The obvious gradation between sentient beings and Buddhas suggests that the same gradation is implied in the mineral metaphors employed for both – i.e. the Cintāmaṇi may be even purer than the *vaiḍūrya*.

The *Ratnagotravibhāga* discusses the example of the Wish-Fulfilling Gem in three different parts of the text. The first time we find a discussion of the Cintāmaṇi it refers to the Buddha's Dharmakāya:

> In respect to the Dharmakāya of the Tathāgata, the similarity to a Wish-Fulfilling Gem should be understood in terms of its own characteristic, which is the nature of having the power to accomplish the wished-for purpose, and so forth. (RGV)[50]

In the second instance, the Cintāmaṇi refers to the Saṁbhogakāya:

> Because it appears as the Dharma, in the form of the enjoyment of varied Dharmas;
>
> Because it is never interrupted in concerning itself with the good of sentient beings, being the outflow of the purity of compassion;
>
> And because, without concepts and without effort, it fulfills wishes,
> The Sāṁbhoga is explained in terms of the perfect power of a Cintāmaṇi.

50 *tatra tathāgatadharmakāye tāvac cintitārthasamṛddhyādiprabhāvasvabhāvatāṃ svalakṣaṇam ārabhya cintāmaṇiratnasādharmyaṃ veditavyam* | (Commentary to I.31)

Its varied coloring is said to be fivefold; in respect to: teaching, being in sight,
Uninterrupted activity, no effort, and appearing with a different nature.

Just like a gem becomes different due to the variety of colors that are near it,
So the Lord becomes different due to the variety of sentient beings that are near him.[51]

The third instance comments more directly on the passage from *The Ornament of the Light of Awareness*, and its focus is probably the Nirmāṇakāya:

He is like a Cintāmaṇi:

Just like the Cintāmaṇi fulfills all wishes for each and every one
Who is staying within its range, although it is free from concepts,

So also people with different inclinations, after going near the Cintāmaṇi that is the Buddha,
Hear a varied Dharmatā; and he does not form concepts about them.

Just like, having no concept, the Gem-Ratna bestows the desired wealth to others, without effort,
So also the Muni, without effort, according to their worthiness,
Accomplishes others' good, continuously, for as long as existence itself.

The Tathāgatas are difficult to obtain:

In this world, obtaining a beautiful gem is extremely difficult,
As people eagerly search for it in the ocean and the netherworld;
Similarly it should be understood that obtaining sight of a Tathāgata is not easy,
In this very unfortunate world, whose mind is seized by many afflictions.[52]

51 *vicitradharmasaṃbhogarūpadharmāvabhāsataḥ* | *karuṇāśuddhiniṣyandasattvārthāsraṃsanatvataḥ* ||2.49||
nirvikalpaṃ nirābhogaṃ yathābhiprāyapūritaḥ | *cintāmaṇiprabhāvarddheḥ sāṃbhogasya vyavasthitiḥ* ||2.50||
deśane darśane kṛtyāsraṃsane'nabhisaṃskṛtau | *atatsvabhāvākhyāne ca citratoktā ca pañcadhā* ||2.51||
raṅgapratyayavaicitryād atadbhāvo yathā maṇeḥ sattvapratyayavaicitryād atadbhāvas tathā vibhoḥ ||2.52||

52 *cintāmaṇivad iti* | *yugapad gocarasthānāṁ sarvābhiprāyapūraṇam* | *kurute nirvikalpo 'pi pṛthak cintāmaṇir yathā* ||4.67||
buddhacintāmaṇiṁ tadvat sametya pṛthagāśayāḥ | *śṛṇvanti dharmatāṁ citrāṁ na kalpayati tāṁś ca saḥ* ||4.68||

To be more precise, as v. 4.95 tells us, the Buddha is even more difficult to obtain than the 'ordinary' Cintāmaṇi.

THE BUDDHIST, AND THE BUDDHA, AS GEMOLOGISTS

Several texts describe the Buddha's teaching as a 'Lion-Roar' that scares non-Buddhist philosophers, who are more timid wild beasts. This powerful-sounding teaching is often said to be about selflessness; and yet, the *Tattvasaṁgraha* of Śāntarakṣita (and Kamalaśīla's glosses), proposes an alternative Lion-Roar that is no other than an invitation to test gold:

> Like experts would do with gold, through heating, cutting, and touchstones, Monks! You should accept my words after examining them, and not due to respect for me as a teacher.[53]

The idea here seems to be that this is a Lion-Roar because it shows, amongst other things, supreme self-confidence (one of the special qualities of a Samyaksambuddha).

However, testing teachings (like testing gems and precious substances) is not an activity only enjoined for student monks. Readers of Buddhist texts are probably used to comparisons between the Buddha and doctors, or with the many royal similes that make it clear that the Buddha is the Supreme Emperor of Dharma (the Dharmacakravartin). The next passage that I translate in this article offers another simile: Buddhas, and Bodhisattvas, are like gemologists; or, more precisely, they are caravan-leaders, expert enough in precious things to ascertain *the* most precious gem of all: Bodhicitta. What follows is Prajñākaramati's commentary to a verse of Śāntideva's *Bodhicaryāvatāra*, part of a larger section devoted to expounding the praiseworthy good qualities of Bodhicitta:

> Even those who wish for prosperity within the different destinies of existence should not have doubts or distorted ideas about [Bodhicitta]: showing this, he says:

---

yathāvikalpaṁ maṇiratnam īpsitaṁ dhanaṁ parebhyo visr̥jaty ayatnataḥ | *tathā munir yatnam r̥te yathārhataḥ parārtham ātiṣṭhati nityam ābhavāt* ||4.69|| *durlabhaprāptabhāvās tathāgatā iti* | *iha śubhamaṇiprāptir yadvaj jagaty atidurlabhā jalanidhigataṁ pātālasthaṁ yataḥ spr̥hayanti tam* | *na sulabham iti jñeyaṁ tadvaj jagaty atidurbhage manasi vividhakleśagraste tathāgatadarśanam* ||4.70||

53 *tāpāc chedāc ca nikaṣāt suvarṇam iva paṇḍitaiḥ*| *parīkṣya bhikṣavo grāhyaṁ madvaco na tu gauravāt* || (TS 3588)

It has been well examined to be very valuable
By those whose intelligence cannot be measured,
The only caravan-leaders of the worlds.
O you, whose habit is to travel out to the merchant-towns of the destinies,
You should firmly seize this Jewel that is Bodhicitta.

[Explanation of 'O you, whose habit is to travel out to the merchant-towns of the destinies (of rebirth)']
The destinies (of rebirth) are themselves the 'merchant-towns.' Here 'merchant towns' means those towns where commodities and goods are bought and sold; similarly, the merchant-towns of the destinies are places where one buys and sells commodities that are vendible through good and bad karma.
Those who have the habit, i.e. the nature, of traveling out, among those, are said to be so. He addresses them: 'O you, who whose habit is to travel out to the merchant-towns of the destinies, You should firmly seize this Jewel that is Bodhicitta!'

[Explanation of 'this Jewel that is Bodhicitta']
Bodhicitta itself is the 'Jewel,' i.e. is like a jewel. Just like the Wish-Fulfilling Gem, the Great Jewel, is the cause of pacifying all poverty and bad destinies; so also is this Jewel that is Bodhicitta.

[The overall sense]
This is what is intended: You are indeed merchants, wishing to obtain the accomplishment of happiness. Therefore, you should seize, with great care, this very Great Jewel. Why?

[Explanation of 'very valuable']
It is 'very valuable'; this expression indicates the reason. *Since* this is invaluable, excelling everything else, as it is the cause for all mundane and supra-mundane accomplishment; *therefore* only this should be seized – that is the sense.

[Explanation of 'well examined']
'And how is that known?' To this he says: 'It has been well examined.' Properly observed, correctly ascertained – this is the meaning.

[Explanation of 'By those whose intelligence cannot be measured']
'By whom?' Thus he says: 'By those whose intelligence cannot be measured.' Those who have an intelligence, an intellect, that cannot be measured, that it is not possible to measure; by them, who have great higher cognition, i.e. the Buddhas and Bodhisattvas. In such a case, in respect to the examination there is also no blunder; thus it is said to be 'well-examined.'

[Explanation of 'The only caravan-leaders of the worlds']
Furthermore, how are they 'The only caravan-leaders of the worlds'?
'They lead the caravan,' thus the word is formed with the suffix *aṇ*. The only caravan-leaders of the worlds are no other than the Buddhas, the Bhagavats and the Bodhisattvas, operating under the control of compassion: by them. Just like for merchants, in respect to obtaining benefit and avoiding the opposite, knowledgeable caravan-leaders, who wish for their benefit, are the guides; thus, in that respect there is no supposition that there could be a disappointment. In the same way it is the case here – this is what is intended. Therefore, the invaluable Jewel that is Bodhicitta should be firmly seized.

[Quote from the Maitreyavimokṣa/Gaṇḍavyūha]
And this has been said in that text itself:
'Just like, Son of Noble Family, as far as the moon and the sun shine with their circular splendor, for that long, all means of enjoyment whatsoever, such as wealth, grains, jewels, gold, silver, flowers, incense, perfume, garlands, ointments or clothes, do not match the value of the Vaśirāja Jewel, the Great Gem;
'In the very same way, as far as in the three times the awareness of the Omniscient One shines upon the domain of the Dharmadhātu, for that long whichever roots of merit of all the deities and humans, of all sentient beings, including Śrāvakas and Pratyekabuddhas, whether with or without outflows, all of those do not match the value of the Vaśirāja Jewel, the Great Gem that is the Arising of Bodhicitta.'[54]

54 *bhavagatiṣu vibhūtikāmair api nātra saṃśayo viparyāso vā kartavyaḥ ity upadarśayann āha- suparīkṣitam aprameyadhībhir bahumūlyaṃ jagadekasārthavāhaiḥ | gatipattanavipravāsaśīlāḥ sudṛḍhaṃ gṛhṇata bodhicittaratnam ||1.11|| gataya eva pattanāni paṇyadravyakrayavikrayanagarāṇīha pattanāni | tadvat śubhāśubhakarmapaṇyadravyakrayavikrayasthānāni gatipattanāni | teṣu vipravāso vipravasanam eva śīlaṃ svabhāvo yeṣāṃ te tathoktāḥ | teṣāṃ saṃbodhanam | he gatipattanavipravāsaśīlāḥ, sudṛḍhaṃ gṛhṇata bodhicittaratnam | bodhicittam eva ratnaṃ ratnam iva | yathā cintāmaṇimahāratnaṃ aprameyadhībhir bahumūlyaṃ, tathā idam api bodhicittaratnam | ayam abhiprāyaḥ-vaṇija eva sukhasaṃpattilābhārthino yūyam | ataḥ idam eva mahāratnaṃ mahatādareṇa gṛhṇata | kutaḥ ? bahumūlyam iti hetupadam etat | yasmād anargham idaṃ sarvātiśāyi laukikalokottarasaṃpattinidānabhūtatvāt, tasmād idam eva grāhyam ity arthaḥ | katham idaṃ jñāyata iti ced āha-suparīkṣitam iti | suṣṭhu nirūpitaṃ samyaṅ nirṇītam ity arthaḥ | kair ity āha-aprameyadhībhiḥ | aprameyā pramātum aśakyā dhīr buddhir yeṣāṃ taiḥ mahāprājñaiḥ buddhabodhisattvaiḥ | etāvatā parīkṣāyāṃ skhalitam api nāsti iti suparīkṣitam ucyate | punar api kiṃbhūtaiḥ ? jagadekasārthavāhaiḥ | sārthaṃ vāhayantīty aṇ | jagatām eka eva sārthavāhāḥ karuṇāvaśavartino buddhā bhagavanto bodhisattvāś ca, taiḥ | yathā khalu vaṇijāṃ hitāhitaprāptiparihārayor hitaiṣiṇo jñānavantaś ca sārthavāhā netāro bhavanti, iti na tatra visaṃvādasaṃbhāvanā, tathā atrāpīty abhiprāyaḥ | tasmād idam eva bodhicittaratnam anarghaṃ sudṛḍhaṃ grāhyam iti | etac ca tatraivoktam-tadyathā*

We should also notice that there is even a more literal sense in which the Buddhas, and Bodhisattvas, should know of gemology. As noted by Finot (RP), the knowledge of *ratnas* is listed by authorities on Kāmaśāstra as one among the arts and crafts that an urbane and refined man should learn. Buddhist texts that describe the previous lives of the Buddha mention that, while still a Bodhisattva, he mastered all arts and crafts (*kalā*). The anonymous commentary on Āryaśūra's *Jātakamālā* preserved in Sanskrit explains the term *prajñā* ('higher cognition,' 'wisdom') as a synonym of the arts and crafts (*kalā*), and offers an extensive list comprising 64 *kalās*, of which 'The Examination of Jewels' (*Ratnaparīkṣā*) is the twelfth.[55] The obvious implication is that the Buddha, during his previous lifetimes, had mastered gemology. We should also notice that the last of the 'Five Topics of Knowledge' (*pañcavidyāsthāna*), which a Bodhisattva needs to master so that they may result in the Omniscience of a Buddha, is also connected to arts and crafts (*śilpakarma*) and is likely to include some amount of gemological expertise. In other words, to know everything that is potentially beneficial, one must know gemology; and Buddhahood includes the knowledge of everything that may bring benefit to sentient beings.

CONCLUSIONS

> What do you think, Subhūti? If some Son of Noble Family or Daughter of Noble Family were to fill this three-throusand-great-thousand world-sphere with seven *ratnas* and give it as an offering to the Tathāgatas, the Arhats, the Samyaksambuddhas, would that Son of Noble Family or Daughter of Noble Family give birth, due to that, to a great heap of merit? Subhūti said: great, Bhavagat, great would be the heap of merit that such Son of Noble Family or Daughter of Noble Family would, due to that, give birth to. And why is it so? Bhagavat! that which the Tathāgata has spoken of as a heap of merit, the Tathāgata has spoken of that as a non-heap.[56]

---

kulaputra yāvac candrasūryau maṇḍalaprabhayā avabhāsete | atrāntare ye kecid dhanadhānyaratnajātarūparajatapuṣpadhūpagandhamālyavilepanaparibhogāḥ, te sarve vaśirājamahāmaṇiratnasya mūlyaṃ na kṣamante, evam eva yāvat triṣv api adhvasu sarvajñajñānaṃ dharmadhātuviṣayam avabhāsayati | atrāntare yāni kānicit sarvadevamanuṣyasarvasattvasarvaśrāvakapratyekabuddhakuśalamūlāni sāsravānāsravāṇi sarvāṇi tāni bodhicittotpādavaśirājamahāmaṇiratnasya mūlyaṃ na kṣamante | [Gaṇḍavyūhasūtra] iti || (BP on 1.11)

55 See JMT, pp. 251, 252.

56 *bhagavān āha- tat kiṁ manyase subhūte yaḥ kaścit kulaputro vā kuladuhitā vā imaṁ trisāhasramahāsāhasraṁ lokadhātuṁ saptaratnaparipūrṇaṁ kṛtvā*

Ratna is a complex idea: extending to anything considered supreme, it involves not just beauty and rarity but also great beneficial power, spanning advantages from medicinal properties to overall good fortune, and even the complete fulfillment of all wishes.

Accordingly, the ancient gemologists' sphere of activity and their role in Sanskrit and Pāli minds are perhaps not entirely or always matched by their modern counterparts. It seems that ancient South Asian gemologists served traditional doctors and adventurous merchants in search of precious things; they brought riches, long-life, health and fortune, not 'just' ornaments in the modern sense.

A *ratna* is the epigraph of past meritorious karma, and an occasion of present merit-making. An environment filled with precious substances is the dependently arisen result of collective wholesome karma, best put to use as the vehicle to create further merit through the many shapes of giving (*dāna*). Gems and jewels too should be understood within the horizon of karma, dependent arising, and repeated lives and deaths, i.e. the basic Buddhist framework that delineates the functioning of *saṁsāra*. Furthermore, gems become symbols of the Ultimate, of the Buddhist path to realize it, and of the realization itself.

Ancient *ratnas*, and their knowers and bearers, help to better appreciate the suggestions and unstated implications of Buddhist texts, ritual activities and sacred spaces, filled as they are with innumerable gems and jewels – starting from the unmatched, rare and most precious Three Jewels.

ABBREVIATIONS AND REFERENCES

P. = Pali Skt. = Sanskrit

AAPṬ *Abhidhammāvatārapurāṇaṭīkā*. In: *Chaṭṭa Saṅgāyana Tipiṭaka* 4.0, Version 4.0.0.15. (1995). Igatpuri: Vipassana Research Institute.

AAV Ārya Vimuktisena: *Abhisamayālaṁkāravṛtti*. Pensa, C. (ed.) (1967). *L'Abhisamayālaṁkāravṛtti di Ārya-Vimuktisena. Primo Abhisamaya*. Roma: Istituto Italiano per il Medio ed Estremo Oriente.

ADṬ *Abhidhānappadīpikāṭīkā*. In *Chaṭṭa Saṅgāyana Tipiṭaka* 4.0, Version 4.0.0.15. (1995). Igatpuri: Vipassana Research Institute.

---

tathāgatebhyo'rhadbhyaḥ samyaksaṁbuddhebhyo dānaṁ dadyāt, api nu sa kulaputro vā kuladuhitā vā tatonidānaṁ bahu puṇyaskandhaṁ prasunuyāt | subhūtir āha-bahu bhagavan, bahu sugata sa kulaputro vā kuladuhitā vā tatonidānaṁ puṇyaskandhaṁ prasunuyāt | tat kasya hetoḥ? yo' sau bhagavan puṇyaskandhas tathāgatena bhāṣitaḥ, askandhaḥ sa tathāgatena bhāṣitaḥ | (Vaj 8)

AH — Vāgbhāṭa: *Aṣṭāṅgahṛdaya*. Hariśāstrī, P.V. (ed.) (1989). *The Aṣṭāṅgahṛdaya*. Bombay: Nirnaya-Sāgar.

AKB — Vasubandhu: *Abhidharmakośabhāṣya*. Pradhan, P. (ed.) (1987). *Abhidharmakośabhāṣya of Vasubandhu*. TSWS 8. Patna: K.P. Jayaswal Research Institute.

Amara(J) — Jātarūpa: *Commentary on the Amarakośa*. Pant, M.R. (ed.) (2000). *Jātarūpa's Commentary on the Amarakośa*. Delhi: Motilal Banarsidass.

Amara(K) — Kṣīrasvāmin: *Amarakośodghāṭana*. Oka, Krishnaji Govind (ed.) (1913). *The Nāmaliṅgānuśānaṁ (Amarakośa) of Amarasiṁha. With the Commentary (Amarakośodghāṭana) of Kṣīrasvāmin*. Poona: Law Printing Press.

Amara(M) — Maheśvara: *Commentary on the Amarakośa*. Thatte, C.S. (ed.) (1882). *Amarakośa, with the Commentary of Maheśvara*. Bombay: Nirṇaya Sāgar.

ASaṁ — Vāgbhāṭa: *Aṣṭāṅgasaṁgrahaḥ*. Chāṅgāṇī, G.Ś. (ed.) (1945). *Aṣṭāṅgasaṁgrahaḥ*. Banaras: Chawkhamba Sanskrit Series.

BBh — *Bodhisattvabhūmi*. Wogihara, U. (ed.) (1930–1936). *Bodhisattvabhūmi: A Statement of Whole Course of the Bodhisattva (being fifteenth section of Yogācārabhūmi)*. Reprint: Tokyo: Sankibo Buddhist Book Store, 1971.

BP — Prajñākaramati: *Bodhicaryāvatārapañjikā*. De La Vallée Poussin, L. (ed.) (1901–1914). *Bodhicaryāvatārapañjikā. Prajñākaramati's Commentary on the Bodhicaryāvatāra of Çāntideva, Edited with Indices* (vols. 983, 1031, 1090, 1126, 1139, 1305 and 1399 of Bibliotheca Indica), Calcutta: Baptist Mission Press.

Deokar 2014 — Deokar, L.M. (ed.) (2014). *Subhūticandra's Kavikāmadhenu on Amarakośa 1.1.1–1.4.8. Together with Si tu Pan chen's Tibetan Translation*. Marburg/Delhi: Indica et Tibetica Verlag.

DN — *Dīghanikāya*.

DSMṬ — *Dhammasaṅganīmūlaṭikā*. In *Chaṭṭa Saṅgāyana Tipiṭaka* 4.0, Version 4.0.0.15. (1995). Igatpuri: Vipassana Research Institute.

GuBh — Raviśrījñāna: *Guṇabharaṇī*. In Sferra, F. (ed., tr.) (2000). *The Ṣaḍaṅgayoga by Anupamarakṣita with Raviśrījñāna's Guṇabharaṇīnāmaṣaḍaṅgayogaṭippaṇī. Text and Annotated Translation*. (Serie Orientale Vol. LXXXV). Roma: Istituto Italiano per l'Africa e l'Oriente.

JMT — Āryaśūra: *Jātakamālāṭīkā*. In Basu, R. (ed.) (1989). *Eine Literatur-Kritische Studie zu Āryaśūras Jātakamālā*. Doctoral Dissertation, Bonn University.

JS — *Jñānālokālaṃkārasūtra*. In Takayasu Kimura et al. (eds)., 'Sarvabuddhaviṣayāvatāra-jñānālokālaṃkāra nāma mahāyāna-sūtra Sanskrit Text.' In The Publishing Committee of the Felicitation Volume for Litt. D. Kichō Onozuka (ed.), *Kōbōdaishi Kūkai's Thought and Culture. In Honour of Litt. D. Kichō Onozuka* on *His Seventieth Birthday*, Vol. 2, pp. 1(596)–89(508). Tokyo: Nomburusha.

KN — Advayavajra: *Kudṛṣṭinirghātana*. Mikkyou-seiten-kenkyūkai (ed.), *Kudṛṣṭinirghātana*. In *Taisho Daigaku Sougou Bukkyou Kenkyūjo Nenpou* 10.

MVBhṬ — Sthiramati: *Madhyāntavibhāgabhāṣyaṭīkā*. Yamaguchi, S. (ed.) (1934). *Sthiramati Madhyāntavibhāgaṭīkā. Exposition Systématique du Yogācāravijñaptivāda*, Tome I-Texte. Nagoya: Librairie Hajinnkaku. Reprint, Tokyo: Suzuki Research Foundation, 1966.

NāSa — Padmaśrī: *Nāgarasarvasva*. Sharma, T. (ed.) (1921). *Nāgarasarvasva of Padmaśrī*. Bombay: Manilal Iccaram Desai.

Ōtani — Suzuki, D.T. (ed.) (1985). *The Tibetan Tripitaka. Peking Edition. Catalogue & Index, Reduced-size Edition*. Kyoto: Rinsen Book Co.

Pagel 2007 — Pagel, Ulrich (2007). *Mapping the Path. Vajrapadas in Mahāyāna Literature*. Tokyo: The International Institute for Buddhist Studies (Studia Philolologica Buddhica).

PP — Nāgārjuna: *Mūlamadhyamakakārikā*. De La Vallée Poussin, L. (ed.) (1903–1913). *Mūlamadhyamakakārikās (Mādhyamikasūtras) de Nāgārjuna avec la Prasannapadā Commentaire de Candrakīrti*. Bibliotheca Buddhica 4.

PTMV — Maitrīpa: *Pañcatathāgatamudrāvivaraṇa*. In Study Group on Sacred Tantric Texts, AICSB 10 (March 1988). *Advayavajrasaṁgraha*, pp. 189–78 (=44–57).

RGV — Maitreya: *Ratnagotravibhāga Mahāyānottaratantraśāstra*. Johnston, E.H. (ed.) (1950). *The Ratnagotravibhāga Mahāyānottaratantraśāstra*. Seen through the press and furnished with indexes by T. Chowdhury. Patna: The Bihar Research Society, 1950.

RP — Buddhabhaṭa: *Ratnaparīkṣā*. Finot, Louis (ed., tr.) (1896). *Les Lapidaires Indiens*. Paris: Libraire Émile Bouillon.

RSV — *Ratanasuttavaṇṇanā*. In *Chaṭṭa Saṅgāyana Tipiṭaka* 4.0, Version 4.0.0.15. (1995). Igatpuri: Vipassana Research Institute.

SSun — Aruṇadatta: *Sarvāṅgasundarā*. Commentary to AH.

Suv — *Suvikrāntavikrāmiparipṛcchā*. In Vaidya, P.L. (ed.) (1961). *Mahāyāna-sūtra-saṃgraha, Part I*. Darbhanga: The Mithila Institute of Post-Graduate Studies and Research in Sanskrit Learning.

Tōhoku — Ui, Hakuju et al. (eds.) (1934). *A Complete Catalogue of the Tibetan Buddhist Canons (Bkaḥ-ḥgyur and Bstan-ḥgyur)*. Sendai: Tōhoku Imperial University.

TS — Śāntarakṣita: *Tattvasaṅgraha*. In *Tattvasaṅgrahapañjikā* 2 = Shastri Dwarikadas (ed.) (1968). *Tattvasaṅgraha of Ācārya Śāntarakṣita with the Commentary 'Pañjikā' of Śrī Kamalaśīla*. 2 vols. Bauddha Bharati Series 1. Varanasi: Bauddha Bharati.

Vaj — *Vajracchedikā*. In Vaidya, P.L. (ed.) (1961). *Mahāyānasūtrasaṁgraha*. Darbhanga: Mithila Insitute.

Vidyābhūṣaṇa 1912	Vidyābhūṣaṇa, S.C. (ed.) (1912). *Amara-ṭīkā-kāmadhenu. The Tibetan Version of Amara-ṭīkā-kāmadhenu, A Buddhist Sanskrit Commentary on the Amarakośa*. Calcutta: Bibliotheca Indica.
VS	*Vyākhyāsudhā*. In Śivadatta, Paṇḍit (ed.) (1929). *The Nāmaliṅgānuśānaṁ (Amarakośa) of Amarasiṁha. With the Commentary Vyākhyāsudhā (or Ramāśramī) of Bhānujī Dīkṣita*. Bombay: Nirṇaya Sāgar.
YGV	*Yaṃkiñcītigāthāvaṇṇanā*, part of the RSV.
YRM	*Yogaratnamālā*. Tripathi, R.S. and Negi, T.S. (eds.) (2006). *Hevajratantram with Yogaratnamālāpañjikā of Mahāpaṇḍita Kṛṣṇapāda*. Sarnath: CIHTS.

Chapter 11

When Earth Comes Alive: Earth-Bodied Beings in Jain Tradition

ANA BAJŽELJ[1]

During a discourse on proper ways of conduct (*ācāra*)[2] for Jain mendicants, which observe the fundamental ethical principle of nonviolence (*ahiṃsā*), the *Ācārāṅgasūtra* of the Śvetāmbara canon states the following:

> A monk or a nun on the pilgrimage, whose road lies through a forest which they are not certain of crossing in one or two or three or four or five days, should, if there be some other place for walking about or friendly districts, not choose the former road for their voyage. The Kevalin says: This is the reason: During the rain (he might injure) living beings, mildew, seeds, grass, water, mud.[3]

Here one may catch a glimpse of just how broad the Jain notion of that which is alive and prone to injury is. Apart from human, heavenly, infernal and animal beings, Jainism understands plants, earth, water, air and fire to be able to possess life as well. That is the reason why treading on mildew, seeds, grass and even water and mud is understood as violent activity that must be avoided. With a specific focus on earth, this chapter will explore Jain understandings of how something that is seemingly only material can be considered to be alive, and what the ethical and practical implications of such an outlook are.

1 Ana Bajželj is Postdoctoral Researcher at the Polonsky Academy, The Van Leer Jerusalem Institute, Jerusalem (Israel).

2 Italicized words between parentheses are given in Sanskrit, even if the original source is Prakrit.

3 ĀS II.3.1.11–12 (as translated by Jacobi 1884: 138–39).

In his *Sarvārthasiddhi*, Pūjyapāda[4] lists four kinds of scriptural references to earth, namely, earth (*pṛthivī*), earth-body (*pṛthivī-kāya*), earth-bodied being (*pṛthivī-kāyika*) and earth-body tending *jīva* (*pṛthivī-jīva*). After mentioning that the word *pṛthivī* is derived from *prathana*, a neuter noun meaning 'extending' or 'spreading out,' he defines the four in the following way:

> That which has no consciousness and has the quality of hardness as its own nature is earth. [...] The earth body is that which has been abandoned by the soul present in it, similar to the dead body of a man. The earth creature is that which has earth for its body (namely the soul that lives in an earth body). The earth soul is that which has acquired the name karma of earth body, and is in transit with the karmic body, but has not actually entered the earth body.[5]

The description of the four different states of earth reveals that earth may exist either independently of or in relation to *jīva*, often translated as soul or self. The first state comprises only earth as such, solely matter, unrelated to anything living, without consciousness (*acetana*) and characterized by the quality of hardness (*kāṭhinya-guṇa*). The other three states disclose a special function that earth can perform, that is, operating as a body (*kāya*) of a *jīva*. A living being that possesses an earth-body is called an earth-bodied being. Right before it enters this kind of body, it is called an earth-body tending *jīva*, and a body that it sheds upon death is called an earth-body. The whole process occurs in a succession of births, deaths and rebirths and is related to acquired *karman*, particularly of the specific kind referred to as name *karman* (*nāma-karman*). The following two sections of this chapter will examine all the four categories, starting with the first category of earth as unconscious matter.

EARTH AS LIFELESS MATTER

According to the Jain doctrine, existence (*sat*) is expressed through substance (*dravya*).[6] There are two main groups of substances, the living (*jīva*) and the non-living (*ajīva*),[7] and matter (*pudgala*) belongs to the latter.[8]

4 Also known as Devanandin, Pūjyapāda, a Digambara Jain mendicant, probably lived in the fifth or sixth century CE. His *Sarvārthasiddhi* is a commentary to Umāsvāti's *Tattvārthasūtra*.

5 SAS II.13 §286 (as translated by Jain 1960: 62).

6 SAS V.29 §582.

7 GK 563.

8 PrS II.35.

Unconsciousness distinguishes matter from *jīvas*, a characteristic that it shares with the other non-living immaterial substances, which are medium of motion (*dharma*), medium of rest (*adharma*), space (*ākāśa*) and, according to some,[9] time (*kāla*).[10] Unlike medium of motion, medium of rest and space, which are singular, matter is plural. It exists either in the form of particles/atoms (*aṇu*, *paramāṇu*) or aggregates thereof (*skandha*), the former being eternal and the latter temporary. Kundakunda[11] describes a material particle as '[t]hat substance, which (is) the beginning, the middle and the end by itself, inapprehensible by the senses, and indivisible [...].'[12] Being independent material units that cannot be further divided, particles represent the most basic level of matter. They may exist either individually or as elemental components of divisible aggregates. When discrete, they are imperceptible and only through their aggregation can matter become visible.[13] The principal dynamics of matter through which aggregates form and disintegrate are aggregation (*saṃghāta*) and disjunction (*bheda*) of particles.[14] As noted, the process of disjunction has a limit; once the particle level is reached, matter can no longer be fragmented.[15] The process of aggregation likewise has a boundary, which lies in the size that aggregates can come to obtain.

Generally speaking, matter qualifies as an extensive kind of substance (*asti-kāya*), that is to say it occupies a multitude of space-points (*pradeśa*).[16] However, with regard to space-points in relation to basic particles, Umāsvāti[17] says that particles do not have space-points.[18] Kundakunda explains: '[A] primary atom is without space-points, because (being a unit) it gives rise to the (measure of) a space-point.'[19] This indicates that

9 Digambara Jains accept time as an independent substance, whereas Śvetāmbaras do not uniformly do so.

10 DS 15; GK 564.

11 Kundakunda was a Digambara Jain mendicant from south India. The dates of his life are uncertain but it is likely that he lived between the second and third centuries CE.

12 NS 26 (as translated by Sain 2006: 37).

13 TAS V.28.

14 TAS V.26.

15 PañS 77.

16 NS 34.

17 Umāsvāti or Umāsvāmin was a Jain mendicant who lived sometime between the second and fifth centuries CE and authored the *Tattvārthasūtra*, the first *sūtra*-style Jain text in Sanskrit, which is recognized as authentic by both Digambaras and Śvetāmbaras.

18 TAS V.11.

19 PrS II.45 (as translated by Upadhye 1984: 396).

a particle itself is not yet extensive (*apradeśa*); it does not possess space-points because it is that which defines a space-point.[20] It is then not material particles but their aggregates that occupy a multitude of space-points and can thereby be considered extensive.[21] All the other substances except time are likewise described as being extensive in character.[22] Covering an infinite (*ananta*) number of space-points,[23] space is the only limitless substance, and albeit undivided and singular, it is depicted as having two parts, these being cosmic space (*loka-ākāśa*) and acosmic space (*aloka-ākāśa*).[24] The edge of the former, which is limited, is delineated by the dimensions of the medium of motion and the medium of rest,[25] which both contain an innumerable (*asaṅkhyeya*) number of space-points.[26] The medium of motion and the medium of rest are depicted as substances, which respectively allow motion and rest in a similar way that water enables fish to swim and earth or shade (of a tree) moving things to rest.[27] No movement is possible beyond them and, consequently, no substance can traverse the border of cosmic space.[28] This is, then, the size-limit of material aggregation; no aggregate can be greater than the size of cosmic space.

The process of aggregation and disjunction of material particles is continuous and has its own logic, which is based on the structural dynamics of matter as a substance. All substances, be they living or non-living, share the same basic structure. Umāsvāti defines it in the following way: 'That which possesses qualities and modes is a substance.'[29] The association between a substance on the one hand and qualities (*guṇa*) and modes (*paryāya*) on the other is constant and can never be broken. This means that there is never a time when a substance is without qualities and modes and, similarly, qualities and modes cannot exist without being tied to a particular substance.[30] Through the qualities they possess, different groups of substances may be distinguished from one another; for example, the qualities of *jīvas* are completely different from the qualities of matter. Qualities, then, provide

20 DS 27.
21 DS 26; GK 586.
22 DS 23–24; GK 620.
23 DS 25; TAS V.9.
24 DS 19.
25 PañS 87; TAS v.13.
26 DS 25; TAS v.7. An innumerable number is a large but finite number.
27 DS 17–18; PañS 85–86.
28 GK 583, 587; PañS 3; PrS II.44.
29 TAS v.37 (as translated by Tatia 2011: 142).
30 PañS 12–13.

substances with general qualifications and they coexist with them. All substances of matter, particles and aggregates alike, have four basic qualities, which are color (*varṇa*), taste (*rasa*), smell (*gandha*) and touch (*sparśa*).[31]

Modes provide substances with specific qualifications and relate to different qualities of substances. Even though every quality must always have a certain mode, these are continuously changing.[32] Particular modes in a series of modal manifestations of qualities are, then, accidental rather than essential characteristics of substances. Since substances are eternal and modes only momentary, all the modal series of substances are potentially infinite. Modes can be either innately produced (*svabhāva-paryāya*), meaning that they arise independently, or produced in relation to an external factor (*vibhāva-paryāya*), meaning that they arise dependently.[33] Like all other substances, substances of matter also constantly undergo modifications of their four qualities. There are said to be eight kinds of touch (hard, soft, heavy, light, cold, hot, viscous and dry), five kinds of taste (bitter, pungent, astringent, acidic and sweet), two kinds of smell (pleasant and unpleasant) and five kinds of color (black, blue, red, yellow and white).[34]

Basic particles can have only a single taste, a single smell, a single color and four kinds of touch (cold, hot, viscous and dry).[35] It is due to the modal changing of their quality of touch along the scale of various degrees of viscosity (*snigdhatva*) and dryness (*rūkṣatva*) that material particles come to join and form aggregates.[36] A difference in the modal manifestation of the quality of touch between two particles is necessary in order for them to be able to combine and they cannot do so if the modal manifestations of this quality are too similar.[37] This means that two particles that are equally dry cannot combine, whereas a very dry and a very viscous particle will form a strong connection. This kind of aggregation is referred to as 'natural' (*vaisrasika*) because its occurrence does not depend on human effort. Another way for particles to combine in a natural way is through such occurrences as lightning, meteor showers, fire and rainbows. The unnatural (*prāyogika*) type of aggregation is that which is produced due to human

31 PrS II.40; TAS V.23.

32 PrS II.4.

33 NS 10–15, 28.

34 SAS V.23 §570; TV V.23.7–10.

35 PañS 81. This means that the characteristics of hardness-softness and heaviness-lightness can apply only to aggregates. The cold-hot, viscous-dry couplets are also referred to as two kinds of touch.

36 GK 609; TAS V.32.

37 TAS V.33–35.

effort, for example by fitting pieces of furniture together. Disintegration, on the other hand, is possible by scratching, grinding, cutting into pieces, breaking into parts, dividing into layers and emitting sparks.[38] Due to diverse combinations of aggregation and disjunction, material particles can form many different aggregates. According to their varying sizes, Kundakunda classifies them into very gross, gross, gross-subtle, subtle-gross, subtle and very subtle kinds.[39] Pūjyapāda emphasizes that the subtlety of subtle (*sūkṣma*) aggregates is relative in contrast to the subtlety of particles which is absolute. The same goes for the grossness of gross (*sthūla*) aggregates, it being absolute in the case of an aggregate that is coextensive with cosmic space, and it being relative in the case of other aggregates, that are large but do not reach cosmic dimensions.[40] Apart from the different kinds of grossness and subtlety that characterize them, some of the other categories of material aggregate formations are sound (*śabda*), darkness (*tamas*), shadow (*chāyā*), light (*uddyota*), heat (*ātapa*), various shapes (*saṃsthāna*)[41] as well as earth (*bhū/pṛthivī*), water (*ambu/āpas/jala*), air (*vāyu*) and fire (*agni/tejas*). Earth, water, air and fire are then not basic material particles, through the varying proportions of which aggregates obtain their different natures, but are already aggregates themselves. In response to the Vaiśeṣika understanding of earth, water, air and fire as basic particles, out of which only earth possesses all the four qualities of color, taste, smell and touch, Akalaṅka[42] writes the following series of questions and answers:

> Q The Vaiśeṣikas postulate (in Vaiśeṣika-sūtra 2.1.1–4) that (i) the earth has colour, taste, smell and touch (ii) the water has colour, taste and touch. It is liquid and smooth (iii) the fire has colour and touch and (iv) the air has touch only. This should be acceptable to the Jainas.
> A This is not correct. All the above four properties are found in all the material[43] realities. [...]
> Q The smell etc. are there in water etc. because of their combination with the earth atoms.
> A There is no special cause for this assumption that these properties belong to the earth atoms and it is due to their combination that they are found

38 SAS v.24 §572.
39 NS 21.
40 SAS v.24 §572.
41 DS 16; TAS v.24.
42 Akalaṅka was an eighth century Digambara Jain philosopher. His *Tattvārtharājavārtika* is a commentary to Umāsvāti's TAS.
43 Originally translated as 'mattergic.'

> in others and that they are not natural properties of water etc. In contrast, the Jainas postulate that these are the properties of water etc. as they are observed there naturally. [...] [T]here is no class distinction in the earth etc. as all of them are material in nature which exists in different forms in nature due to specific aggregation of atoms.[44]

A product of integration of basic particles, earth is, then, a material aggregate that is not conscious and possesses all the four material qualities, just like water, air and fire. How does this lifeless aggregate of earth come to possess life?

EARTH AS A SEAT OF LIFE

Apart from the various classifications of material aggregates already mentioned, Jain texts further divide them into those that may be karmically bound (*yogya*) and those that may not (*aprayogya*).[45] When *jīvas*' activities are informed by passions (*kaṣāya*) of anger, pride, deceit and greed,[46] the karmically bondable material aggregates attain karmic nature. These aggregates, the modes of which are externally affected, are referred to as karmic matter.[47] They are drawn to *jīvas* and stick to them like dust to moist ground. Karmic matter is bound to *jīvas* for a limited period of time (*sthiti*). Following a dormant phase, it bears fruit (*anubhāva*), after which it falls off and is then normally replaced by newly drawn *karman*. The nature (*prakr̥ti*) of bound *karman* is molded after the kind of activity that generated its bondage, and the volitional level that governed the performed activity determines its amount (*pradeśa*). The duration of its binding and the nature of its effect are both decided by the extent to which the *karman*-producing activities were influenced by passions.[48] The doctrine that *karman* is material is unique to Jainism. Like most other indigenous traditions of South Asia, Jainism is based on the recognition that living beings have been trapped in the cycle (*saṃsāra*) of births, deaths and rebirths since beginningless time. It agrees with them in the idea that the central factor driving the succession and determining the nature of recurring embodiments is *karman*, a residue of bodily, verbal and mental activities of

44 TV ii.20.4 (as translated by Jain 1999: 117).
45 PrS ii.76.
46 GK 284–87; SAS ii.6 §265.
47 PAS 12.
48 DS 33.

living beings. As long as beings are bound by *karman*, they remain trapped in *saṃsāra* and after bodily death in one particular life-form, they are reborn in another. It likewise shares their belief that liberation (*mokṣa*) from *saṃsāra* can be attained only when *karman* is completely destroyed.[49] Understanding *karman* to be material in character, however, distinguishes Jainism from the rest of the South Asian traditions. Like a weight,[50] *karman* physically binds immaterial *jīvas* and prevents them from their natural movement upwards. Upon destruction of all karmic matter, disembodied and thus weightless *jīvas* follow this natural course to the top of cosmic space where they remain forever.[51] Throughout the series of rebirths, *jīvas* can be born into many different birth-states (*gati*). Which life-form they are born as, is decided by the specific kind of karmic matter called the *nāma-karman*, already mentioned at the beginning of this chapter. There are said to be 148 kinds of *karman* and these are divided into two basic types, destructive (*ghātiyā*) and non-destructive (*aghātiyā*). The destructive kinds of karmic matter hinder the functioning of *jīvas*' essential qualities (consciousness, bliss and energy), the result of which is the impairment of their modal manifestations. The non-destructive kinds of karmic matter do not exert any influence on the *jīvas*' qualities, but they generate the state and conditions of the specific embodiments of *jīvas*. There are four different types of non-destructive karmic matter: (1) *vedanīya-karman*, which determines whether the experience of living beings is characterized by happiness or unhappiness; (2) *āyu-karman*, which decides specific lifespans and is always related to the particular birth-states that living beings are born into;[52] (3) *gotra-karman*, which determines the status or environment of living beings, especially with regard to them being either conducive or non-conducive for spiritual progress; and (4) *nāma-karman*, which brings about the birth-state, the subclass (*jāti*) in that birth-state and the formation of the specific kind of body (*śarīra*) that *jīva* comes to occupy. The last includes the particular sense-faculties that living beings possess, the specific ways in which their bodily parts are formed, their mobility and so forth. Apart from the gross physical body (*audārika-śarīra*) other kinds of bodies are generated as well, for example, the luminous body

49 TAS x.3.

50 GK 202.

51 TAS x.5.

52 Unlike other kinds of karmic matter this specific karmic type is bound only once and is reactivated in each successive existence.

(*taijasa-śarīra*) and the karmic body (*kārmaṇa-śarīra*).[53] The first sustains the temperature of living beings, whereas the second represents all of the subtle karmic matter that adheres to *jīvas*. Upon death, these two bodies function as vehicles which ensure the instantaneous passing of *jīvas* from one existence to another (Jaini 1998: 115–27). These transitions between different life-forms in no way affect the essential nature of *jīvas*. Like matter, *jīvas* are considered to be extensive in nature. Because of that they can adapt to all sorts of bodies without suffering any consequences, not unlike a piece of cloth that remains the same piece of cloth despite being folded in numerous ways.[54]

This complex karmic theory explains how a particular living being and the conditions of its embodied existence come to be, as well as accounts for there being such a great diversity of existent life-forms despite all *jīvas* being essentially alike. There are said to be in total 8,400,000 possible birth-states, which are classified in various ways. One of the most common classifications sorts them into four main groups, which include human beings (*manuṣya*), heavenly beings (*deva*), infernal beings (*nāraka*) and animals, plants and 'elemental' beings (*tiryañc*).[55] Another divides them with respect to how many sense-faculties (*indriya*) they possess, the most highly developed beings experiencing the world through the five senses of touch (*sparśana*), taste (*rasana*), smell (*ghrāṇa*), sight (*cakṣus*) and hearing (*śrotra*) and the lowest through the tactile faculty only.[56] Among further criteria (*mārgaṇā-sthāna*) for classification are also their being mobile (*trasa*) or immobile (*sthāvara*),[57] their biological gender (*liṅga*), the color of their *jīvas* (*leśyā*) and their possession of the capacity for reflection (*manas*) or lack thereof. According to the latter, living beings are divided into those that can reason (*saṃjñin*) and those that cannot (*asaṃjñin*).[58] Another distinction is between developed (*paryāpta*) and undeveloped (*aparyāpta*) beings, according to which bodily organs and capacities of some life-forms are not fully developed (von Glasenapp 1999: 199).

53 There are additionally two bodies that may be formed: the transformational body (*vaikriya-śarīra*) of heavenly and infernal beings and the translocational body (*āhāraka-śarīra*), which allows one to travel to those places in the cosmos where Tīrthaṅkaras (lit. ford-makers, also referred to as Jinas) teach.

54 DS 10; Jaini 1998: 102; Jaini 2010: 123.

55 SAS II.6 §265; TAS IV.28.

56 TAS II.20; GK 167; SAS II.22 §304.

57 TAS II.12.

58 *Jīvas* that possess this capacity are referred to as 'those with a mind' (*sa-manaska*) and those that do not as 'those without a mind' (*a-manaska*). TAS II.11.

Earth-bodied (*pṛthivī-kāyika*) beings belong to the group of what are often referred to as 'elemental' beings. These are *jīva*s that due to the operation of their respective *nāma-karman*s occupy bodies of different 'elements,' which in this context means the material aggregates of earth, water, air and fire.[59] With regard to which material conglomerates they occupy, they are referred to as earth-bodied, water-bodied (*ambu-kāyika/āp-kāyika/jala-kāyika*), fire-bodied (*tejo-kāyika/agni-kāyika*) and air-bodied (*vāyu-kāyika*) kinds. Wiley explains the formation of the body of elemental beings, which are all one-sensed (*eka-indriya*), in the following way:

> Birth as a one-sensed being is attained by the fruition (*udaya*) of those *karma*s that, at the time of death of its current physical body, cause the transition of the soul to its next place of birth where the soul begins to form a new physical body through the fruition of the *nāma karma* that forms a body with one sense (*ekendriya-śarīra-nāma karma*). If the soul is to be earth-bodied, a specific set of subvarieties (*uttara-prakṛti*s) of *nāma karma*s comes into fruition simultaneously and causes the formation of a separate body (*pratyeka-śarīra-nāma karma*) by attracting particles of earth, transforming them, and binding them together to form a body of a specific size and shape. Until the time of death, certain of these *nāma karma*s will continue to rise, causing the continuous influx of matter to maintain the body. (2006: 40–41)

Pūjyapāda's distinction between four states of earth, which were mentioned at the beginning of the chapter, is now perhaps more clear. Earth is just a material aggregate, which is not conscious. An earth-bodied being is a conscious *jīva* with a body that is a material aggregate of earth. An earth-body is a material body of an earth-bodied being that has been 'vacated' and cast off after the *jīva* that previously occupied it transferred to a new body. An earth-body tending *jīva* is a *jīva* that has left the previous body it occupied and is now in transit with the karmic body in order to inhabit an earth-body.

What are the characteristics of earth-bodied beings? Although some texts deem fire- and air-bodied beings to be mobile, earth- and water-bodied beings are always regarded as immobile, immobility being defined as an inability to move away from objects of fear.[60] Apart from possessing the one sense faculty of touch (*sparśana-indriya*),[61] Pūjyapāda describes

59 GK 182.

60 TAS II.13.

61 DS 11.

all elemental beings as having another three life-principles or vitalities (*prāṇa*). He says: 'How many life-principles or vitalities do these possess? They possess the four vitalities of the sense-organ of touch, strength of body or energy, respiration and life-duration.'[62] Earth-bodied beings, then, have the sense of touch, strength of body (*kāya-bala*), inhaling/exhaling (*ucchvāsa-niśvāsa*)[63] and life-duration (*āyus*). These are imperceptible (Wiley 2006: 41), which is one of the principal reasons for earth-bodied beings being discredited as non-living. Earth-bodied beings are further described as having four instincts (*saṃjñā*), which are: desire of food (*āhāra-saṃjñā*),[64] fear (*bhaya-saṃjñā*), desire to reproduce (*maithuna-saṃjñā*) and desire to accumulate things for later use (*parigraha-saṃjñā*).[65] Like other elemental beings, earth-bodied beings belong to a third-gender category (*napuṃsaka-liṅga*),[66] since they do not have sexual organs (von Glasenapp 1999: 199). The colors of *jīvas* as stains of *karman*[67] indicate the level on the scale of purification (*guṇa-sthāna*) they have reached and *jīvas* of earth-bodied beings adopt either black, blue or grey colors, which are a sign of low levels of purity.[68] They are described as beings 'with false belief' (*mithyādṛṣṭi*), which represents the very lowest stage of spiritual development (von Glasenapp 1942: 76; cf. ibid. 54). Earth-bodied beings do not possess the capacity for reflection, which, however, does not mean that they are not conscious; even if the quality of consciousness is very heavily burdened, as in the case of earth-bodied beings, it never ceases to function.

It is already in the early Jain doctrine that earth-bodied beings are also described as being able to experience pleasure and pain. With regard to the latter, the *Ācārāṅgasūtra* explains:

62 SAS II.13 §286 (as translated by Jain 1960: 62–63). The other six vitalities are the sense of taste, faculty of speech, sense of smell, sense of sight, sense of hearing and the mind. GK 130; SAS II.14 §288.

63 In the *Bhagavatīsūtra*, Indrabhūti Gautama, the first chief disciple of Mahāvīra, asks his teacher if one-sensed beings breathe, even though one cannot observe their breathing. Mahāvīra pronounces that they do. BhS, *śataka* 2, *uddeśaka* 1, *sūtras* 3–7.

64 For earth-bodied beings nourishment (*āhāra*) is an involuntary activity that involves the assimilation of matter through the whole surface of their existing body (Wiley 2006: 42).

65 GK 134–38.

66 GK 275.

67 GK 489, 493.

68 GK 692. Von Glasenapp points out that developed gross one-sensed living beings can have also the yellow or fiery color if their previous existence was in the form of a heavenly being (1942: 53–54).

> As somebody may cut or strike a blind man (who cannot see the wound), as somebody may cut or strike the foot, the ankle, the knee, the thigh, the hip, the navel, the belly, the flank, the back, the bosom, the heart, the breast, the neck, the arm, the finger, the nail, the eye, the brow, the forehead, the head, as some kill (openly), as some extirpate (secretly), (thus the earth-bodies are cut, struck, and killed though their feeling is not manifest).[69]

This excerpt indicates that earth-bodied beings undergo pain and suffering even though they possess only a single sense faculty and despite the fact that they are unable to express what they feel. Moreover, when being hurt, they cannot voluntarily move away from the agent that violates them, which makes them all the more exposed and vulnerable. In his *Triṣaṣṭiśalākāpuruṣacaritra*, Hemacandra[70] describes the many afflictions that earth-bodied beings undergo:

> [T]hey are divided by implements such as plows; they are crushed by horses, elephants, etc.; they are submerged by streams of water; and are burned by forest-fires. They are pained also by water – salt-water, rice-water, etc., and when they have become salt, they are boiled in hot water. They are cooked by potters, etc., who have turned them into bricks for pots, etc., and they are piled up in walls when they have reached the form of mud. Some are ground by grindstones by persons after they have heated them with layers of saline soil; some are split by chisels and burst by mountain streams.[71]

Earth-bodied beings live solitary (*pratyeka*) rather than communal (*sādhāraṇa*) lives, which means that a single earth-body is always inhabited by just one *jīva*. Individual earth-bodies can be either gross (*bādara/sthūla*) or subtle (*sūkṣma*).[72] The former are obstructive (*ghāta*) and can be obstructed by other material forms, whereas the latter are neither obstructive nor obstructed, freely passing through any kind of matter.[73] Consequently, the motion of gross-bodied *jīvas* is restricted to only a portion of cosmic space, while subtle-bodied *jīvas* can move all throughout it without any obstruction.[74] The non-obstructability of subtle earth-bodied beings also means their indestructability. They cannot be killed and die

69 ĀS I.1.2.5 (as translated by Jacobi 1884: 4–5); cf. Wiley (2006: 43).

70 Hemacandra was a Śvetāmbara Jain mendicant leader who lived between 1089 and 1172 CE.

71 TC II.4.100–02 (as translated by Johnson 1937: 295–96).

72 US XXXVI.71.

73 GK 183.

74 GK 184; US XXXVI.79.

only when their lifespan-determining *karman* runs its course. All subtle earth-bodies are the same in character, but there are many varieties of gross earth-bodies.[75] They may first be divided into smooth and rough types. There are said to be seven varieties of smooth gross earth-bodies and the *Uttarādhyayanasūtra* lists the following: 'black, blue, red, yellow, white, pale dust, and clay.'[76] The different kinds of rough gross earth-bodies are thirty-six in number. These include soil, gravel, sand, stones, rocks, rock-salt, iron, copper, tin, lead, silver, gold, diamond, orpiment, vermilion, realgar, antimony, coral, hyacinth, natron, crystal, emerald, sapphire, red chalk, sulfur, lapis lazuli and so forth.[77]

According to the *Bhagavatīsūtra*, earth-bodied beings can live anywhere from forty-eight minutes to 22,000 years.[78] This holds for the developed kinds of earth-bodied beings. The undeveloped kinds, which do not acquire the necessary vitalities for keeping alive in the form of an earth-bodied being, live only the length of one-eighteenth of a pulse beat.[79] The *Uttarādhyayanasūtra* further says that a *jīva* can keep being reborn in the form of an earth-bodied being for an immeasurable amount of time but that it can also take an endless amount of time for it to return to the earth kind of body after having left it.[80] In line with this, it is possible to infer that there may be *jīvas* that left an earth-body and were never reborn as an earth-bodied being again. It is likewise possible to assume that there may be *jīvas* that have always been born in a state of an earth-bodied being. However, even if that is so, the prospects of the latter *jīvas* are not completely gloomy. In fact, earth-bodied beings are supposed to not only be able to be reborn in the one-sensed bodily forms but even as human beings (von Glasenapp 1942: 51–54). Jainism considers the human life-form to be the most auspicious of all, since it is only as a human being that any *jīva* can attain *mokṣa*. Although great jumps from one level of existence to another are not frequent, they are possible and the development of *jīvas*' lives towards liberation need not be linear (Wiley 2006: 40). How can earth-bodied beings affect this development or, better, can they affect it

75 US XXXVI.78.

76 US XXXVI.72–73 (as translated by Jacobi 1895: 213).

77 US XXXVI.74–77. I have only listed the ones which Jacobi identifies with certainty. He includes all, in fact, thirty-nine kinds of rough gross earth-bodies, and writes Sanskrit names for those the identity of which is ambiguous or unclear (US, tr. Jacobi 1895: 213n3, 214n2).

78 BhS, *śataka* 1, *uddeśaka* 1, *sūtra* 27.

79 Wiley (2006: 56n29); cf. US XXXVI.81.

80 US XXXVI.82–83.

at all? Earth-bodied beings are described as having the passions of anger, pride, deceit and greed, which means that they are capable of attracting new *karman*. As a matter of fact, it is just by breathing that they are supposed to be able to commit three to five kinds of actions, the five being those caused with one's body, those caused with an instrument, those caused with animosity, those that cause pain and those that cause death to other living beings (ibid.: 42, 57n34). This indicates that earth-bodied beings are not only passive recipients of previously accumulated *karman*, waiting for it to be exhausted before they can move on to another kind of body, but have some (very limited) sort of autonomy. The kind of understanding of earth-bodied beings that has been described reverberates with complex implications for practical life of both mendicants and laity. These will be the subject of the last section of this chapter.

ENCOUNTERING EARTH-BODIED BEINGS

It was not only curiosity that led Jain authors into such meticulous analyses of the world as described in the previous sections. The incentive for drawing a very clear line between that which is living and that which is not was inspired by the aspiration for spiritual development, guided by the highest ethical imperative of nonviolence which has, ideally speaking, overseen all of Jain practice since the early developments of the tradition. The *Ācārāṅgasūtra* reads:

> [A]ll breathing, existing, living, sentient creatures should not be slain, nor treated with violence, nor abused, nor tormented, nor driven away. This is the pure, unchangeable, eternal law, which the clever ones, who understand the world, have declared [...].[81]

If no breathing, existing, living, sentient beings are to be killed, it is crucial to know what in the world matches those criteria, for only by knowing that which can be harmed, may one restrain from harming it. This kind of examination led Mahāvīra, the twenty-fourth ford-maker (Tīrthaṅkara) of Jainism,[82] to recognize life in most unobvious places, embedded in seemingly merely material forms. Upon this recognition, he resolved not to violate that life:

81 ĀS I.4.1.1–2 (as translated by Jacobi 1884: 36).

82 According to the traditional dating, Mahāvīra or Vardhamāna Jñātṛputra, the last ford-maker of our era, lived between the sixth and fifth centuries BCE.

> Thoroughly knowing the earth-bodies and water-bodies and fire-bodies and wind-bodies, the lichens, seeds, and sprouts, he comprehended that they are, if narrowly inspected, imbued with life, and avoided to injure them [...].[83]

Mahāvīra realized that the diverse occurrences of life, the whole range from a human examiner to an earth-bodied being, are ontologically related[84] and share similar wants. The *Ācārāṅgasūtra* says that all of them are 'fond of life, like pleasure, hate pain, shun destruction, like life, long to live. To all life is dear.'[85] As noted previously in this chapter, in violating these wants of a living being with bodily, mental or verbal activities driven by passions, the executor not only intervenes in a life of another but also damages itself by attracting *karman* through the passions which inform its performed actions.[86]

During initiation (*dīkṣā*), every Jain novice mendicant undertakes to follow the five great vows (*mahā-vrata*) of nonviolence, truthfulness (*satya*), restraint from stealing (*asteya*), celibacy (*brahmacarya*) and non-possession (*aparigraha*), the principal one among which is the vow of nonviolence. The *Ācārāṅgasūtra* conveys it in the following way: 'I renounce all killing of living beings, whether subtile or gross, whether movable or immovable. Nor shall I myself kill living beings (nor cause others to do it, nor consent to it).'[87] This resolve to non-discriminatorily avoid injuring life affects both the inward and the outward practice of Jain mendicants. A fundamental aspect of inward practice is the vow of equanimity (*sāmāyika*), which includes the development of an equanimous attitude towards all living beings in order to avoid injurious activities. With regard to it, Kundakunda says: 'I have equanimity towards all living beings, and I have no ill-feeling towards any of them [...].'[88] An obligatory daily practice (*āvaśyaka*) for every Jain mendicant is also a kind of confession (*pratikramaṇa*), which is a recollection of past wrongdoings and vow transgressions. *Pratikramaṇa* is concluded with a sentence *micchā mi dukkaḍaṃ*, a Prākrit phrase, which not only implores forgiveness for all the faults done but also implores them to have been done in vain, namely, not to have attracted new *karman*. The following *pratikramaṇa*, which addresses also earth-bodied beings, calls to mind harm caused by walking:

83 ĀS I.8.1.11–12 (as translated by Jacobi 1884: 80–81).
84 DVS X.5.
85 ĀS I.2.3.4 (as translated by Jacobi 1884: 19).
86 PAS 43–48.
87 ĀS II.15.i.1 (as translated by Jacobi 1884: 202).
88 NS 104 (as translated by Sain 2006: 82).

> I want to make *pratikramaṇa* for injury on the path of my movement, in coming and in going, in treading on living things, in treading on seeds, in treading on green plants, in treading on dew, on beetles, on mould, on moist earth, and on cobwebs; whatever living organisms with one or two or three or four or five senses have been injured by me or knocked over or crushed or squashed or touched or mangled or hurt or affrighted or removed from one place to another or deprived of life – may all that evil have been done in vain. (Williams 1963: 204)

The outward practice of nonviolence towards all living beings is reflected in the appearance and conduct of Jain mendicants. With a specific regard to the encounter with earth-bodied beings, it should be mentioned that Jain mendicants carry special whisk brooms (*piñchī*, *rajoharaṇa*), with which they sweep the ground in order to remove minute life-forms, such as earth-beings, so as not to hurt them. With the same intention, Jain mendicants frequently and carefully inspect their clothing. As noted at the very beginning of this chapter, they also have a four-month restriction of movement during the time of the monsoon, in order not to harm flourishing lives, including earth-bodied beings, such as mud. Because ploughing and digging into earth with any purpose, including farming, injures earth-bodied beings, Jain mendicants do not grow their own food. In fact, they do not prepare it but are offered it by lay people. There are further restrictions on whom they can obtain the food from and what kind of food they may take. The *Daśavaikālikasūtra* prohibits taking food from laypeople whose hands are wet or dusty.[89] This instruction rests on the resolve to refrain from injuring water- and earth-bodied beings and holds also for the food that mendicants intend to take from lay people. In the words of the *Ācārāṅgasūtra*:

> When a male or a female mendicant, having entered the abode of a householder with the intention of collecting alms, recognises food, drink, dainties, and spices as affected by, or mixed up with, living beings, mildew, seeds or sprouts, or wet with water, or covered with dust – either in the hand or the pot of another – they should not, even if they can get it, accept of such food, thinking that it is impure and unacceptable.[90]

If the food that is being offered is free of even the tiniest living beings, it may be taken. Once it is, Jain mendicants who use alms bowls ought to pay

89 DVS v.1.33.

90 ĀS ii.1.1.1 (as translated by Jacobi 1884: 88).

special attention to where they set their bowls and thoroughly examine them before eating, again in order to avoid injury to miniscule life-forms.

It is clear that in accordance with the early Jain doctrine, any act of violence attracts *karman* and thereby contributes to the agent's bondage in *saṃsāra*. However, the strictness of this notion, which demands the practice of most rigorous asceticism, slightly loosened over time. Harm caused by performing one's occupation (Jaini 2006: 144) and accidental injury that occurs despite carefully executed actions with correct motivation were eventually deemed to draw less *karman* (Dundas 2006: 110). Similarly, karmic matter attached due to carrying out religious duties was considered to burn off immediately.[91] Another difference, significant for the topic of this chapter, was a newly drawn distinction between harm caused to mobile and immobile living beings, the former being proclaimed to attract much more *karman* than the latter. This distinction coincided with the development of laity, of whom only abstention from injuring mobile rather than all beings was required (Johnson 1995: 25). Although some lay people occasionally adopt certain mendicant practices, such as using whisk brooms and mouth-shields during ritual activities, they generally follow limited versions of the five mendicant vows, referred to as the five lesser vows (*aṇu-vrata*). It seems that the loosening of the restrictions was a necessary development, since unlike mendicants who renounce worldly life, Jain lay people continue to be householders and active participants in the religious community. With respect to their role in the community, the distinction between harming mobile and immobile beings, for example, 'allowed' them to erect monumental temples made of stone, a practice that was further justified and endorsed by the claim of bringing about the 'greater good,' namely, facilitating the temple worship of Jinas, which outweighs the harm involved in the construction of the temple (Dundas 2006: 107–10). An analogy has been drawn between the necessity of building a temple for veneration and that of digging a well for survival, despite the great number of lives that are hurt throughout both processes (ibid.: 109). Apart from following the five lesser vows, Jain laity may also observe three specialized vows (*guṇa-vrata*), which aim at a further restriction of causing harm to living beings. The *anartha-daṇḍa-vrata* looks to restrain purposeless activities, such as prodding the ground with no particular reason. The *dig-vrata* is a spatial restraint on movement, with which a lay person's activities and harm produced through them are restricted to a limited space only (Williams 1963: 99–102):

91 Cf. PAS 79–80.

> Like a heated iron sphere the layman will inevitably, as a result of *pramāda*, bring about the destruction of living creatures everywhere, whether he is walking, or eating, or sleeping, or working. The more his movements are restricted, the fewer *trasa-jīvas* and *sthāvara-jīvas* will perish. (ibid.: 100)

The *bhogopabhoga-parimāṇa-vrata* sets stricter dietary restrictions as well as proscribes eating and cooking at night-time in order not to unknowingly harm small creatures. Additionally, some occupations for lay people are preferred to others. Whereas trade in precious stones, for example, is accepted as one of the permissible occupations (Jaini 2006: 144), some of the undesired ones, which relate specifically to earth-bodied beings, are trade in ploughs and spades, livelihood from producing and selling charcoal, working with bricks, pottery and metal as well as hewing and digging, which includes cultivation of soil, excavating and quarrying and shaping rocks (Williams 1963: 117–21). The implications of the ontology of earth-bodied beings therefore carry weight not only for mendicants but for laity as well.

Despite practical differences in individual endeavors to avoid injury to earth-bodied beings, there is no doubt that Jain texts identify these kinds of beings as living. Conscious and capable of attracting new *karman* through passion-motivated activities, they are independent agents in *saṃsāra* and on their possible path to liberation. According to the ideals of Jain ethical theory, it is not only for the reason that their very next life could be in a human form, but simply because they are living and thereby prone to harm, that like any other life their life matters and should not be violated.

REFERENCES

ĀS — *Ācārāṅgasūtra.* Jacobi, H. (tr.) (1884). *Gaina Sûtras.* Part I. *The Âkârâṅga Sûtra; The Kalpa Sûtra*, pp. 1–213. Oxford: Clarendon Press.

BhS — *Bhagavatīsūtra.* Lalwani, K.C. (tr.) (1999). *Sudharma Svāmī's Bhagavatī Sūtra.* Vol. I. Śatakas 1–2. Calcutta: Jain Bhawan.

DS — Nemicandra: *Dravyasaṃgraha.* Balbir, N. (tr.) (2010). *Dravyasaṃgraha*. Mumbai: Hindi Granth Karyalay.

Dundas 2002 — Dundas, P. (2002) *The Jains*. London and New York: Routledge.

Dundas 2006 — Dundas, P. (2006). 'The Limits of a Jain Environmental Ethic.' In Chapple, C.K. (ed.), *Jainism and Ecology. Nonviolence in the Web of Life*, pp. 95–117. Delhi: Motilal Banarsidass.

DVS — *Daśavaikālikasūtra.* Lalwani, K.C. (tr.) (1973). *Ārya Sayyambhava's Daśavaikālika Sūtra*. Delhi: Motilal Banarsidass.

GK	Nemicandra: *Gommaṭasārajīvakāṇḍa*. Jaini, J.L. (tr.) (1927). *Gommatsara Jiva-Kanda (The Soul) by Shri Nemichandra Siddhanta Chakravarti*. Assisted by Brahmachari Sital Prasada. Lucknow: The Central Jaina Publishing House.
Jaini 1998	Jaini, P.S. (1998). *The Jaina Path of Purification*. Delhi: Motilal Banarsidass.
Jaini 2006	Jaini, P.S. (2006). 'Ecology, Economics, and Development in Jainism.' In Chapple, C.K. (ed.), *Jainism and Ecology. Nonviolence in the Web of Life*, pp. 141–56. Delhi: Motilal Banarsidass.
Jaini 2010	Jaini, P.S. (2010). '*Karma* and the Problem of Rebirth in Jainism.' In Jaini, P.S. (ed.), *Collected Papers on Jaina Studies*, pp. 121–45. Delhi: Motilal Banarsidass.
Johnson 1995	Johnson, W.J. (1995). *Harmless Souls. Karmic Bondage and Religious Change in Early Jainism with Special Reference to Umāsvāti and Kundakunda*. Delhi: Motilal Banarsidass.
KS	*Kalpasūtra*. Jacobi, H. (tr.) (1884). *Gaina Sûtras*. Part I. *The Âkârâṅga Sûtra; The Kalpa Sûtra*, pp. 215–311. Oxford: Clarendon Press.
NS	Kundakunda: *Niyamasāra*. Sain, U. (tr.) (2006). *Niyamasāra of Āchārya Kundakunda*. Assisted by Brahmachari Sital Prasad. Delhi: Bharatiya Jnanpith.
PañS	Kundakunda: *Pañcāstikāyasāra*. Chakravarti Nayanar, A. (tr.) (2001). *Ācārya Kundakunda's Pañcāstikāya-Sāra (The Building of the Cosmos)*. Delhi: Bharatiya Jnanpith.
PAS	Amṛtacandra Sūri, *Puruṣārthasiddhyupāya*. Prasada, A. (tr.) (1933). *Purushartha-Siddhyupaya (Jaina-Pravachana-Rahasya-Kosha) by Shrimat Amrita Chandra Suri*. Lucknow: Central Jaina Publishing House.
PrS	Kundakunda: *Pravacanasāra*. Upadhye, A.N. (tr.) (1984). *Śrī Kundakundācārya's Pravacanasāra (Pavayaṇasāra)*. Agas: The Parama-Śruta-Prabhāvaka Mandal, Shrimad Rajachandra Ashrama.
SAS	Pūjyapāda: *Sarvārthasiddhi*. (1) Shastri, P. (ed. and tr.) (1997). *Āchārya Pūjyapāda's Sarvārthasiddhi. The Commentary on Āchārya Griddhapiccha's Tattvārtha-sūtra*. New Delhi: Bharatiya Jnanpith. (2) Jain, S.A. (tr.) (1960). *Reality. English Translation of Shri Pujyapada's Sarvarthasiddhi*. Calcutta: Vira Sasana Sangha.
SKS	*Sūtrakṛtāṅgasūtra*. Jacobi, H. (tr.) (1895). *Gaina Sûtras*. Part II. *The Uttarâdhyayana Sûtra; The Sûtrakritâṅga Sûtra*, pp. 233–435. Oxford: Clarendon Press.
TAS	Umāsvāti: *Tattvārthasūtra*. Tatia, N. (tr.) (2011). *Tattvārtha Sūtra. That Which Is*. New Haven and London: Yale University Press.
TC	Hemacandra: *Triṣaṣṭiśalākāpuruṣacaritra*. Johnson, H.M. (tr.) (1937). *Triṣaṣṭiśalākāpuruṣacaritra or The Lives of Sixty-Three Illustrious Persons*. Vol. II. Books II and III. Baroda: Oriental Institute.

TV	Akalaṅka: *Tattvārtharājavārtika.* (1) Jain, N.L. (tr.) (1999). *Biology in Jaina Treatise on Reals [Biology in Tattvārtha-Sūtra]. English Translation with Notes on Chapter Two of Tattvārtha-Rāja-Vārtika of Akalaṅka.* Varanasi: Pārśvanātha Vidyāpīṭha and Chennai: Śri Digambar Jain Samāj. (2) Jain, N.L. (tr.) (2000). *The Jaina World of Non-Living [The Non-Living in Tattvārthasūtra]. English Translation with Notes on Chapter Five of Tattvārtha-Rāja-Vārtika of Akalaṅka.* Varanasi: Pārśvanātha Vidyāpīṭha and Plano TX., USA: Pradyuman Zaveri.
US	*Uttarādhyayanasūtra.* Jacobi, H. (tr.) (1895). *Gaina Sûtras.* Part II. *The Uttarâdhyayana Sûtra; The Sûtrakritâṅga Sûtra*, pp. 1–232. Oxford: Clarendon Press.
von Glasenapp 1942	von Glasenapp, H. (1942). *The Doctrine of Karman in Jain Philosophy.* Trans. G. Barry Gifford. Bombay: The Trustees, Bai Vijibai Jivanlal Panalal Charity Fund.
von Glasenapp 1999	von Glasenapp, H. (1999). *Jainism. An Indian Religion of Salvation.* Delhi: Motilal Banarsidass.
Wiley 2006	Wiley, K.L. (2006). 'The Nature of Nature: Jain Perspectives on the Natural World.' In Chapple, C.K. (ed.), *Jainism and Ecology. Nonviolence in the Web of Life*, pp. 35–59. Delhi: Motilal Banarsidass.
Williams 1963	Williams, R. (1963). *Jaina Yoga. A Survey of the Mediaeval Śrāvakācāras.* London: Oxford University Press.

Index

Ācārāṅgasūtra (Skt.)/*Āyāraṁgasutta* (Pkt.): xx, xx fn. 34, 255, 265, 268–69, 270
adharma: 257
Ādi Granth: *see Gurū Granth Sāhib*
aggregates (*skandha*): xii, 241, 257–59, 259 fn. 35, 260–61, 264; subtle (*sūkṣma*): 260; gross (*sthūla*): 260
Akāl (Purukh): 184
alchemy: xxiii, 32, 61–62 fn. 39, 65 fn. 52, 73, 75, 97, 111, 112 fn. 63
alum: 206
aluminum: 151, 162–64, 164 fn. 21, 165, 167, 169, 188
Amar Dās (Gurū): 177–79
Amarakośa (=*Nāmaliṅgānuśānaṁ*): 221, 221 fn. 6–7, 222, 222 fn. 8–9, 226, 234, 234 fn. 30, 235 fn. 32 and 34
Amitābha: xxvi, 220
ammonite: 15, 16 fn. 17, 22
amulets: xxiv, 63, 74, 190 fn. 35
Anasūyā: xxv, 202, 207–09, 213–14
animate (Skt. *cetana*; Tib. *rgyuba*): ix, xi–xiii, xiii fn. 15, xiv–xxii, 123–24, 134
antimony: 267
aṇu, *paramāṇu*: xiii, 56, 257; *see* particle
aphrodisiac (*vājīkaraṇa*): 61, 97, 101 fn. 23, 109–10, 215
Arjan Dev (Gurū): 177–78, 180
Arthaśāstra: 31, 62 fn. 42, 89
asceticism: x, 65, 145, 149, 170–72, 271
āśram: 144–45, 149–50, 152, 159, 162, 165, 167, 172
Aṣṭādhyāyī: xvi–xvii, xvii fn. 26 and 28, xviii
Aṣṭāṅgahṛdayasaṃhitā: xxiv, 81, 94, 104, 104 fn. 32–36, 105, 105 fn. 38
Aṣṭāṅgasaṃgraha: 81, 104, 104 fn. 32–33 and 36, 113, 232 fn. 26, 252
Atharvaveda: xi, xi fn. 8, xiv fn. 17
ātman: x, xii–xiv, xiv fn. 17, xvi, xvi fn. 23, xviii, xx, 64 fn. 48, 67, 67 fn. 57; *anātman* (Skt.)/ *anatta* (P.): xii
atom: *see aṇu*
avatāra: 6, 10, 16, 52, 52 fn. 5, 183
Āyurveda: ix fn. 3, xv, 21, 53 fn. 10, 61, 73–75, 81, 95, 98, 98 fn. 13, 99, 99 fn. 16, 100 fn. 20, 101–03, 105–06, 112, 122, 149, 152, 152 fn. 17, 153–54, 153 fn. 18, 230, 232–33

Baluchistan: xxiii, 28–30, 34–35, 39, 40, 43–44
bāṇaliṅga: 3, 5–6, 9–11, 11 fn. 9, 12–14, 13 fig. 1.2, 14 fn. 13, 16, 20–23
Bhagavadgītā: 179 fn. 12, 183
Bhagavatīsūtra: 265 fn. 63, 267, 267 fn. 78
Bhairava: 63–64, 64 fn. 49
Bhaiṣajyaratnāvalī: 111, 111 fn. 59–60, 112, 112 fn. 64
Bhartṛhari: xvi–xviii, xviii fn. 30, xxii
Bhāvaprakāśa: 97, 97 fn. 8, 98, 98 fn. 9, 101, 101 fn. 21–22, 110 fn. 52, 111, 112 fn. 64
bhūta: xviii–xx, 99–100, 105, 106 fn. 40, 123 fn. 12, 152
Bhuvneshwari Puri: 144; *see* Guru Ma
Bībī Nānī: 29–30, 36, 40, 44
Bihar: 3, 11, 19
bitumen: 122

blacksmith: 60, 177 fn. 7, 180–81
body (*śarīra*; *rūpa*): xii, xiv, xvi, xvi fn. 23, xxi fn. 36, xxii, xxiv, xxv, xxvi, 6, 7, 23 fn. 27, 38–39, 49, 51 fn. 3, 66, 74–77, 79, 80, 84, 96–101, 101 fn. 23, 104–06, 106 fn. 40, 108 fn. 46, 109, 144, 146, 146 fn. 9, 152, 154, 157–64, 169–71, 179, 191, 202–03, 207–08, 215–16, 219, 226, 230, 238, 243, 256, 262, 262 fn. 53, 263–68; physical body (*audārika śarīra*) 262; transformational body (*vaikriya śarīra*): 262 fn. 53; luminous body (*taijasa śarīra*): 263; translocational body (*āhāraka śarīra*): 263 fn. 53
brass: 4, 126, 188–89
breath: xiv, xiv fn. 17, xv–xvi, xxiii, 6, 98, 98 fn. 11; five breaths: xiv; *see prāṇa*
Bṛhajjātaka: 77 fn. 10, 78
Bṛhaspati (Jupiter): 51 fn. 3, 52 fn. 5, 53–54, 77, 88, 90–91
Bṛhatpārāśarahorāśāstra: 52, 52 fn. 6, 53, 64 fn. 48, 65
Bṛhatsaṃhitā: xxi, 63, 63 fn. 43, 76–77, 77 fn. 8–9, 78 fn. 11, 84, 84 fn. 30–31, 86
bronze: 4–5, 116, 126–27
Buddha: 52 fn. 5, 206, 219–20, 222–30, 233, 238, 241, 243–50
Buddhaghosa: xix fn. 31
Buddhism: xviii, 123 fn. 12, 146 fn. 8, 147, 147 fn. 10
Budha (Mercury): 51 fn. 3, 52 fn. 5, 53–54, 77, 90

cakra: 15–17, 192; Kāla Cakra: 55–56, 66; Graha Cakra: 56–57
Cakradatta: xxiv, 94, 106, 106 fn. 43, 107 fn. 44 and 45, 108, 110 fn. 52, 111
Cakrapāṇidatta: xxiv, 94, 106–08, 108 fn. 47, 111, 111 fn. 60
cakravartin (P. *cakkavatti*): 77, 222, 225, 247
Candra (Moon): 35, 41, 41 fn. 28, 42–44, 51 fn. 3, 52 fn. 3 and 5, 53–54, 57, 59 fn. 28, 60, 77, 85, 90, 104–05, 150–52, 154, 156, 160–61, 172, 226, 249
Carakasaṃhitā: xiii fn. 13, xv, 74–75, 75 fn. 3 and 4, 98 fn. 12, 99, 99 fn. 14 and 17, 100, 100 fn. 17–20, 101, 101 fn. 23 and 24, 102–03, 103 fn. 29 and 30, 105 fn. 39, 106 fn. 40, 108–09
caste: 34, 39, 53–54, 57, 58 fn. 24, 60, 144, 150, 167, 169, 181 fn. 15, 192
cat's eye (*vaiḍūrya*): 54, 62, 77, 79–80, 85, 90, 245
Chennai: 3, 4 fn. 2, 5, 9, 9 fn. 7, 11, 19, 19 fn. 19, 20–22, 22 fn. 26, 24, 26, 27
China: 119–21, 125, 125 fn. 16, 130
Chittagong: 37
Cikitsāsārasaṃgraha: xxiv, 95, 108, 108 fn. 46 and 48, 109 fn. 49
cinnabar (vermilion, Skt. *sindūra*): xxiii, xxiv, 30, 30 fn. 5, 31, 31 fn. 6 and 8, 32, 109, 117, 117 fn. 3, 120, 125, 125 fn. 15 and 17, 126–31, 134–35, 215, 267
cintāmaṇi (wish-fulfilling gem): 242–43, 243 fn. 46–47, 244, 244 fn. 48, 245, 245 fn. 50, 246, 246 fn. 51–52, 247, 247 fn. 54
clay: xiv, xxiii, 9, 143–44, 162–64, 267
conch (*śaṅkha*): 7, 62 fn. 41, 82 fn. 26
copper: xi, xxvi, 21, 54, 61, 62 fn. 43, 63, 96, 102, 102 fn. 28, 103, 106–07, 110–11, 126–27, 152, 162–64, 164 fn. 21, 167–69, 178, 178 fn. 10, 189, 204, 220, 267
coral (*pravāla*): 54, 62, 77, 79–80, 80 fn. 26, 85–88, 90, 221, 221 fn. 8, 232, 267
cosmetics: xxv, 201–09, 211–17
crystal: 3, 8, 18–22, 79, 82 fn. 26, 220, 232, 235, 267
Crystal Garland (Tib. *'Khrungsdpedri med shelgyi me long*): 121, 127, 131
Crystal Rosary (Tib. *Shel greng*): 117, 119 fn. 7, 125–27, 130–132, 132 fn. 30
Cullavagga: 206

Dalai Lama: 129–30
Devī Stotra: 157
Devī: xxiii, 3, 5–6, 6 fn. 3, 8–9, 16–17, 20, 29, 32, 33 fn. 13, 34, 34 fn. 16, 184
Devībhāgavatapurāṇa: 33, 33 fn. 13
Devīmāhātmya: 44, 64, 150 fn. 15

dhammas: xii
Dharamsala: 116 fn. 1, 122 fn. 10, 126 fn. 19, 129 fn. 24, 130 fn. 25, 132 fn. 29, 133, 133 fn. 33,
dharma: xv, 52, 66, 146 fn. 8, 229, 238, 238 fn. 38, 241, 241 fn. 43, 243, 243 fn. 46, 245, 245 fn. 50, 246 fn. 51 and 52, 247, 250 fn. 54, 257; Buddhist Dharma/ Dhamma: 146 fn. 8, 147 fn. 10, 223, 223 fn. 13, 224, 225; Sikh Dharma: 176, 191
diamond (*vajra*; *hīraka*): xxv, xxvi, 18, 54, 62, 75–77, 78 fn. 11, 79–81, 82 fn. 26, 84–86, 90–91, 110, 143, 143 fn. 2, 144, 158–59, 220–21, 231–36, 238–43, 267
Diamond Vehicle (Vajrayāna): xxvi, 220, 243
dīkṣā: 269
direction (*diś*): xii, 23, 77, 82, 82 fn. 96, 90, 153, 243
disease(s): xv, 55 fn. 12, 61, 61 fn. 39, 73, 76, 84–85, 94–100, 103–04, 107–08, 108 fn. 46, 109, 111–12, 128, 145 fn. 6, 146, 151–54, 156–57, 161, 164, 177 fn. 6, 204, 206, 211
doṣa: 53, 74, 83, 98, 98 fn. 12, 108, 145, 152–53, 155–60, 163, 170, 230–31
Draupadī: xxv, 201–02, 208–15
Durgā: 33, 64, 150, 150 fn. 15, 184–85
dust: xxvi, 56, 178, 261, 267, 270

earth (*pṛthivī*): ix, xi, xii, xii fn. 9, xiii, xiv, xviii, xix, xx, xxv, xxvi, 9, 23 fn. 27, 26, 39, 44, 46, 63 fn. 43, 80, 99, 105, 107, 121, 121 fn. 9, 125, 150, 152, 154, 158, 161, 171, 171 fn. 22, 179, 183, 187, 211, 233, 255–56, 258, 260–65, 265 fn. 64, 267, 267 fn. 77, 268–72
eight elements/minerals (Tib. *khamsbrgyad*): 126, 130 fn. 26
emerald (*marakata*; *gārutmata*; *tārkṣya*): 54, 62, 75–77, 79–80, 82, 82 fn. 26, 84, 90, 158, 220–21, 267
equinox: 56, 144, 150, 182
Essential Drop of Nectar (Tib. *Sdedge Drungyig Guru 'phel*): 130 fn. 27, 131 fn. 28

fossil: 3, 10, 15, 16 fn. 17, 22, 23 fn. 27
Four Treatises (Tib. *Rgyudbzhi*): 120–21, 121 fn. 8, 123–28

Gandhasāra: 214
Gaṇeśa (or Gaṇapati): xxiii, 3, 5–6, 8–9, 16–17, 19, 58 fn. 22
Garuḍapurāṇa: 80, 80 fn. 20–21
gem (gemstone) (*maṇi*): ix, xxiv, xxv, xxvi, 18–19, 22, 61, 61 fn. 38 and 39, 62, 62 fn. 42, 63 fn. 43, 68, 73–91, 124, 143–45, 152, 152 fn. 17, 161, 220, 227, 229, 231–35, 243, 247, 251; gemstone powder (*cūrṇa*; *piṣṭa*): 75
glass: 62 fn. 41, 110, 178
Gobind Siṅgh (Gurū): xxv, 174–75, 180–84, 184 fn. 22, 185, 187, 188 fn. 32, 193
gold: xii fn. 9, xxvi, 20–21, 43, 54, 60, 62 fn. 43, 63, 65 fn. 52, 75, 89–90, 96, 102, 102 fn. 28, 103, 110, 110 fn. 52, 111 fn. 56, 116, 123 fn. 11, 126–27, 139, 152, 158, 160, 175, 177–78, 178 fn. 10, 179, 189, 205, 220–21, 225, 227, 229, 235, 247, 249, 267; goldsmith: 160
Golden Annotations (Tib. *Gser mchan rnam bkra gan mdzod*): 129
Gondwana: 15
Gorakṣanātha: 39, 39 fn. 26
graha ('grasper'): 52, 52 fn. 3–4, 77, 83; *navagraha*: 52, 76, 78, 81, 82 fn. 26, 90; *see* planets
gravel (*saśarkara*): 84, 267
Gujarat: 34, 53 fn. 11, 145, 145 fn. 5, 150
guṇa: xiii, xviii, 37, 53, 57, 83, 85, 99, 224, 228, 229–31, 231 fn. 23, 238 fn. 38, 256, 258, 265, 271
Gurū Granth Sāhib: xxv, 175, 177–80, 180 fn. 13, 184, 189, 190 fn. 34, 193
Guru Ma: 143–44, 145 fn. 5, 146, 149ff., 155 fig. 7.1

Hanumān: 63, 63 fn. 46, 64; *Hanumān Cālīsā*: 64
Hargobind (Gurū): 174–75, 180–82
Haribhadra: xx, xxi fn. 36
Hemacandra: xxiii, 266, 266 fn. 70

Herodotus: 176
hessonite (*gomeda*): 54, 62; *see* zircon
Himalaya (Himavanta): 15, 33, 37, 80
hiṅgula (cinnabar/vermilion): 30, 30 fn. 5, 31, 31 fn. 6 and 7
Hiṅgulā (Hiṅglāj): xxiii, xxiv, 28ff.; Hiṅglāj Devī: 40 fig. 2.1
hyacinth: 267

inanimate (Skt. *acetana*; Tib. *mi rgyuba*): ix, xi–xiii, xiii fn. 15, xiv–xv, xv fn. 20, xvi–xxiii, 52, 117, 123, 123 fn. 12, 124, 128, 134
incense (*dhūpa*): 59 fn. 27, 60, 187, 215, 249
insentience/insentient: xii, xiv, xvii–xviii, xix fn. 31, xxii–xxiii, 229, 244
iron (*loha*): xi, xxiv, xxv, 20–21, 54–55, 55 fn. 12, 58, 59 fn. 26, 60, 60 fn. 33, 61, 61 fn. 39, 62, 62 fn. 43, 63, 65 fn. 52, 66 fn. 55, 91, 101–02, 102 fn. 28, 103–04, 107–08, 110, 110 fn. 52, 111, 123, 126–27, 130 fn. 26, 174–75, 175 fn. 3, 176, 176 fn. 5, 177, 177 fn. 6 and 7, 178–82, 184, 184 fn. 22, 185 fn. 23, 186–88, 188 fn. 32, 189–90, 190 fn. 35, 191–93, 193 fn. 39, 204, 220, 225, 267, 272; pure iron (Pan. *sarabloh*): xxv, 174–75, 183–84, 184 fn. 22, 188, 188 fn. 32, 189–190, 193; iron sulfide: 20–21; black iron; 55, 58; iron cooking pot (Pan. *deg*): 174, 174 fn. 2, 186, 189–90, 190 fn. 34, 193
Īśvara: xii, xxii, 66
Īśvarapratyabijñākārikā: xxii

Jain, Jainism: xii, xix, xx, xxi, xxi fn. 36, xxiii, xxvi, 79, 255–56, 256 fn. 4, 257 fn. 9, 11 and 17, 260, 260 fn. 42, 261, 261 fn. 66, 265, 265 fn. 62, 266 fn. 70, 267–72
jewel (Skt.: *ratna*; *maṇi*; *see* gem): xii, fn. 9, xxv, xxvi, 63, 78 fn. 15, 83, 90, 129 fn. 23, 143, 143 fn. 2, 144, 144 fn. 3, 159–60, 178, 178 fn. 11, 179, 205, 212, 219–21, 223, 223 fn. 13, 224, 226, 229, 232–33, 237–38, 240, 242, 244–45, 248–50
jewelry: x, 167, 202, 205–07, 210, 212–13
jīva: xx, xxi fn. 36, 256–58, 261–63, 263 fn. 58, 264–67, 272–73; *jīvājīva*: xx; *jivātman*: 52 fn. 6, 64 fn. 48; *ajīva*: xx, 256
Jyotiṣa (astrology): 52–53, 53 fn. 9, 55, 57, 67 fn. 57, 68–69, 73, 137, 154; Jyotiṣaśāstra: 57 fn. 21; *jyotiṣī* (astrologer): 58, 58 fn. 24, 61–62

Kabīr: 178ff.
Kailash Das: 165, 166 fig. 7.3, 167, 171
kajjalī (mixture of mercury and sulfur): 108, 110
Kālacakra: 243
Kālī: 63–64
Kāmasūtra: 205–06
Kathmandu: 4 fn. 2, 16, 16 fn. 17, 133, 133 fn. 32
Kerala: 203
Ketu (Cauda Draconis): 51 fn. 3, 52 fn. 5, 53 fn. 8, 54–55, 59 fn. 28, 67 fn. 57, 77, 90
Khajuraho: 33, 46,
Khālsā: xxv, 174ff., 177 fn. 8, 180ff., 182 fn. 16, 185, 185 fn. 23, 186–89, 193
Koh-i-Noor: 144
kohl: 203–04, 207, 214
kuṁkuṁ (vermilion): 206
Kundakunda: 257, 257 fn. 11, 260, 269
Kūrmapurāṇa: 56 fn. 17, 57 fn. 20

lapis lazuli: 75, 77, 77 fn. 7, 204, 267
lead: xxvi, 54, 62 fn. 43, 63, 96, 102 fn. 28, 108, 112 fn. 64, 116, 126–27, 130 fn. 26, 151, 178, 186, 204–05, 267; lead sulfide: 130 fn. 26
Lhasa: 130

Madhya Pradesh: 3, 12, 12 fn. 11, 13, 19, 20, 33–34, 59 fn. 26, 150
Madhyamaka: xviii, 240–41
Mahābhārata: xxv, xix fn. 31, 24, 31, 31 fn. 6, 201ff., 214
Maharashtra: 34, 59, 59 fn. 26, 145 fn. 5, 181 fn. 15, 187
makeup: xxv, 201–05, 207, 211–12, 214–16
Makran: 28, 30–38, 43–44

malachite: 204
Maṅgala (Mars): 51 fn. 3, 52 fn. 5, 53–54, 77, 85, 87, 90
maṇi (gem, jewel; see *ratna*): 73, 221–22, 231 fn. 21, 232, 242
mantra: xv fn. 20, 6–8, 58, 60, 60 fn. 32, 61, 68, 88, 124, 157, 230, 232, 240
gāyatrīmantra: 7; *mantrapuṣpa*: 8; *brahmamantra*: 16, 16 fn. 16; *siddhamantra*: 232; Mantranaya: 238, 240
Manusmṛti: xiv
Mārkaṇḍeyapurāṇa: 44, 64
matter: ix, x, xii, xiii, xviii, xviii fn. 30, xxiii, xxvi, xxix, 52, 65 fn. 52, 97, 120, 144, 256–59, 262, 262 fn. 52, 263–64, 265 fn. 64, 266, 271
māyā: 67 fn. 57, 179, 215
medicine: xxiv, 21, 52, 61 fn. 39, 75, 94ff., 99, 101, 101 fn. 23, 102, 105–06, 106 fn. 40, 107, 109, 112, 112 fn. 63, 116ff., 124–26, 129–30, 134, 143–46, 146 fn. 7 and 8, 147, 147 fn. 9, 148–51, 152–73, 230, 232–33
mercury (Skt. *pārada*; Tib. *ngülchu*): xxiii–xxiv, 30–32, 94–98, 98 fn. 11, 100–01, 101 fn. 22, 103–04, 104 fn. 35, 106, 106 fn. 42, 107–10, 110 fn. 52, 111, 111 fn. 56, 112, 112 fn. 64, 116–20, 124 fn. 14, 125, 125 fn. 16, 126, 126 fn. 18, 127–30, 133, 133 fn. 31, 134, 152; sulfide: 30–31, 125, 130, 133; liquid mercury: 117–18, 128, 133, 135
metallurgy: x fn. 5, 123 fn. 11, 175–76, 205
metals: ix, ix fn. 3, x–xii, xxiv, xviii, xxiii–xxvi, 3, 7, 9, 31, 53–54, 62 fn. 43, 63, 65 fn. 52, 75, 77, 95–97, 101–02, 102 fn. 28, 103–04, 104 fn. 27, 108–11, 116, 121, 123, 123 fn. 11, 124, 126, 129, 133, 133 fn. 31, 134, 143–45, 151–52, 152 fn. 17, 158, 163–64, 167, 176–77, 220–21
mica: 106–08
Mīmāṃsā: xv, xviii
Mīmāṃsāsūtra: xv, xv fn. 20, xvi, xvi fn. 23–25
minerals: ix, ix fn. 3, x, xviii, xix, fn. 31, xxiii–xxvi, 18, 22, 30–31, 37, 43, 61 fn. 39, 62 fn. 43, 75–76, 84, 95, 102, 104 fn. 37, 110, 121–22, 126, 143, 202–05, 212, 215–16, 223, 229, 245
mines; mining: x fn. 5, 18, 80, 158, 170
Muktinath: 3, 4 fn. 2, 14, 16, 17, 38
Mūlamadhyamakakārikā: 241, 241 fn. 43, 242 fn. 44
mūrti: xxiii, 28, 55 fn. 14, 58–59, 64 fn. 49, 206; *svayambhūmūrti*: 28
myth, mythology: xxiii, xxiv, xxviii, 14–16, 20, 29, 29 fn. 3, 56, 58, 63 fn. 43, 66, 79–80, 124, 135, 183, 189, 203

Nāgārjuna: 241–242
nakṣatras (lunar mansions): 55–56, 87, 87 fn. 45
Nāmdev: 178
Nānak (Gurū): 177–80, 192
Narmada: 3, 11–14, 20, 25
Nātha Siddha: 39, 39 fn. 26, 41, 145 fn. 4
natron: 267
Nepal: xxiv, xxvi, 3, 4 fn. 2, 11, 14, 16, 31, 38, 125, 133
Nyāya: xiii, xiii fn. 13 and 14, xiv–xv, xviii
Nyāyasūtra: xiii fn. 14, xiv

Omkareshwar: 3, 12, 14, 22
orpiment: 267

pañcāyatanapūjā: xxiii, 3–6, 5 fig. 1.1, 6 fn. 3, 8 fn. 5, 9–11, 11 fn. 9, 16–25, 21 fig. 1.4, 27
Pāṇini: xvi, xvii fn. 26
pāras: *see* philosopher's stone (s.v. stone)
particle: 257–61, 264
pativratā: xxv, 202
Patna: 4 fn. 2, 19, 19 fn. 20, 20, 20 fn. 21
pearl (Skt. *muktā*, *mauktika*; H. *motī*): xxiv, 54, 62, 75, 75 fn. 3, 76–77, 79–80, 82 fn. 26, 85, 88, 90–91, 158–59, 160, 232, 240
pebbles: xxvi, 161
pharmacopoeia: 106, 117–18, 119 fn. 7, 120–23, 126–27, 134–35
pigment: 31, 204
planets (*graha*): 51–52, 52 fn. 3 and 6, 53, 53 fn. 7 and 9, 54–57, 57 fn. 20, 59, 59

fn. 26, 63 fn. 46, 64–65, 67 fn. 57, 74–78, 78 fn. 13 and 15, 79, 81, 87–89
Plato: 56 fn. 18
poison, poisoning: 61, 75, 83–84, 102–03, 112 fn. 64, 117, 124, 124 fn. 13 and 14, 125, 127, 130, 132, 135, 138, 151, 169–70, 203, 215, 217, 230, 232
polar star (*dhruva*): 55–56
pradakṣiṇā: 8, 29
Prājāpati: x, xi, 53, 66
prāṇa: xiv, xiv fn. 17, xv, xvi, xvi fn. 23, xxiii, xxviii, 88, 264; *aprana*: xiv; *prāṇapratiṣṭhā*: xxiii fn. 37, 9; *prāṇāyāma*: 6
prasād: 151, 186, 190 fn. 34, 224
pūjā: xxiii, xxviii, 3–27, 58 fn. 24, 59 fn. 29, 60, 60 fn. 33, 61, 63, 63 fn. 47, 64, 65 fn. 50, 68, 150, 151, 187–88, 196, 206; *pūjārī* (ritualist): 58 fn. 24
pyrite (*suvarṇamukhi*): 3, 5–6, 9, 11, 20, 20 fn. 22, 21–22, 103–04, 110 fn. 52, 130 fn. 26

quartz (*sphaṭika*): 17–20, 22, 80, 157–59
quicksilver (*rasakarman*): 83
Qur'ān: 177

Rāhu (Caput Draconis): 51 fn. 3, 52 fn. 5, 53 fn. 8, 54–55, 59 fn. 28, 67 fn. 57, 77, 89, 90
Rajasthan: xxv, 18, 31, 39, 143–44, 150, 153 fn. 18, 154 fn. 20, 165
Rām Dās (Gurū): 177–79
Rāmacaritmānasa: 64
Rāmānuja: xxii
Rāmāyaṇa: xxv, 201–02, 207–08, 214
Rasahṛdayatantra: 96, 96 fn. 2–4, 97–98, 112
Rasaratnasamuccaya: 75, 75 fn. 5, 81–86, 81 fn. 22–24, 83 fn. 29, 84 fn. 32–35, 85 fn. 36–42, 96 fn. 3–5, 104 fn. 36
Rasārṇava: 98, 98 fn. 10 and 11, 104 fn. 36
Rasaśāstra: 73, 75, 81, 82
rasāyana: xxiv, 62 fn. 39, 65 fn. 52, 83, 94–95, 96 fn. 5, 97, 98 fn. 9, 101, 101 fn. 21, 23 and 24, 102, 102 fn. 26, 103–04, 104 fn. 35, 105–06, 106 fn. 40, 107–08, 108 fn. 46, 109–10, 110 fn. 52, 111, 111 fn. 59, 112–13
ratna: xxv, 61 fn. 38, 73, 77, 83, 90, 219–24, 229–34, 238, 243–44, 246, 250–51; *strīratna* (woman-gem): 222; *uparatna* (semi-precious stones): 62, 80, 91; *navaratna*: 62, 76–78, 78 fn. 15, 79, 81–82, 84, 87 fn. 43, 89–90, 90 fig. 4.1; *see* jewel
Ratnaśāstra: 73, 221, 221 fn. 8, 222–24, 229, 229 fn. 19, 230–31, 233, 236, 238, 240
Ravidās: 177–78, 178 fn. 10
realgar (Tib. *ldongros*): 131, 131 fn. 28, 267
red chalk: 267
red jasper (*śoṇabhadra*): 3, 5–6, 9, 11, 19–22
Ṛgveda: ix–x, x fn. 4, xiv fn. 17
Ṛgvidhāna: xxiii
ritual: x, xxiii, xxiv, 3–4, 6, 6–8, 10–11, 11 fn. 9, 15, 23–24, 29, 51, 55, 55 fn. 15, 58–59, 59 fn. 26, 60 fn. 30, 32 and 33, 63, 66 fn. 55, 74, 78, 87 fn. 15, 82, 88, 144–45, 145 fn. 6, 146 fn. 7, 150–51, 156–57, 164, 170, 186, 188, 203, 208, 251, 271
rock: 18, 28–29, 36–38, 40, 40 fig. 2.1, 41, 41 fn. 28, 43, 59, 125, 130, 161; rock resin, shilajit (*śilājatu*): 103; rock-salt: 110, 267; cinnabar rock: 120, 125
ruby (*māṇikya*; *padmarāga*): 54, 62, 77, 79, 81, 81 fn. 26, 82, 82 fn. 26, 84, 86–87, 90–91, 221, 221 fn. 8, 232, 232 fn. 25 and 26
rūpa: *see* body

Śābarabhāṣya: xv, xv fn. 20, xvi fn. 23 and 25
sacrifice (*yajña*): x, xi, xv, 52, 60 fn. 32, 64, 80, 80 fn. 21, 81, 189
sadhātu (mineral substances): 84, 108, 215
sādhu: vi, xxv, 65, 143–44, 144 fn. 4, 145, 145 fn. 6, 146, 146 fn. 7, 8 and 9, 147, 149, 156–57, 159, 161–62, 165, 167, 169–72

Śaivism, Śaiva: xvii fn. 27, 11, 23, 64–65, 144, 144 fn. 4, 145 fn. 5
Śakti: 5, 16, 28–30, 32, 34 fn. 16
śakti: xv; *yathāśākti*: 89, 91; *prāṇ-śakti*: 160–61
Śaktism, Śākta: 33, 37, 44, 64; Śākta *pīṭha*: xxiii, 29, 29 fn. 3, 33, 37 fn. 21, 38
śālagrāma: 3, 5–6, 8, 8 fn. 6, 9–11, 11 fn. 9, 14–18, 17 fig. 1.3, 20–23, 23 fn. 27
Sāṃkhya: xiii fn. 13, xv fn. 19, xviii–xix
Sāṃkhyakārikā: xviii, xix
saṃskāra: xii, 97, 97 fn. 7, 108
sand: xxvi, 107, 267; red sand (Tib. *cu gshag*): 125 fn. 15
Śani (Saturn): xxiv, 51, 51 fn. 2, 52 fn. 5, 53–54, 55 fn. 16, 65, 65 fn. 51, 69, 77, 85, 90–91, 177 fn. 7; Śani Pūjā: 59 fn. 29, 63, 65 fn. 50; *Śani Cālīsā*: 65 fn. 51
Śaṅkara: xxi, 3, 10, 20 fn. 24, 23, 65, 145 fn. 4
sapphire: xxiv, 51, 54, 58, 60–61, 61 fn. 39, 62, 62 fn. 39, 40, 42 and 43, 63 fn. 44, 77, 79–82, 82 fn. 26, 85–86, 90–91, 221, 232, 245, 267; blue sapphire (*nīla, indranīla*): xxiv, 54, 58, 60–62, 62 fn. 39–40, 63, 63 fn. 44, 77, 85, 90; yellow sapphire (*puṣparāga, pītāśmā*): 54, 62
sarabloh (Pan. all iron; pure iron): xxv, 174–75, 183–84, 188, 188 fn. 32, 189–90, 193
śarīra: 294, 294 fn. 53, 262–63, 262 fn. 53, 264; *see* body
Śārṅgadharasaṃhitā: 109, 109 fn. 50–51, 110, 110 fn. 53, 111, 111 fn. 54
sentience/sentient: xii, xiv, xvi–xix, xxii, 52, 123 fn. 12, 161, 171, 119, 225, 227–29, 231, 238, 240, 244–46, 249–50, 268
Sikh, Sikhism: vi, xxv, 174–75, 175 fn. 3, 176–77, 177 fn. 8, 179–181, 181 fn. 15, 184–87, 187 fn. 28, 189, 189 fn. 33, 190, 190 fn. 35, 191–93, 193 fn. 41
silver: xxiv, xxvi, 5, 21, 54, 62 fn. 43, 75, 91, 96, 102, 102 fn. 28, 103, 110, 116–17, 126, 151–52, 157–58, 163–65, 167, 189, 205, 220–21, 225, 227, 229, 249, 267; silver ash: 127; silver ore (Tib. *rdodngul*): 126, 130; silver foil: 151, 157; silversmith: 160
Sindh: 29, 34, 39
Sītā: xxv, 63 fn. 46, 64, 64 fn. 48, 145 fn. 4, 201–02, 207–09, 213–15
soil: xix fn. 31, xxvi, 38, 124, 132–33, 133 fn. 31, 266–67, 272
śoṇabhadra: *see* red jasper
space (*ākāśa*): xiii, xix, 53, 123, 227, 257–58, 262, 271; space-points (*pradeśa*): 257–58; cosmic space (*loka-ākāśa*): 258, 260, 266; acosmic space (*alokaākāśa*): 258
sphaṭika (crystal): 3, 5–6, 8–9, 11, 18, 21–22, 82 fn. 26; *see* quartz
Stainless Rosary: 127–28, 128 fig. 6.1, 131
steel (*tegh*): 12, 150, 174–75, 175 fn. 3, 176, 176 fn. 5, 177, 179–82, 184, 187–88, 191, 195, 197
stone: ix–xi, xiv, xxiii, xviii, xix fn. 31, xxiii–xxiv, 3–12, 12 fn. 11, 13–15, 18–24, 28–29, 38, 62–63, 75–77, 77 fn. 7, 78, 80 fn. 21, 83–91, 133, 144, 149, 157–62, 164, 167, 170–71, 191, 236, 267, 271; artificial stone: 13, 91; *bāṇaliṅga* stones: 12–14; five stones: xxiii, 3–7, 8 fn. 5, 9–11, 11 fn. 9, 16, 21; memorial stones: 41; natural/authentic stone: 3, 13, 24, 91; philosopher's stone (*lapis philosophorum*): xv, 175, 178, 178 fn. 10 and 11, 189; precious stone (*ratna, ratnāni*): xxiv, 61–62, 62 fn. 41, 63 fn. 44, 66, 73–76, 76 fn. 6, 77, 79, 80, 82–85, 88–89, 91, 171, 212, 220, 240, 272; pressing stone/stone mortar: x–xi, 107; sacred stone: 12; *śālagrāma* stone: 14–18; semi-precious stone (*uparatnāni*): 76, 80, 80 fn. 21, 83, 91; silver stone (Tib. *dngulrdo*): 131; *śoṇabhadra* stone: 19–20; *sphaṭika* stone: 18–19; stone medicine: 121, 121 fn. 9, 126, 165; stone pillar: 59 fn. 26; stone therapy: 74, 143, 145–46, 146 fn. 8, 147, 149–50, 152–53, 156–62, 164–65, 166 fig. 7.3, 167, 168 fig. 7.5, 169–72; sun-stone (*sūryakānta*): 82, 91; *suvarṇamukhi* stone: 20–22; Vallam stone: 18

Śukra (Venus): 51 fn. 3, 52 fn. 5, 53–54, 77, 85, 90
sulfur: xxiii, 30–32, 35, 37, 103, 106–07, 109–11, 127, 267
Sunārs: 165, 167
Sūrya (Sun): 17, 18, 41, 41 fn. 28, 51 fn. 3, 52 fn. 5, 53, 53 fn. 8, 54, 56 fn. 19, 57 fn. 20, 59, 64 fn. 48, 67, 67 fn. 57, 77, 82, 82 fn. 26, 87, 90, 154, 160–61, 163, 226, 249
suvarṇamukhi: *see* pyrite
sword: xvi, xxv, 63, 174–75, 180–86, 186 fn. 25, 190 fn. 34, 191, 193

temple: 4, 7, 9, 10, 14, 16–17, 19–20, 20 fn. 23, 23, 23 fn. 27, 28, 30, 33–34, 37–39, 39 fn. 27, 42, 46, 55, 58 fn. 24, 59, 59 fn. 26, 60, 60 fn. 33, 63–64, 64 fn. 49, 65, 78, 78 fn. 15, 88, 145, 147 fn. 10, 150–51, 154, 165, 190 fn. 34, 191, 206, 235, 271
Thanjavur: 18–19
Three Jewels (*triratna*; *ratnatraya*): xxv, 220, 223–24, 229, 251
Tibet: xxiv, 31, 47, 94, 116 fn. 1, 117, 117 fn. 4, 120, 123 fn. 11, 125, 125 fn. 15, 16 and 17, 126, 128–30, 130 fn. 27, 132, 137
time (*kāla*): xii, 257
tin: 62 fn. 43, 96, 102 fn. 28, 108, 112 fn. 64, 126, 267
Tīrthaṅkaras: 263 fn. 53, 268
tiryañc: 263–65
topaz (*pītāśmā*, *puṣparāga*,*puṣyarāga*): 77, 79–81, 84, 88, 90–91, 221, 232
Tulsi Giri: 167, 168 fig. 7.5
turmeric: 7, 206

Udaipur: 153 fn. 18, 154 fn. 20, 159, 170
Umāsvāti: 256 fn. 4, 257 fn. 17, 258, 260 fn. 42
Upaniṣad: x, xii, xiv fn. 17; *Bṛhadāraṇyaka Upaniṣad*: xi, xi fn. 6; *Chāndogya Upaniṣad*: xi fn. 6, xiv; *Maitrī Upaniṣad*: x fn. 5, xiv; *Śvetāśvatara Upaniṣad*: xii; *Taittirīya Upaniṣad*: xiv
Ursa Major (Sapta Ṛṣi): 55–56
Utpaladeva: xvii fn. 27, xxii
Uttar Pradesh: 59 fn. 26, 145

Vāgbhaṭa: 81, 83, 104, 104 fn. 32, 106, 232
Vaiśeṣika: xii–xiii, xiii fn. 12–15, 15, 15 fn. 19, xviii, 260
Vaiśeṣikasūtra: xiii, xiv
Vaiṣṇavism, Vaiṣṇava: 11, 17, 23, 23 fn. 27, 144, 144 fn. 4, 145 fn. 4
Vajratirtham: 18
Vākyapadīya: xviii
Vallam: 3, 11, 18–19
Vāmanapurāṇa: 32, 32 fn. 10, 33
Varāhamihira: 76–77, 84
Veda: xiv fn. 5, xv, xvii, 24, 52
Vedāṅga: 52
Vedānta: 24; Advaita Vedānta: xxi, 4, 20 fn. 24, 23–24; Viśiṣṭādvaita Vedānta: xxii; Vedāntin: xxi, 22
vermilion: *see cinnabar*
Viṣṇu: xxiii, 3, 5–6, 6 fn. 3, 8, 8 fn. 6, 10, 14–17, 23, 23 fn. 27, 52, 54, 56, 144 fn. 4, 184, 208
Visuddhimagga: xix fn. 31

weapon: 180–81, 181 fn. 15, 182, 185, 185 fn. 23, 186–88, 190, 193; Indra's weapon: 63 fn. 43
weaponry: x, 174, 179, 181–83, 187

yajña: *see* sacrifice
yajñopavīta: 7, 192
Yama: 53, 55 fn. 15, 59 fn. 27, 61, 66, 66 fn. 55, 67, 77
Yoga: xiii fn. 13, 66, 240
Yogācāra: 240
Yogaratnākara: 111, 111 fn. 55–56 and 58
Yogaratnamālā: 240 fn. 41, 241, 241 fn. 43

zircon (*gomeda*): 77, 79–80, 82 fn. 26, 85–86, 89–90; *see* hessonite
zodiac: 51 fn. 2, 53, 55–56, 56 fn. 19, 57, 87

CPSIA information can be obtained at www.ICGtesting.com
Printed in the USA
BVOW02*1942061016

464365BV00005B/7/P

9 781781 791288